Enzymes in the Valorization of Waste

Enzymes in the Valorization of Waste: Next-gen Technological Advances for Sustainable Development of Enzyme-based Biorefinery focusses on key enzymes which are involved in the development of integrated biorefinery. It highlights the modern next-gen technologies for promoting the application of sustainable and greener enzymatic steps at industrial scale for the development of futuristic and self-sustainable "consolidated/integrated biorefinery/enzyme-based biorefinery." It also deals with technological advancement for improvement of enzyme yield or specificity, conversion capability, such as protein and metabolic engineering and advances in next generation technologies, and so forth.

Features:

- Explores all modern-day technologies that can potentially be used in enzyme-based biorefinery conversion of wastes to value-added products.
- Covers technological, economic, and environmental assessments of enzyme-based biorefinery prospects.
- Deliberates all possible products that can be generated from wastes including biofuel and essential chemicals.
- Illustrates techniques for enhanced yield and properties to be used in various industrial applications.
- Reviews advanced information of relevant sources and mechanism of enzymes.

This book is aimed at graduate students, researchers and related industry professionals in biochemical engineering, environmental science, wastewater treatment, biotechnology, applied microbiology, biomass-based biorefinery, biochemistry, green chemistry, sustainable development, waste treatment, enzymology, microbial biotechnology, and waste valorization.

Novel Biotechnological Applications for Waste to Value Conversion

Series Editors: *Neha Srivastava, IIT BHU Varanasi, Uttar Pradesh, India and Manish Srivastava, IIT BHU Varanasi, Uttar Pradesh, India*

Solid waste and its sustainable management is considered as one of the major global issues due to industrialization and economic growth. Effective solid waste management (SWM) is a major challenge in the areas with high population density, and despite significant development in social, economic and environmental areas, SWM systems is still increasing the environmental pollution day by day. Thus, there is an urgent need to attend to this issue for green and sustainable environment. Therefore, this proposed book series is a sustainable attempt to cover waste management and their conversion into value added products.

Utilization of Waste Biomass in Energy, Environment and Catalysis
Dan Bahadur Pal and Pardeep Singh

Nanobiotechnology for Safe Bioactive Nanobiomaterials
Poushpi Dwivedi, Shahid S. Narvi, Ravi Prakash Tewari and Dhanesh Tiwary

Sustainable Microbial Technologies for Valorization of Agro-Industrial Wastes
Jitendra Kumar Saini, Surender Singh and Lata Nain

Enzymes in the Valorization of Waste
Enzymatic Pre-treatment of Waste for Development of Enzyme based Biorefinery (Vol I)
Pradeep Verma

Enzymes in the Valorization of Waste
Enzymatic Hydrolysis of Waste for Development of Value-added Products (Vol II)
Pradeep Verma

Enzymes in the Valorization of Waste
Next-Gen Technological Advances for Sustainable Development of Enzyme based Biorefinery (Vol III)
Pradeep Verma

Biotechnological Approaches in Waste Management
Rangabhashiyam S, Ponnusami V and Pardeep Singh

Agricultural and Kitchen Waste
Energy and Environmental Aspects
Dan Bahadur Pal and Amit Kumar Tiwari

For more information about this series, please visit: www.routledge.com/Novel-Biotechnological-Applications-for-Waste-to-Value-Conversion/book-series/NVAWVC.

Enzymes in the Valorization of Waste

Next-gen Technological Advances for Sustainable Development of Enzyme-based Biorefinery

Edited by

Pradeep Verma

CRC Press is an imprint of the
Taylor & Francis Group, an **Informa** business

First edition published 2023
by CRC Press
6000 Broken Sound Parkway NW, Suite 300, Boca Raton, FL 33487-2742

and by CRC Press
4 Park Square, Milton Park, Abingdon, Oxon, OX14 4RN

CRC Press is an imprint of Taylor & Francis Group, LLC

ISBN: 978-1-032-03517-8 (hbk)
ISBN: 978-1-032-03518-5 (pbk)
ISBN: 978-1-003-18772-1 (ebk)

DOI: 10.1201/9781003187721

Typeset in Times
by SPi Technologies India Pvt Ltd (Straive)

Contents

Preface

The modern era can be called an era of waste. Rapid population growth has increased the needs of people and accommodating these requirements of a modern, fast-paced life has led to the generation of huge amounts of waste. Different anthropogenic activities result in generation of waste from non-biodegradable plastics to biodegradable organic or biological products (for example from agriculture, animal product processing, forest product processing, marine and municipal waste, and food, feed, textile, pulp, leather processing industries etc.). Waste disposal has been considered one of the most critical issues in current times. Nevertheless, humankind always surprises with a great ability to turn a crisis into an opportunity.

With this aim in the past few decades, several processes have been developed to convert these wastes into value-added compounds such as: biobutanol, bioethanol, and other high-value essential compounds. Several physical and chemical methods have been developed to facilitate these conversions and resource recoveries from waste. But the cost involved, environmental concerns and the non-sustainable nature of physical and chemical methods has led the focus to shift towards biological methods. Among several biological methods, the role of enzymes in these important steps of bioconversion of waste to value has been critical.

Waste-based biorefineries focus on an integrated system for complete valorization of wastes by generation of value-added compounds. The waste biomass consists of lignin, hemicelluloses, cellulose, starch, pectin, lipids, and so on that can be used as precursor compounds for generation of several building block chemicals, as well as fuels. The production of suitable liquid or gaseous fuel and value-added compounds occurs through a series of steps. The role of different enzymes is important to the biorefinery concept with a critical role in the two most rate-limiting steps of biorefinery: pretreatment and hydrolysis. These enzymes, by their action on different components of the waste, can lead to generation of specific oligomers and monomers. These oligomeric and monomeric compounds can be subsequently converted to biofuels, or can be used as the building blocks for commercially important chemicals. In order to understand the role of enzymes in these two critical steps, their source, structural and mechanistic properties, and technological advancements in the field of enzymology for their application in waste-to-value generation need to be understood. Thus, this book entitled *Enzymes in the Valorization of Waste: Next-gen Technological Advances for Sustainable Development of Enzyme-based Biorefinery* as a part of the book series Novel Biotechnological Applications for Waste-to-Value Conversion consists of ten chapters under the thematic areas associated with next-generation advances in enzyme-based biorefinery. This book provides insight into a wide range of recent next-generation advances in enzyme-mediated biorefinery, starting with the basics of enzyme-based biorefinery for conversion of biowaste to value-added products. This book also provides insight into creation of biowealth from biowaste using advanced biotechnological and bioprocessing technologies, with special emphasis on advanced enzyme purification strategies, immobilizations strategies, nanotechnological approaches, and cell surface engineering. Further, this book presents a detailed overview of next-generation approaches available for integration of bioinformatics to waste biomass degradation and valorization. Additionally, this book provides insight into the economics of biochemical conversion based biorefinery. This book will be helpful to researchers, students, academicians, scientists, and professionals in planning their future strategies for waste biomass valorization.

Acknowledgments

First, I would like to convey my gratitude to the series editors, Dr. Neha Srivastava and Dr. Manish Srivastava for inviting us to submit this book entitled *Enzymes in the Valorization of Waste: Next-gen Technological Advances for Sustainable Development of Enzyme based Biorefinery* in the book series Novel Biotechnological Applications for Waste-to-Value Conversion. I am thankful to CRC Press for accepting my proposal to act as editor for the current book volume. The current volume of this book series is only possible because of the support from all the researchers and academicians who contributed to this book, therefore I am thankful for their contributions. I would also like to thank my PhD Scholar, Dr. Bikash Kumar, currently working as Post Doctoral Researcher at Indian Institute of Technology, Guwahati. For providing me with editorial assistance and technical support during all stages of book development. I am also thankful to the Central University of Rajasthan (CURAJ), Ajmer, India for providing infrastructural support and a suitable teaching and research environment. The teaching and research experience at CURAJ has provided the necessary understanding of the needs of academicians, students, and researchers. This was greatly helpful during the development of this book. I am also thankful to the Department of Biotechnology for providing me with funds through sponsored projects (Grant No. BT/304/ NE/TBP/2012 and BT/PR7333/PBD/26/373/2012), for setting up my laboratory: the Bioprocess and Bioenergy Laboratory. I am always thankful to God and my parents for their blessings. I also express my deep sense of gratitude to my wife and children for their support, both during the development of this book and in life.

Notes on the Editor

Prof. Pradeep Verma

Prof. Verma completed his PhD at Sardar Patel University Gujarat, India in 2002. In the same year he was selected as UNESCO fellow and joined the Czech Academy of Sciences, Prague, Czech Republic. He later moved to Charles University, Prague to work as Post-doctoral Fellow. In 2004, he joined the UFZ Centre for Environmental Research, Halle, Germany as a visiting scientist... He was awarded a DFG fellowship which provided him another opportunity to work as a Postdoctoral Fellow at Gottingen University, Germany. He moved to India in 2007 where he joined Reliance Life Sciences, Mumbai and worked extensively on biobutanol production, which led to the attribution of a few patents to his name. Later, he was awarded the JSPS Post-doctoral Fellowship Program and joined the Laboratory of Biomass Conversion, Research Institute of Sustainable Humanosphere (RISH), Kyoto University, Japan. He is also a recipient of various prestigious awards such as the Ron Cockcroft award by Swedish society, UNESCO Fellow ASCR, Prague. He has been awarded Fellow of Mycological Society of India (MSI-2020); the Prof. PC Jain Memorial Award, Mycological Society of India 2020; and Fellow of Biotech Research Society, India (2021).

Prof. Verma began his independent academic career in 2009 as a reader and founder head at the Department of Microbiology at Assam University. In 2011, he moved to the Department of Biotechnology at Guru Ghasidas Vishwavidyalaya (A Central University), Bilaspur, and served as an associate professor until 2013. He is currently working as professor (former head and dean, School of Life Sciences) at the Department of Microbiology, CURAJ. He is a member of various national and international societies/academies. He has completed two collaborated projects worth 150 million INR in microbial diversity and bioenergy.

Prof. Verma is a group leader of the Bioprocess and Bioenergy laboratory at the Department of Microbiology, School of Life Sciences, CURAJ. His area of expertise involves microbial diversity, bioremediation, bioprocess development, lignocellulosic- and algal-biomass-based biorefinery. He also holds 12 international patents in the field of microwave-assisted biomass pretreatment and biobutanol production. He has more than 62 research articles in peer-reviewed international journals and has contributed to several book chapters (36 published; ten in press) in different edited books. He has also edited four books for international publishers such as Springer, Taylor and Francis, CRC Press, and Elsevier. He is a guest editor for several journals such as *Biomass Conversion and Biorefinery* (Springer), *Frontiers in Nanotechnology* (Frontiers), and the *International Journal of Environmental Research and Public*

Health (mdpi). He is also an editorial board member for the journal *Current Nanomedicine* (Bentham Sciences). He is acting as reviewer for more than 40 journals in different publishing houses such as Springer, Elsevier, RSC, ACS, Nature, Frontiers, and mdpi.

List of Contributors

Cecil Antony
School of Biotechnology
National Institute of Technology Calicut
Kozhikode, Kerala 673601, India

Tarun Kanti Bandyopadhyay
Department of Chemical Engineering
National Institute of Technology Agartala
Jirania, Tripura, India-799046

Akansha Bhatia
Department of Civil Engineering (CED)
Indian Institute of Technology Roorkee
(IITR)
Roorkee-247667, Uttarakhand, India

Biswanath Bhunia
Department of Bioengineering
National Institute of Technology Agartala
Jirania, Tripura, India-799046

Anupama Binoy
Discipline of Chemistry
Indian Institute of Technology Palakkad
Kerala-678557, India

Manswama Boro
Department of Microbiology
Sikkim University
Gangtok-737102, Sikkim, India

Judith M. Braganca
Department of Biological Sciences
Birla Institute of Technology and
Science Pilani
KK Birla Goa Campus, NH 17B,
Zuarinagar, Goa-403726, India

Dixita Chettri
Department of Microbiology
Sikkim University
Gangtok-737102, Sikkim, India

Bhaskar Jyoti Deka
Department of Hydrology
Indian Institute of Technology Roorkee
(IITR)
Uttarakhand, India-247667

Sumit Dhawane
Department of Chemical Engineering
Maulana Azad - National Institute of
Technology (MANIT)
Bhopal, 462003 (M.P.) India

Irene J. Furtado
Department of Microbiology
Goa University
Goa, India-403206

Praveen Kumar Ghodke
Department of Chemical Engineering
National Institute of Technology Calicut
Kozhikode, Kerala-673601, India

Bhaskar Kalita
School of Agro and Rural Technology
Indian Institute of Technology Guwahati
Assam, India

A.A. Kazmi
Department of Civil Engineering
(CED)
Indian Institute of Technology Roorkee
(IITR)
Roorkee-247667, Uttarakhand, India

Muthusivaramapandian Muthuraj
Department of Bioengineering
National Institute of Technology
Agartala
Jirania, Tripura, India-799046

Devika Nagar
Department of Biological Sciences
Birla Institute of Technology and
Science Pilani
K K Birla Goa Campus, NH 17B,
Zuarinagar, Goa-403726, India

Pinku Chandra Nath
Department of Bioengineering
National Institute of Technology
Agartala
Jirania, Tripura, India-799046

Sanjukta Patra
Department of Bioscience and
Bioengineering
Indian Institute of Technology Guwahat

John M. Pisciotta
Biology Department
West Chester University
West Chester, PA 19383, USA

Ankur Rajpal
Department of Civil Engineering (CED)
Indian Institute of Technology Roorkee
(IITR)
Roorkee-247667, Uttarakhand, India

Biplab Roy
Department of Chemical Engineering
National Institute of Technology Agartala
Jirania, Tripura, India-799046

Sushabhan Sadhukhan
Discipline of Chemistry
Indian Institute of Technology Palakkad
Kerala-678557, India

Azar E. Saikali
Biology Department
West Chester University
West Chester, PA 19383, USA

Jata Shankar
Genomics Laboratory, Department of
Biotechnology and Bioinformatics
Jaypee University of Information
Technology
Solan, Himachal Pradesh 173234, India

Satyam Singh
Department of Biosciences and
Biomedical Engineering
Indian Institute of Technology Indore
Madhya Pradesh 453552, India

Chhavi Thakur
Genomics Laboratory, Department of
Biotechnology and Bioinformatics
Jaypee University of Information
Technology
Solan, Himachal Pradesh 173234, India

Saravanakumar Thiyagarajan
Department of Biochemistry and
Molecular Biology
Michigan State University
East Lancing, Michigan – 48824, USA

Vinay Kumar Tyagi
Environmental Hydrology Division
National Institute of Hydrology
Roorkee-247667, Uttarakhand, India

Anil Kumar Verma
Department of Microbiology
Sikkim University
Gangtok-737102, Sikkim, India

List of Abbreviations

AFEX	ammonia fiber explosion
API	active pharmaceutical ingredients
ARCD	aromatic ring cleavage dioxygenases
ARHD	aromatic ring hydroxylating dioxygenases
AST	aspartate transaminase
ATCC	american tissue cell culture
ATP	adenosine triphosphate
ATPS	aqueous two-phase system
ATR-FTIR	attenuated total reflection-fourier transform infrared
BGLs	β-glucosidases
BNICE	biochemical network integrated computer explorer
CAMD	computer-aided molecular design
CBD	cellulose-binding domain
CBH II	cellobiohydrolase II
CBMs	carbohydrate binding modules
CBN	carbon-based nanomaterials
CBP	consolidated bioprocessing
CBS	consolidated bio-saccharification
CD	circular dichroism
CdTe	cadmium tellurium
CFU	colony forming units
CLEAs	cross-linked enzyme aggregates
CMCase	carboxymethylcellulase
CPW	citrus processing waste
CTD	comparative toxicogenomics database
CYP450	cytochrome P450
DHA	docosahexaenoic acid
EG	endoglucanase
EPA	eicosapentaenoic acid
EPS	extracellular polymeric substances
ESCY	ethanol substrate conversion yield
ETA	ethanol tolerance assay
FAO	food and agriculture organization
FAOSTAT	Food and Agriculture Organization Corporate Statistical Database
FMM	from metabolite to metabolite
GA	glucoamylase
GH	glycoside hydrolase
GLIMMER	gene locator and interpolated markov modeler
GlpD	glycerol-3-phosphate dehydrogenases
GPI	glycosylphosphatidylinositol
GRAS	generally regarded as safe
GRB	green biorefinery

HFCS	high fructose corn syrup
HFI	hydrogen fuel injection
HMF	5-hydroxymethyl-2-furaldehyde
HMF	hydroxymethylfurfural
HRT	hydraulic retention time
IL	ionic liquid
IONP	iron oxide nanoparticle
IUCLID	International Uniform Chemical Information Database
KEGG	Kyoto Encyclopedia of Genes and Genomes
LAC	laccases
LCB	lignocellulose feedstock biorefinery
LiP	lignin peroxidases
MEC	microbial electrolysis cell
MeO-NP	metal alloy nanoparticle
MICs	minimum inhibitory concentrations
MnP	manganese peroxidases
MSW	municipal solid waste
MTBE	methyl tert-butyl ether
MTs	million tons
MW	megawatts
NADH	nicotinamide adenine dinucleotide hydrogen
NADP	nicotinamide adenine dinucleotide phosphate
NGS	next-generation sequencing
NRCM	National Research Centre on Mushrooms
OD	optical density
OLR	organic loading rate
PGDBs	pathways and genomes databases
PHB	poly(3-hydroxybutyrate)
PM-IRRAS	polarization modulation infrared reflectance absorption spectroscopy
PNS	purple non-sulfur
POME	palm oil mill effluent
PSSF	prehydrolysis simultaneous saccharification and fermentation
PUFA	polyunsaturated fatty acids
PVP	polyvinylpyrrolidone
QSAR	quantitative structure-regulatory activity relationship
ROS	reactive oxygen species
SBB	sugar-based biorefinery
SCFAs	short-chain fatty acids
SDS-PAGE	sodium dodecyl sulphate-polyacrylamide gel electrophoresis
SERS	surface enhanced raman spectroscopy
SFG	sum frequency generation
SHF	separate hydrolysis and fermentation
SmF	submerged fermentation
SOD	super oxide dismutase
SPR	surface plasmon resonance

SSCF	simultaneous saccharification and co-fermentation processes
SF	saccharification and fermentation
SSF	simultaneous saccharification and fermentation
SsF	solid state fermentation
SSI	single spore isolates
TBCC	tribasic copper chloride
TOF-SIMS	time-of-flight secondary ion mass spectrometry
TPP	three phase partitioning system
TRI	toxics release inventory
TS	total solids
UM-BBD	University of Minnesota Biocatalysis/Biodegradation Database
URN	universal reaction network
VFA	volatile fatty acids
WCB	whole crop biorefinery
XDH	xylitol dehydrogenase
XK	xylulokinase
XR	xylose reductase
XYN	xylanase
YPD	yeast extract peptone dextrose

1 Fundamentals of Enzyme-based Biorefinery for Conversion of Waste to Value-added Products

Pinku Chandra Nath, Biplab Roy,
Tarun Kanti Bandyopadhyay, Biswanath Bhunia,
and Muthusivaramapandian Muthuraj
National Institute of Technology Agartala, Jirania, India

CONTENTS

DOI: 10.1201/9781003187721-1

1.1   INTRODUCTION

With a worldwide population of over 7.9 billion, a demographic expansion of up to 28% is expected by 2057, according to Worldometer projections (Worldometer, 2021). This record high demographic expansion, augmenting daily needs and associated industrialization creates the significant challenges that disturb sustainability and environmental stability. Rapid depletion of natural resources and fossil fuel reserves have prompted the pursuit of alternate resources that can fulfill the material and energy demands of the growing population (Weber et al., 2020; Goswami et al., 2022a). In addition, with an ever-expanding global population, the amount of wastes produced is steadily increasing. For instance, the US alone generates municipal solid wastes up to ~292 million tons per year, and about 50% of the global municipal wastes usually end up in landfills and dumping sites. In the present scenario, dependence on a bioeconomy that relies on renewable biomass resources remains feasible to attain sustainability and environmental stability attributed to the potentials in obtaining the paradigm of multiple products from the broad range of biomass resources. Thus, biorefining based on biological resources is one of the potential aspects of the emerging bioeconomy.

Biorefinery aims to fractionate the biomass resources into lipids, polysaccharides, proteins, pigments, fibers, and other polymers and further generates a marketable product (value-added bioproducts and bioenergy) via biochemical fermentation process or through a thermochemical process. Furthermore, achieving commercial-scale feasibility with sustainability and high-efficiency product generation remains one of the significant hurdles. However, few biorefinery operations at commercial-scale are reported in biofuel production, food, paper and pulp industries, with few other innovative biorefinery operations from different resources under the pipeline. The bottlenecks spread across the process from raw material procurement to final product purification and marketing, which determine the overall economic feasibility (Matharu et al. 2016). In the case of raw material procurement, the utilization of waste materials has gained significant interest as a feasible alternative. In general, the substantial waste piles generated by the US alone could be a significant resource for biorefinery operations. Waste agricultural residues, municipal solid materials, and industrial wastes are predominantly utilized in biorefinery operations to synthesize value-added chemicals and bioenergy molecules.

Bioenergy is an expensive source strategically promoted for climate change mitigation (Wang et al., 2018). The feedstock for bioenergy, lignocellulosic biomass, is also the only renewable and cost-effective resource of valuable bioproducts (Nikodinoska et al., 2017; Kumar & Verma 2020). Nonfood biomass from metropolitan administrations and cultivating exercises is reasonable feed for creating bioenergy without influencing food safety (Alexander et al., 2015).

Pretreatment, saccharification, fermentation, and separation are all unit processes in biomass bioconversion (Cai & Zhang, 2005). The unit process design criteria are directly associated to the types of biomass. The biomass obtained from agricultural waste includes wheat straw (Salapa et al., 2017), rice straw (Kapoor et al., 2017; Bhardwaj et al., 2020), corn stover (da Costa Sousa et al., 2016), and rice husk (Ruiz et al., 2013). Biomass can come from various plant parts, for example leaves, bark, trunk, and stems. Each part comprises of different structures and measures of components of the building block (i.e., hemicelluloses, cellulose, and lignin) combined into a sophisticated 3D structure (Leu & Zhu, 2013; Kumar & Verma 2020). The plant cell wall should be disintegrated before additional preparation to amplify the yields of the value-added products. Chemical and physical treatment strategies involve hazardous molecules and are less efficient, respectively as compared to enzymatic processes. Thus, the disintegration of complex cellular systems rely on enzymes to achieve high yields and efficiency. Although biorefineries have a lot of potentiality, they have not been fully exploited yet, which is strongly linked to enzyme technologies. This chapter has therefore explored the influence and practical applications of enzymes in biorefineries and biotechnological processes.

1.2 SOURCES AND VARIOUS TYPES OF WASTE

Waste is described as undesired and unusable items, as well as a substance that is no longer useful. The wastes may generally be classified as follows: Liquid wastes, solid wastes, and gaseous wastes.

1.2.1 WASTEWATER

Approximately 90% of the world's wastewater is released untreated into oceans, rivers, and lakes, endangering the environment and human health (Khan et al., 2017). Water treatment facilities utilizing new advancements is a pressing need and the ideal opportunity for accomplishing manageability, especially in nonindustrial nations (Shannon et al., 2010). The waste biorefinery concept, which produces hydrogen from wastewater, is a viable solution to remediate wastewater while also providing sustainable energy. There are a variety of strategies for producing H_2, but most of them are too expensive and emit too much CO_2 to be practical on a wide scale (Miandad et al., 2016; Nair et al., 2022).

For instance, Khan et al. (2017) researched the chance of a microbial electrolysis cell (MEC) for delivering hydrogen from metropolitan wastewater in Saudi Arabia. MEC innovation utilizes wastewater as a substrate by means of the reactant activities of organisms within the sight of an electric momentum yet not without oxygen. In contrast with water electrolysis and aging methodologies, MEC has a high hydrogen age rate (80–100%) and can eliminate up to 60% chemical oxygen demand (COD). In Saudi Arabia (KSA), around 1.18 billion m^3 of homegrown and 0.39 billion m^3 of mechanical wastewater are delivered, separately, with no office to treat or change the wastewater over to energy. It has been estimated that if MEC innovation is applied to KSA modern and homegrown wastewater, a sum of 434 MW each year could be created, with the possibility to increase to 615 MW by 2025 and 769 MW by 2035 (Khan et al., 2017). Within waste biorefinery thinking, MEC innovation is more suitable and practical than other H_2 creating advancements. Low energy proficiency; ohmic, fixation, and conductivity misfortunes; microbial populaces and contending measures; and the development of H_2 foragers are largely troubles that should be addressed to develop cycle execution. Khan et al. (2017) discovered that applying MEC innovation in Saudi Arabia and other Gulf countries will fundamentally develop wastewater treatment plant supportability, including ecological, mechanical, and monetary efficiencies (Pasupuleti et al., 2015).

1.2.2 FOOD WASTE

Agrofood ventures are those that depend on agrarian products as crude materials. Food waste is generated when organic raw agrarian materials are processed into foodstuffs, which is usually accomplished by collecting the nutritionally useful part of the raw materials (Poggi-Varaldo et al., 2014). This generates a huge assorted waste ranging from edible to non-edible. The unused residues are mostly organic material, but their application as a food source is restricted due to their low nutritional value or the presence of inedible components. Efforts to reduce food waste while maintaining product quality may be done by enhancing manufacturing efficiency; nevertheless, the waste reduction potential is quite limited (Russ & Schnappinger, 2007). However, unused residues comprising edible parts may act as a source material for fodder industries and related sectors. These include plant-based food squanders, got from vegetables and organic products, and grain materials, as well as animal food squanders, created from dairy items and meat items (Mirabella et al., 2014). Table 1.1 summarizes the features of food waste.

1.2.3 AGRICULTURAL WASTE

There are numerous agricultural products that extends from simple organic compounds, vegetables, to complex meat, poultry, and dairy items. The waste generated after recovering these useful commodities generates a huge waste which predominantly comprise biomass. Agriculture is one of the major biological sectors with the greatest biomass output, making it an important bioeconomy input. The huge agricultural residue output generated from agricultural sector contain material that is helpful to people, even though their monetary value may not justify the expense of assortment, transportation, and handling for advantageous use. They may be fluids, slurries, or solids and their organization will differ contingent upon the framework and type of horticultural action. Rural wastes also comprises animal dungs, wastes from slaughter houses, food handling waste (just 20% of maize is canned, leaving the other 80% to be squandered), crop squander, and perilous and harmful

TABLE 1.1

Features of Various Types of Food Waste

Sources	pH	Moisture	Total Solid	Volatile Solid	Total Sugar	Starch	Cellulose	Lipid	Protein	Ash	References
Municipal solid waste (MSW)	7.3	85.1	14.8	88.6	–	32.7	15.4	8.7	6.8	11.3	(Rao & Singh, 2004)
Dining hall	5.6	82.8	17.2	85.0	62.7	46.1	2.3	18.1	15.6	–	(Ma et al., 2008)
Food court	6.6	64.3	35.5	87.3	–	37.5	–	8.9	4.7	18	(Cekmecelioglu & Uncu, 2013)
Restaurant	6.2	81.5	18.5	94.1	55.0	24.0	16.9	14.0	16.9	5.9	(Vavouraki et al., 2014)
Eatery	5.8	81.6	18.4	87.6	35.4	–	–	24.2	14.3	–	(He et al., 2012)
Residents	4.9	80.8	19.2	92.7	–	15.5	–	20.4	15.7	–	(Pan et al., 2008)
Canteen	6.5	80.2	19.5	95.7	59.6	–	1.5	15.4	21.9	1.5	(Tang et al., 2008)

farming waste. Farming waste assessments are uncommon, yet essentially are considered to contribute a significant amount to the waste generated worldwide. In Asian regions, it is expected that over 998 MTs of agricultural waste are produced each year (Agamuthu, 2009).

1.3   ENZYMES INVOLVED AND THEIR ROLE IN BIOREFINERY

Proteins are molecular driving factors found in all living things that initiate complex responses; by bringing down the initiation energy, they can stay unaltered until the response is finished. As they stay unaltered, they can be reconditioned for various responses in lesser amounts. They are used for the viable recuperation of bioproducts like biodiesel and biogas from the garbage in eco-accommodating green innovations (Battista et al., 2020; Ng et al., 2020). Plants, animals, and microbes are all examples of enzyme resources that are used for waste valorization. However, there are several disadvantages in the employment of plant- and animal-based enzymes for waste recovery, such as inadequate accessibility, changes to the action of the enzymes, and their particularly destructive character. Microbial enzymes, on the other side, have a significant market presence due to their high stability, robustness, and broad spectrum of activities when compared to other resources. The transformation of waste streams into value-added products including sweets, bioplasm, biofuels, and prebiotics is an active part of microbial enzymes such as amylases, cellulases, and xylanases. Traditional techniques to these alterations require a substantial energy intake and chemical catalysts are of restricted reactions and generate byproducts, especially in complex matrices, such as waste streams (Basso & Serban, 2019).

1.3.1   Cellulases

The term "cellulases" refers to a group of enzymes that includes β-1, 4 endoglucanase (EC 3.2.1.4), β-glucosidase (EC 3.2.1.21), and β-1, 4 exoglucanase (EC 3.2.1.91). These enzymes synergistically depolymerize cellulose efficiently (Lynd et al., 2002; Zhang et al., 2006). While endoglucanases target β-1, 4 links within glucan chain in amorphous areas, cellobiohydrolases separate cellobioses from reduction and non-reduction ends, and β-glucosidases transform cellobioses into glucose monomers (Kiran et al., 2014; Menon & Rao, 2012; Bhardwaj et al., 2021a). *Aspergillus niger, Cellulomonas flavigena, Trichoderma reesei,* and *Clostridium thermocellum* are just a few of the microorganisms that produce cellulases. Depending on microorganism and strain, the enzymatic pattern can differ significantly (Lynd et al., 2002; Alam et al., 2021).

Cellulases from fungi and bacteria with a modular structure may have a cellulose-binding domain (CBD) along with a 30–200 amino acid sequence that belongs to the carbohydrate binding modules (CBMs) category (Howard et al., 2003). CBDs assume a significant part in cellulase affinity and selectivity for lignocellulose fibers (Howard et al., 2003; Agrawal & Verma 2021). After cleansing and sliding one glycosidic bond, CBDs inhibit cellulase from detaching from the substratum, permitting effective adsorption by the enzyme to cellulose and processive hydrolyze (Levy & Shoseyov, 2002). On the contrary, β-glucosidase is not regarded as CBD, that iss why cellulose is less closely linked afterwards to cellobiohydrolases and endoglucanases (Gomes et al., 2015). CBDs have been discovered in esterases, mannanases, pectate-lyases, and xylanases in addition to cellulases (Levy & Shoseyov, 2002).

1.3.2 HEMICELLULASES

Parasites, microbes, yeast, marine green growth, protozoans, snails, shellfish, bugs, seeds, and different living beings produce hemicellulases (Gírio et al., 2010). Filamentous fungi and bacteria including *Bacillus*, *Actinomycetales*, and *Clostridiales* are the most important species in terms of productivity (Rabemanolontsoa & Saka, 2016). The enzyme levels of filamentous fungi have been reported to be substantially greater than bacteria and yeast. Among the significant species employed as indus-trial sources of commercialized xylanases are *T. fusca, T. niger, Humicola insolens, Trichoderma koningii*, and *Trichoderma reesei* (Howard et al., 2003).

1.3.3 AMYLASES

The amylase bunch comprises two primary classes: α-amylase (EC 3.2.1.1) and glucoamylase (GA) (EC 3.2.1.3). By cleaving the 1,4-α-D-glucosidic connections between consecutive glucose units in the linear amylose chain (Pandey et al., 2000), α-amylase hydrolyzes starch into maltose, glucose, and maltotriose, whereas glu-coamylase hydrolyzes amylose and amylopectin to glucose at nonreduction end-points (Anto et al., 2006). In food, fermentation, textiles, and paper, amylases have been extensively used (Sundarram & Murthy, 2014). They also serve to boost the bioproducts yield in future operations for the pretreatment of agro-industrial and organic products. As a result, there is a growing interest in producing amylases from low-cost feedstock (Wang et al., 2008).

1.3.4 LIGNOCELLULOLYTIC ENZYMES

Lignocellulolytic enzymes are mostly produced by fungi and consist of cellulases, xyl-anases, and ligninases that decompose lignocellulosic materials. Cellulases are utilized in an assortment of enterprises, including food, animal feed, blending and wine produc-tion, farming, biomass refining, mash and paper, fabric materials, and washing (Kuhad et al., 2011). Endoglucanases (EC 3.2.1.4) perform alone on insoluble and soluble cellulose chains, whereas exoglucanases (cellobiohydrolases; EC 3.2.1.91) release cel-lular biosections from reduction and nonreduction endpoints of cellulose chains, and β-glucosidases (EC 3.2.1.91) (Jørgensen et al., 2007; Kumar et al., 2018). With or without the assistance of cellulases, xylanases have a wide scope of employment in the food, feed, mash and paper, blending, winemaking, and fabrics (Khandeparkar & Bhosle, 2006; Bhardwaj et al., 2021b). Lignin is an unfavorable polymer for biofuel generation because it hinders plant-derived polysaccharides from being accessible (Kumar et al., 2020). The lignin components obtained can, however, be utilized in the development of important products including scattering, detergents, flocculants, coagulants, surfactants and mud thinner, adhesives, plastics, grafted polymers includ-ing polyurethanes, polyesters, polyamines, and epoxies (Menon & Rao, 2012).

1.3.5 Pectinolytic Enzymes

Pectinolytic compounds, otherwise called pectinases, depolymerize gelatin polymers consecutively and synergistically. Endo- and exo-acting polygalacturonases and gelatin and pectate lyases, that cuts the rhamnogalacturonan chain, are needed for complete gelatin breakdown (Kashyap et al., 2001). Pectinases are commonly employed for juice, wine, and many other typical industrial processes, such as textiles, processing of tea, fiber plants, oil extraction, coffee, and wastewater treatment for industrial purposes (Pedrolli et al., 2009). The majority of pectinases are produced by solid-state fermentation (SSF) of fungal organisms, specifically *Aspergillus* strains (Heerd et al., 2012). Pectinases may be produced for commercial usage from wastes containing pectin including citrus and orange trash (Giese et al., 2008).

1.3.6 Lipases

Lipases (EC 3.1.1.3) rank third in terms of total sales volume (Contesini et al., 2010). They are widely used in a variety of industries including pharmaceutical, food, detergent, cosmetic, and organic industries. They catalyze triacylglycerol hydrolysis into di-, mono-acylglycerols, fatty acids, and glycerol (Alkan et al., 2007; Li et al., 2009; Vaseghi et al., 2013). Under specific conditions, they can also catalyze alcoholysis, acidolysis, aminolysis, esterification, and transesterification (Saxena et al., 2003). Phospholipases are subclasses of lipases that catalyze hydrolysis of the glycerophospholipids esters and phosphodiesters. They differ in the site of action on phospholipids which may be utilized to modify/produce novel phospholipids for certain petroleum refining, healthcare, food production, milk, and cosmetics applications (Ramrakhiani & Chand, 2011). The majority of the study has been on the generation of high activity extracellular lipase via submerged fermentation (SmF) and solid-state fermentation employing a diverse range of microorganisms like fungus, bacteria, yeast, and actinomyces (Ramrakhiani & Chand, 2011; Vaseghi et al., 2013). Several commercial lipase-producing fungi, such as *Yarrowia, Rhizomucor, Aspergillus, Rhizopus, Geotrichum*, and *Penicillium* species are quite prominent (Colen et al., 2006).

1.3.7 Proteases

Proteases are enzymes that help proteins hydrolyze their peptide linkages. It is considered to be the most economically important enzyme family due to its wide assortment of utilizations in the food, drug, cleanser, dairy, and calfskin areas (Gupta et al., 2012). *Aspergillus, Penicillium, Rhizopus* fungi, and *Bacillus* genus bacteria have been recognized as active protease producers (Khosravi-Darani et al., 2008). According to researchers (Banerjee et al., 2010), enzymes that are not substantially active in covalent bonding may be significant in lignocellulose breakdown. These are fungus and bacteria loosening and expansions, as well as several enzymes that break down non-glycosidic wall segments to allow access to glycosyl hydrolases.

1.4 CONFIGURATIONS OF WASTE BIOREFINERIES

1.4.1 WASTE BIOREFINERIES WITH MULTIPLE PLATFORMS

Anaerobic digestion, the well-known biological phenomenon, is now used on a wide scale to treat highly heterogeneous waste, which may play a prominent role in upcoming biorefinery systems (Roni et al., 2019). A proposed multiplatform biorefinery process plan combining anaerobic digestion and different biochemical, chemical, and thermochemical processing unit is depicted in Figure 1.1. The suggested design comprises initial segregation of various waste feed components and specific treatments for each constituent, to maximize the production of biomolecules and biofuels (Asunis et al., 2019). The separation methods themselves are intrinsically dependent on the contents and properties of feed waste and may include operations including extraction and washing (Ao et al., 2020), enzyme usage (Arbige et al., 2019), and separation of the membrane (Abels et al., 2013). In some circumstances, the bioproducts incorporated (see dashed line in Figure 1.1) in the suggested design are also considered as alternatives such that the different treatment steps can be adapted according to the immediate requirements of the intended end products.

Implementing a multi-platform, multi-product strategy in its entirety, as illustrated in Figure 1.1, entails addressing bottlenecks linked with conversion procedures involving heterogeneous and low-purity materials like organic wastes. Therefore, before the growth of the entire process chain, it is essential to go through a transitional phase. When transitioning from one stage of implementation to another, the biorefinery idea can be extended and formulated by using simplified layouts focusing on techniques that are already being established and proved at a large scale, to reduce complexities about process performance during the early implementation phases.

This is intended to serve as the foundation for a processing system whose uncertainty can be gradually extended as new, increasingly advanced options became readily accessible for execution. These types of arrangements have the potential to evolve into a fully integrated high-performance strategy in the long run. The development of simplified configurations portraying treatment train services with even a short to moderate application frontier can be accomplished in this respect. These simplified configurations, which are regarded as having a greater likelihood of becoming more comfortably executed within the waste management sector, can be successfully implemented.

1.4.2 WASTE BIOREFINERY OUTPUT

In light of the announced market costs for every result of revenue (Moscoviz et al., 2018), the monetary worth of the items that can be gotten from food waste (FW) in a biorefinery application was assessed as follows: Hydrogen costs 0.25–15.70 €/t FW (normal: 4.8), CH_4 costs 1.8–11.7 €/t FW (normal: 7.4), ethanol costs 9.2–538 €/t FW (normal: 232), and ethanol costs 23–4495€/t FW (normal: 1490). Subsequent to deducting the plant's development and functional expenses, the income produced from biowaste in a biorefinery would be determined. Regardless, given the measure of food squander created (in Europe, 45.7 and 42.3 MT/year from civil and

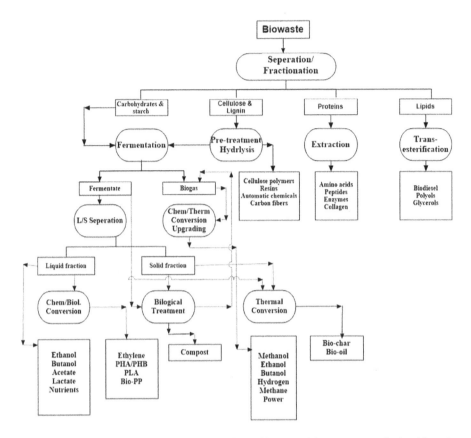

FIGURE 1.1 Multiplatform biorefinery design with anaerobic process producing biomolecules and biofuels.

modern sources, separately), just as the impetuses for the creation of green synthetic substances and energy, the extended food squander and the board framework are required to produce critical monetary advantages (Caldeira et al. 2020).

1.5 SCOPE/BOUNDARY CONDITIONS OF WASTE BIOREFINERIES

Several economic, social, environmental, and political advantages can be derived by developing and executing waste biorefinery approach:

- Encourage the involvement of neighboring communities for the promotion and implementation of environmentally friendly waste management solutions by providing incentives.
- Provide a commercially viable substitute option for waste management in urbanized regions that are experiencing rapid growth.
- Encourage the execution of circular economy ideas; minimize the impact on nonrenewable resources.

- Contribute to the diversification of strategic supply sources and the reduction of dependency on imported materials.
- Encourage the use of distributed production processes and the preservation of regional and rural growth.
- Providing valuable products and mitigating the consumption of fossil carbon can both help to alleviate the effects of climate change

1.5.1 'WASTE BIOREFINERIES' UNDERLYING CONCEPTS

In biorefineries, the cascade principle is frequently used. This entails the adaptable and progressive combination of diverse thermal, chemical, and biological techniques aiming at obtaining a blend of biomolecules and biofuels in order to enhance manufacturing yields and profits (Task, 2016). To this goal, both direct as well as reverse cascade techniques can be applied, relying on whether the synthesis of bioenergy is downstream or upstream. The combination of both methods is based on technological viability, financial viability, bazaar trends, ecological and local demands, and limits and results in a specific range of biomaterials and biofuels (Maina et al., 2017). The expansion of the variety of output goods is projected to have an influence on the reduction of waste recovery accomplished since it will minimize organic waste from being dumped in landfills or open dumps. The flows from the landfill must meet biorefinery-specific quality and management specifications. Therefore, a waste biorefinery in comparison with a normal biorefinery would have an additional complicated aspect due to heterogeneity, low purity, and variability of waste products rather than biomasses (Ubando et al., 2020).

1.5.2 TECHNICAL AND ECONOMIC SUSTAINABILITY

In terms of both technological and economic challenges, the most significant issues are as follows: (i) managing the potential consequences of variations in waste components and features on the variety of processes used in a biorefinery; (ii) putting together an extensive overview of eligible waste resources to use as feedstock in order to enhance the production and superiority of finished product; (iii) determination of the system's finest size, from multifeedstock high-performance to decentralized, more specialist systems with smaller platforms (Galanopoulos et al., 2020); (iv) connecting the process with many other sectors in order to increase the flow of energy and materials throughout the process; (v) making provisions for changeable marketplace conditions and product price volatility.

1.5.3 SUSTAINABILITY IN THE ENVIRONMENT

Waste management strategies are distinguished by environmental consequences of the actions and technologies employed within the process, that is, the collection, transport, and manufacturing of waste products. The generated outputs from wastes help environmental benefits by compensating the needs for other resources. In order to make a waste biorefinery environmentally viable, its "value" must be larger than the "effort" involved in the delivery of outputs. More precisely, it iss vital to determine whether using organic waste mostly as precursor material necessitates fewer resources than the production of similar products from virgin materials (Cristóbal et al., 2018). A biorefinery's environmental performance will rely on regional conditions and whether simpler options like anaerobic digestion or composting have comparable or higher environmental advantages. Thus, a broad range of variables is essential when analyzing a waste biorefineries environmental sustainability, for example:(i) feedstock accessibility, content, characteristics, and fluctuation that may result in greater quantities of feedstock which needs to be disposed of; (ii) in comparison to easier and more adaptable composting or digesting plants, logistical concerns like as transportation distance and storage space are a concern; (iii) more complex process designs, including the requirement for sophisticated pretreatments, (iv) framework circumstances and interaction with "neighboring" waste management industries; and (v) management of the refinery chain coproducts and side streams.

1.5.4 Market Potential

The waste-biorefinery sector has continuously increased despite the recent decade's economic crisis. International organic chemical production makes for a significant portion of the overall chemical industry, with estimates indicating that it amounts to more than 300 metric tons per year (excluding fuels). During the period 2009 to 2014, the market had a total value of approximately USD 6 billion and had grown at an average annual rate of 8%. In the forecast period from 2019 to 2025, it is predicted to expand at an annual compound growth rate of approximately seven toeight percent, reaching USD 16 billion (Alibardi et al., 2020).

Waste biorefineries, in particular, have the capability to unlock the enormous potential contained in the about 130–151 MTs/year of biowaste that is expected to be created in the EU by 2020. According to the most recent figures provided by (Ragazzi et al., 2020) the real overall (municipal + industrial) output capability of organic waste for the EU28 (European Union) in 2016 was around 230 MTs/year, consisting primarily of agricultural and vegetable waste accounting for 42 percent of total waste, 26 percent of municipal solid waste's organic portion, waste of wood (20 percent), and 9 percent of sludge from water treatment plants that is not hazardous.

1.6 IMPLEMENTATION OF WASTE BIOREFINERY FRAMEWORKS

In contrast to conventional biorefineries, waste biorefineries do not have the same economic and technological prospects as traditional biorefineries, and vice versa. Waste materials have a variable composition and may contain contaminants or other undesirable components that are difficult to remove from the waste stream (Bisinella et al., 2017). Organic waste pretreatment is regarded as a key stage in the biorefining system to address waste materials complexity. A pretreatment step is necessary to eliminate undesired components, modify physical characteristics of the solid matrix to enable faster downstream processes, and make important components more accessible for subsequent treatments (Tonini & Astrup, 2012). Three necessary steps to be followed for the pretreatment process include selectiveness of pretreatment approaches, produced quantity of rejected fraction, and the severity with which pretreatment steps are conducted (Budzianowski & Postawa, 2016).

A waste biological refinery is designed most simply to recover low-value products, such as energy or biofuels transporters, soil enhancers, and fertilizers. There is a need for greater complexity to produce clean streams of platform chemicals for producing biomaterials in which increasingly strict technical demands are met. The practicality of high-quality biorefinery is contingent upon the accessibility and quality of feedstock residues, economic conditions and required product consumptions, and the capability of integrating a waste biorefinery into the current industrial system (Shahzad et al., 2017). According to widespread consensus, in order to develop high outputs along with minimizing environmental impact, the process must be implemented to include two, and sometimes more, platforms (Naik et al., 2010). As per IEA-Bioenergy, (de Jong et al., 2012 platforms are intermediaries that serve as a link between raw materials and finished products. The combined manufacturing of numerous platforms would assure that the specific precursors were recovered from the raw material in the most efficient manner. Production of numerous platforms includes the combination of a variety of processes that are based on the properties of waste feed to be used and the finished products. From this perspective, selectivity, precision, and separation efficiency play an important role in fully implementing multiplatform biorefineries.

The minimal economically feasible size of complicated biorefinery installations, suitable waste feed, and sustainable products generated through waste biorefineries remain a matter of conversation. Generally, traditional biorefineries require large facilities with a minimum capacity of approximately 500,000–700,000 T/annum to assure financial viability (Alibardi et al., 2020). Biorefineries can be configured in a variety of ways, from massive, high-performance installations to decentralized, reduced layout systems. In order to reap the benefit of cost-cutting measures, larger facilities must generate biocommodities that can be sold to huge markets. More complicated process designs are therefore expected to be included, integrating numerous platforms and various processes to diversify, functionalize and maximize resources and energy recovery. Limited scope biorefineries have less sophisticated treatment schemes with lesser capital and operational expenses, because of fewer platforms and fewer finished products. Decentralized medium to small factories will employ a limited number of feedstocks that will probably be available locally. The possible spread of pollutants and contaminants connected with generated waste or formed during processing due to its side reactions and inclusion of existing chemical substances is a key threat linked with waste materials. This factor should be taken into consideration when developing any waste management or recycling strategy (Astrup et al., 2018). Compared to typical bioprocesses including anaerobic digestion and composting, the properties of finished residues from complicated biorefinery systems will be distinct, and this must be taken into account when assessing the viability of biorefinery designs (Cattle et al., 2020).

1.7 VALUE-ADDED PRODUCTS GENERATED

1.7.1 AGRICULTURAL PRODUCTS

The regeneration of value-added substances from agricultural waste biomass has significant commercial viability, with applications in the cosmetics, pharmaceutical, and food industries. The appeal of the various added-value chemicals stems from their numerous fitness benefits, which include antiproliferative, antihypertensive, cardiovascular, and antioxidant actions. The economic value of fully recovered added-value substances derived from agricultural biomass varies according to agricultural waste types, the created volume, and the concentrations of desired substances.

1.7.1.1 Lignocellulosic Agricultural Byproducts

Agricultural byproducts derived from lignocellulosic include wheat straw, ensiled sorghum feed, maize stalks, corn straw, sunflower, rice straw, barley, and cotton gin waste. These products differ in their content of lignin, hemicellulose, and cellulose, as well as trace amounts of wax, pectin, protein, and inorganic substances. A wide range of goods including chemicals, biofuels, and other biomass-derived products with excellent potentiality, can be generated from lignocellulosic agricultural byproducts by integrating clean technologies with the production process (Abraham et al., 2016). Table 1.2 represents the methane output and biodegradability of several lignocellulosic byproducts

1.7.2 ORGANIC ACIDS

Grass, alfalfa, clover, and undeveloped cereals are examples of raw wet biomass that can be processed by green biorefinery (GRB). The first step in the preparation of wet biomass is dewatering, which results in the production of two distinct intermediates: a nutritional juice called "Organic Solutions" and a lignocellulosic press cake rich in fiber. Both intermediates serve as starting places for a variety of value-adding processes and pathways.

Based on biomass characteristics in feedstock, the organic solution provides beneficial elements including proteins, minerals, carbohydrates, hormones, free amino acids, and enzymes (de Jong et al., 2012). Anaerobic digestion produces lactic acid and its constituents, along with bioethanol, amino acids, energy, and proteins, which are the most valuable finished products from the organic solution. The organic (liquid and solid) wastes may be easily converted into biogas, which can then be utilized to generate heat and energy.

1.7.3 ENERGY/FUEL

Anaerobic digestion of manure, biosolids, and waste stream (food processing industries) leads to the production of biogas. Methane is produced as a byproduct of anaerobic digestion, which is normally cleaned and extensively applied for its energy value (Chynoweth et al., 2001; Goswami et al., 2021a). It is possible to incorporate biogas

TABLE 1.2
Different Types of Lignocellulosic Agricultural Byproducts

Waste	Pretreatment	Methane yield (mLCH$_4$/g Volatile Solids)	Biodegradability (in Volatile Solids)	References
Grass silage (GS)	Pretreatments:100°C and 1%, 2.5%, 5% and 7.5% loading rates (NaOH) by volatile solids mass in GS	359.5 / 401.8 / 449.5 / 452.5	76.8% / 85.2% / 95.3% / 96.8%	(Xie et al., 2011)
Wheat Straw	Pretreatment: Temperature: 160–200°C Time (min):10–20	Raw: 276 Pretreated: 331 (at 180°C for 15 min)	Maximum volatile solids removal efficiency: 46.3% at 200°C for 10 min.	(Bauer et al., 2009)
Switchgrass (summer and winter harvested)	Pretreatment: Grinding +Alkalinization +Autoclaving.	298 (summer harvested) 140 (winter harvested)	Not available	(Frigon et al., 2012)
Peanut hull	Mechanical (grinding into less than 20 mm) + thermal (at 80°C for 70 min) + 3 g NaOH/100 g TS (for 24 h, at 55°C).	Raw: 112 Pretreated:182	Raw: 26% Pretreated: 47%	(Dahunsi et al., 2017)
Salix	Pretreatment: Temperature:170–230°C Time(min):5–15	Raw Salix: 161 Pretreated Salix: 234 (at 230°C for 10min)	Not available	(Estevez et al., 2012)
Grass (Pennisetum hybrid) Hay	Pretreatment (Grass): 30 min (water vapor) Pretreatment (Hay):175°C for 10 min.	Grass(raw): 190 Hay (raw): 236 Grass (pretreated): 198 Hay (pretreated): 281	Not available	(Rusanowska et al., 2018)
Sunflower stalks	Thermochemical pretreatments (H$_2$O$_2$, NaOH, HCl, Ca(OH)$_2$, and FeCl$_3$) and thermal (55 and 17°C)	Pretreated sunflower stalks: 259±6 at 55°C (4% NaOH, 24 h) sunflower stalks (raw): 192	Not available	(Monlau et al., 2012)
Wheat straw (WS) Ensiled sorghum forage (ESF)	Pretreatments (thermal/thermo-alkaline):100°C, and 160°C for 30 min	ESF (raw): 269, raw WS: 204 pretreated ESF: (361, 10% NaOH, 100°C), pretreated WS: (341, 10% NaOH, 100°C)	WS: 10% and Pretreated (ESF): 84-85%	(Sambusiti et al., 2013)

production into long-term renewable biochemicals and biofuels production strategies since it can provide value to wet streams. By maximizing methane production and the cost efficiency of biogas generation as well as deducing nutritional content from digestion streams, this value can be raised.

Numerous fuels including biohydrogen, biogas, biobutanol, and bioethanol are involved in the formation of energy (Srirangan et al., 2012; Goswami et al., 2021b). Energy production through consecutive methane and ethanol production has been shown to be greater than straight biogas production (Baskar et al., 2012). Furthermore biogas, biohydrogen, and bioethanol include two to three times the amount of energy that is required to generate ethanol from a conventional source. Likewise, the coproduction of several fuels from one particular feedstock also boosts overall energy efficiency and process economy. Apart from integrating two or even more processes, it has been observed that integrating bioprocesses with thermal processes results in increased energy output.

1.7.4 INDUSTRIAL PRODUCTS

The paper and pulp industries are regarded as being one of the principal producers of environmental resources and traditional fuels. The water consumption per ton of paper is estimated to be between 300–2600 m^3 while the resulting sludge had a moisture level of 70% (Gottumukkala et al., 2016). Furthermore, up to 50 kg of sludge is created for every ton of paper, calculated on a dry weight basis. Sludge contains a high proportion of carbohydrates, making it a prospective supplier of cellulose-based fuels and value-added substances in industrial waste biorefineries. There are many benefits to the use of paper sludge which comprises little or no preprocessing of cellulose fibers by extensive pulping, resulting in cellulose fibers being removed from lignin by chemical or biological catalysts (Gottumukkala et al., 2016).

Cassava is the third biggest source of carbohydrate for human utilization. Zhang et al. (2016) studied the possibilities for industrial waste derived from cassava using the waste biorefinery approach. With 720×1012 kg/day, cassava became one of the most affordable sources of dietary carbohydrate energy, ranking fourth behind sugarcane, maize, and rice. 60% of cassava is eaten as flour or fermented goods, 33% as animal feed, and 7% in the paper, textile, fermentation, and food sectors. Additionally, due to its low price and widespread accessibility, cassava has emerged as a critical source of carbohydrate for the manufacturing of biobased compounds and bioethanol. In China, for instance, cassava-based alcohol firms create more than 400,000 tons of ethanol annually. As a result, large amounts of waste are created from cassava sectors (Zhang et al., 2016).

The citrus manufacturing industry develops millions of tons of solid waste, including pulp, peel, and water-containing seeds of 75–80 wt percent. These wastes are composed of soluble sugars, structural polysaccharides, and hemicelluloses, which are composed of lignin-derived substances including essential oils and flavonoids. Due to the high concentration of these major elements in citrus processing wastes (CPW), extensive studies have been undertaken for converting

CPW to energy, bioethanol, fermentable sugars, and value products via acid and enzymatic hydrolysis (Duan et al., 2020).

1.7.5 ALGAE-BASED PRODUCTS

Microalgae are widely involved in the manufacturing of numerous fuels and animal feeds because they contain polyunsaturated fatty acids (PUFAs), carotenoid pigments, polyamines, and polysaccharides (Rathore et al., 2016; Goswami et al., 2022b). Microalgae's significant lipid content makes them an excellent source of biogas, bioethanol, and biodiesel. In some algal biorefineries, however, the fundamental limitation is the lack of freshwater as a large volume of water is required for growth. As a result, the utilization of naturally produced microalgae in marine water appears to be a viable alternative, particularly in coastal regions of developing countries with access to seawater (Nizami et al., 2016).

To circumvent the difficulty of providing high water requirements for algae growth in algal-based biorefineries, wastewater may be utilized as a growing medium. The microalgae utilize the generated carbon dioxide as both a carbon and energy resource for their development and minimize CO_2 emissions from the atmosphere, a process called biofixation. Integrating wastewater and biofixation not just reduces total production costs, but is also considered to be among the most efficient ways of minimizing pollution from agricultural waste. Hadiyanto and Hendroko (2014) have designed a new pilot plant in Indonesia depending on the integrated POME (palm oil mill effluent) biogas microalgae concept. Owing to its high COD concentration, POME is considered to be one of the most significant contributors to water contamination. Every year, Indonesia generates over 16 million tons of crude palm oil, making it the country's leading producer. As such, the country faces a significant risk of POME contamination. According to reports, the pilot plant will be built soon and will be capable of generating 1 MW of electricity from the biogas it produces (Hadiyanto & Hendroko, 2014).

1.7.6 PHARMACEUTICAL PRODUCTS

Many methodologies have been applied in recent decades for converting biomass into high-esteem items. The identification of potential mechanisms for the conversion of biomass into medicinal compounds, on the other hand, has received relatively little attention. The first phase is to research possible and viable medicinal products from palm-based biomass, in order to establish a comprehensive approach for selecting the optimal conversion route. The palm-based biomass consists mostly of hemicellulose, cellulose, and lignin base molecules, which are constituents of diverse C, H_2, and O_2 compounds and structures. Paracetamol, aspirin, vitamin B_{12}, and entacapone are some of the pharmaceutical compounds that are manufactured in this multifunctional biorefinery (Ng et al., 2017). Figure 1.2 depicts the synthesized design, which reflects the preferred processing pathways as well as the ultimate products produced by the system.

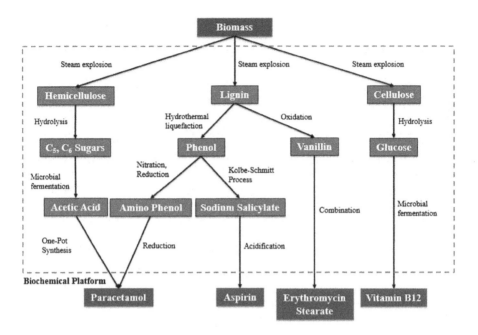

FIGURE 1.2   General flow diagram of pharmaceutical products produced from biomass.

1.8   CONCLUSION

Biorefineries are a sort of integrative thinking that focuses on producing as many high-esteem items as factually and financially conceivable. It is a promising technique to mitigate climate change while not damaging food security. The importance and potential of biomass waste employing a biorefinery concept are highlighted in this chapter. The variety and availability of biomass waste compositions provide prospects for the development of useful chemical platforms and value products. These items are recoverable because they have applications in the synthetic, drug, corrective, and food enterprises. As a result, the purpose of this review is to go over the most important enzymes and their applications in biotechnological processes.

ACKNOWLEDGMENTS

The authors express their gratitude to the Director, National Institute of Technology Agartala for his encouragement, and support.

REFERENCES

Abels, C., Carstensen, F., Wessling, M. 2013. Membrane processes in biorefinery applications. *Journal of Membrane Science*, **444**, 285–317.

Abraham, A., Mathew, A.K., Sindhu, R., Pandey, A., Binod, P. 2016. Potential of rice straw for bio-refining: An overview. *Bioresource Technology*, **215**, 29–36.

Agamuthu, P. 2009. Challenges and opportunities in agro-waste management: An Asian perspective. *Inaugural meeting of first regional 3R forum in Asia*. pp. 11–12.

Agrawal, K., Verma, P., 2021. Fungal metabolites: A recent trend and its potential biotechnological applications. In: Singh, J., Gehlot, P. (eds) *ew and future developments in microbial biotechnology and bioengineering Recent Advances in Application of Fungi and Fungal Metabolites: Current Aspects*. Elsevier, USA, pp. 1–14.

Alam, A., Agrawal, K., Verma, P., 2021. Fungi and its by-products in food industry: An unexplored area. In: Arora, P., (eds) *Microbial products for health, environment and agriculture*. Springer, Singapore, pp. 103–120.

Alexander, P., Rounsevell, M.D., Dislich, C., Dodson, J.R., Engström, K., Moran, D. 2015. Drivers for global agricultural land use change: The nexus of diet, population, yield and bioenergy. *Global Environmental Change*, **35**, 138–147.

Alibardi, L., Astrup, T.F., Asunis, F., Clarke, W.P., De Gioannis, G., Dessì, P., Lens, P.N., Lavagnolo, M.C., Lombardi, L., Muntoni, A. 2020. Organic waste biorefineries: Looking towards implementation. *Waste Management*, **114**, 274–286.

Alkan, H., Baysal, Z., Uyar, F., Dogru, M. 2007. Production of lipase by a newly isolated Bacillus coagulans under solid-state fermentation using melon wastes. *Applied Biochemistry and Biotechnology*, **136**(2), 183–192.

Anto, H., Trivedi, U., Patel, K. 2006. Glucoamylase production by solid-state fermentation using rice flake manufacturing waste products as substrate. *Bioresource Technology*, **97**(10), 1161–1166.

Ao, T., Luo, Y., Chen, Y., Cao, Q., Liu, X., Li, D. 2020. Towards zero waste: A valorization route of washing separation and liquid hot water consecutive pretreatment to achieve solid vinasse based biorefinery. *Journal of Cleaner Production*, **248**, 119253.

Arbige, M.V., Shetty, J.K., Chotani, G.K. 2019. Industrial enzymology: The next chapter. *Trends in Biotechnology*, **37**(12), 1355–1366.

Astrup, T.F., Pivnenko, K., Eriksen, M.K., Boldrin, A. 2018. Life cycle assessment of waste management: Are we addressing the key challenges ahead of us? *Journal of Industrial Ecology*, **22**(5), 1000–1004.

Asunis, F., De Gioannis, G., Isipato, M., Muntoni, A., Polettini, A., Pomi, R., Rossi, A., Spiga, D. 2019. Control of fermentation duration and pH to orient biochemicals and biofuels production from cheese whey. *Bioresource Technology*, **289**, 121722.

Banerjee, G., Scott-Craig, J.S., Walton, J.D. 2010. Improving enzymes for biomass conversion: A basic research perspective. *Bioenergy Research*, **3**(1), 82–92.

Baskar, C., Baskar, S., Dhillon, R.S. 2012. *Biomass conversion: The interface of biotechnology, chemistry and materials science*. Springer Science & Business Media, New York.

Basso, A., Serban, S. 2019. Industrial applications of immobilized enzymes—A review. *Molecular Catalysis*, **479**, 110607.

Battista, F., Frison, N., Pavan, P., Cavinato, C., Gottardo, M., Fatone, F., Eusebi, A.L., Majone, M., Zeppilli, M., Valentino, F. 2020. Food wastes and sewage sludge as feedstock for an urban biorefinery producing biofuels and added-value bioproducts. *Journal of Chemical Technology & Biotechnology*, **95**(2), 328–338.

Bauer, A., Bösch, P., Friedl, A., Amon, T. 2009. Analysis of methane potentials of steam-exploded wheat straw and estimation of energy yields of combined ethanol and methane production. *Journal of Biotechnology*, **142**(1), 50–55.

Bhardwaj, N., Agrawal, K., Verma, P. 2021b. Xylanases: An overview of its diverse function in the field of biorefinery. In: Srivastava, M., Srivastava, N., Singh, R., (eds) *Bioenergy research: Commercial opportunities & challenges*. Springer, Singapore, pp. 295–317.

Bhardwaj, N., Kumar, B., Agrawal, K., Verma, P. 2020. Bioconversion of rice straw by synergistic effect of in-house produced ligno-hemicellulolytic enzymes for enhanced bioethanol production. *Bioresource Technology Reports*, **10**, 100352.

Bhardwaj, N., Kumar, B., Agrawal, K. & Verma, P., (2021a). Current perspective on production and applications of microbial cellulases: A review. *Bioresources and Bioprocessing*, **8**(1), 1–34.

Bisinella, V., Götze, R., Conradsen, K., Damgaard, A., Christensen, T.H., Astrup, T.F. 2017. Importance of waste composition for Life Cycle Assessment of waste management solutions. *Journal of Cleaner Production*, **164**, 1180–1191.

Budzianowski, W.M., Postawa, K. 2016. Total chain integration of sustainable biorefinery systems. *Applied Energy*, **184**, 1432–1446.

Cai, J., Zhang, L. 2005. Rapid dissolution of cellulose in LiOH/urea and NaOH/urea aqueous solutions. *Macromolecular Bioscience*, **5**(6), 539–548.

Caldeira, C., Vlysidis, A., Fiore, G., De Laurentiis, V., Vignali, G., Sala, S. 2020. Sustainability of food waste biorefinery: A review on valorization pathways, techno-economic constraints, and environmental assessment. *Bioresource Technology*, **312**, 123575.

Cattle, S.R., Robinson, C., Whatmuff, M. 2020. The character and distribution of physical contaminants found in soil previously treated with mixed waste organic outputs and garden waste compost. *Waste Management*, **101**, 94–105.

Cekmecelioglu, D., Uncu, O.N. 2013. Kinetic modeling of enzymatic hydrolysis of pretreated kitchen wastes for enhancing bioethanol production. *Waste Management*, **33**(3), 735–739.

Chynoweth, D.P., Owens, J.M., Legrand, R. 2001. Renewable methane from anaerobic digestion of biomass. *Renewable Energy*, **22**(1–3), 1–8.

Colen, G., Junqueira, R.G., Moraes-Santos, T. 2006. Isolation and screening of alkaline lipase-producing fungi from Brazilian savanna soil. *World Journal of Microbiology and Biotechnology*, **22**(8), 881–885.

Contesini, F.J., Lopes, D.B., Macedo, G.A., da Graça Nascimento, M., de Oliveira Carvalho, P. 2010. *Aspergillus* sp. lipase: Potential biocatalyst for industrial use. *Journal of Molecular Catalysis B: Enzymatic*, **67**(3–4), 163–171.

Cristóbal, J., Caldeira, C., Corrado, S., Sala, S. 2018. Techno-economic and profitability analysis of food waste biorefineries at European level. *Bioresources Technology*, **259**, 244–252.

da Costa Sousa, L., Jin, M., Chundawat, S.P., Bokade, V., Tang, X., Azarpira, A., Lu, F., Avci, U., Humpula, J., Uppugundla, N. 2016. Next-generation ammonia pretreatment enhances cellulosic biofuel production. *Energy & Environmental Science*, **9**(4), 1215–1223.

Dahunsi, S., Oranusi, S., Efeovbokhan, V.E. 2017. Optimization of pretreatment, process performance, mass and energy balance in the anaerobic digestion of Arachis hypogaea (Peanut) hull. *Energy Conversion and Management*, **139**, 260–275.

de Jong, E., Higson, A., Walsh, P., Wellisch, M. 2012. Bio-based chemicals value added products from biorefineries. *IEA Bioenergy, Task42 Biorefinery*, **34**, 1–33.

Duan, Y., Pandey, A., Zhang, Z., Awasthi, M.K., Bhatia, S.K., Taherzadeh, M.J. 2020. Organic solid waste biorefinery: Sustainable strategy for emerging circular bioeconomy in China. *Industrial Crops and Products*, **153**, 112–568.

Estevez, M.M., Linjordet, R., Morken, J. 2012. Effects of steam explosion and co-digestion in the methane production from Salix by mesophilic batch assays. *Bioresource Technology*, **104**, 749–756.

Frigon, J.-C., Mehta, P., Guiot, S.R. 2012. Impact of mechanical, chemical and enzymatic pretreatments on the methane yield from the anaerobic digestion of switchgrass. *Biomass and Bioenergy*, **36**, 1–11.

Galanopoulos, C., Giuliano, A., Barletta, D., Zondervan, E. 2020. An integrated methodology for the economic and environmental assessment of a biorefinery supply chain. *Chemical Engineering Research and Design*, **160**, 199–215.

Giese, E.C., Dekker, R.F., Barbosa, A.M. 2008. Orange bagasse as substrate for the production of pectinase and laccase by *Botryosphaeria rhodina* MAMB-05 in submerged and solid state fermentation. *BioResources*, **3**(2), 335–345.

Gírio, F.M., Fonseca, C., Carvalheiro, F., Duarte, L.C., Marques, S., Bogel-Łukasik, R. 2010. Hemicelluloses for fuel ethanol: A review. *Bioresource Technology*, **101**(13), 4775–4800.

Gomes, D., Rodrigues, A.C., Domingues, L., Gama, M. 2015. Cellulase recycling in biorefineries—is it possible? *Applied Microbiology and Biotechnology*, **99**(10), 4131–4143.

Goswami, R.K., Agrawal, K., Verma, P. 2021a. Microalgae-based biofuel-integrated biorefinery approach as sustainable feedstock for resolving energy crisis. Springer, pp. 267–293.

Goswami, R.K., Agrawal, K., Verma, P. 2021b. Microalgae *Dunaliella* as biofuel feedstock and β-carotene production: An influential step towards environmental sustainability. *Energy Conversion and Management: X*, **13**, 100154.

Goswami, R.K., Mehariya, S., Karthikeyan, O.P., Gupta, V., Verma, P. 2022a. Multifaceted application of microalgal biomass integrated with carbon dioxide reduction and wastewater remediation: A flexible concept for sustainable environment. *Journal of Cleaner Production*, **339**, 130654.

Goswami, R.K., Mehariya, S., Karthikeyan, O.P., Verma, P. 2022b. Influence of carbon sources on biomass and biomolecule accumulation in *Picochlorum* sp. Cultured under the mixotrophic condition. *International Journal of Environmental Research and Public Health*, **19**(6), 3674.

Gottumukkala, L.D., Haigh, K., Collard, F.-X., Van Rensburg, E., Görgens, J. 2016. Opportunities and prospects of biorefinery-based valorization of pulp and paper sludge. *Bioresource Technology*, **215**, 37–49.

Gupta, R.K., Prasad, D., Sathesh, J., Naidu, R.B., Kamini, N.R., Palanivel, S., Gowthaman, M.K. 2012. Scale-up of an alkaline protease from *Bacillus pumilus* MTCC 7514 utilizing fish meal as a sole source of nutrients. *Journal of Microbiology and Biotechnology*, **22**(9), 1230–1236.

Hadiyanto, H., Hendroko, R. 2014. Integrated biogas-microalgae from waste waters as the potential biorefinery sources in Indonesia. *Energy Procedia*, **47**, 143–148.

He, M., Sun, Y., Zou, D., Yuan, H., Zhu, B., Li, X., Pang, Y. 2012. Influence of temperature on hydrolysis acidification of food waste. *Procedia Environmental Sciences*, **16**, 85–94.

Heerd, D., Yegin, S., Tari, C., Fernandez-Lahore, M. 2012. Pectinase enzyme-complex production by *Aspergillus* spp. in solid-state fermentation: A comparative study. *Food and Bioproducts Processing*, **90**(2), 102–110.

Howard, R., Abotsi, E., Van Rensburg, E.J., Howard, S. 2003. Lignocellulose biotechnology: Issues of bioconversion and enzyme production. *African Journal of Biotechnology*, **2**(12), 602–619.

Jørgensen, H., Kristensen, J.B., Felby, C. 2007. Enzymatic conversion of lignocellulose into fermentable sugars: Challenges and opportunities. *Biofuels, Bioproducts and Biorefining*, **1**(2), 119–134.

Kapoor, M., Soam, S., Agrawal, R., Gupta, R.P., Tuli, D.K., Kumar, R. 2017. Pilot scale dilute acid pretreatment of rice straw and fermentable sugar recovery at high solid loadings. *Bioresource Technology*, **224**, 688–693.

Kashyap, D., Vohra, P., Chopra, S., Tewari, R. 2001. Applications of pectinases in the commercial sector: A review. *Bioresource Technology*, **77**(3), 215–227.

Khan, M., Nizami, A., Rehan, M., Ouda, O., Sultana, S., Ismail, I., Shahzad, K. 2017. Microbial electrolysis cells for hydrogen production and urban wastewater treatment: A case study of Saudi Arabia. *Applied Energy*, **185**, 410–420.

Khandeparkar, R., Bhosle, N. 2006. Isolation, purification and characterization of the xylanase produced by *Arthrobacter* sp. MTCC 5214 when grown in solid-state fermentation. *Enzyme and Microbial Technology*, **39**(4), 732–742.

Khosravi-Darani, K., Falahatpishe, H., Jalali, M. 2008. Alkaline protease production on date waste by an alkalophilic *Bacillus* sp. 2-5 isolated from soil. *African Journal of Biotechnology*, **7**(10).

Kiran, E.U., Trzcinski, A.P., Ng, W.J., Liu, Y. 2014. Enzyme production from food wastes using a biorefinery concept. *Waste and Biomass Valorization*, **5**(6), 903–917.

Kuhad, R.C., Gupta, R., Singh, A. 2011. Microbial cellulases and their industrial applications. *Enzyme Research*, **2011**, 1–10.

Kumar, B., Bhardwaj, N., Alam, A., Agrawal, K., Prasad, H., Verma, P. 2018, Production, purification and characterization of an acid/alkali and thermo tolerant cellulase from *Schizophyllum commune* NAIMCC-F-03379 and its application in hydrolysis of ligno-cellulosic wastes. *AMB Express*, **8**(1), 173: 1–16.

Kumar, B., Bhardwaj, N., Verma, P. 2020. Microwave assisted transition metal salt and orthophosphoric acid pretreatment systems: Generation of bioethanol and xylo-oligosaccharides. *Renewable Energy*, **158**, 574–584.

Kumar, B., Verma, P. 2020. Enzyme mediated multi-product process: A concept of bio-based refinery. *Industrial Crops and Products*, **154**, 112607.

Leu, S.-Y., Zhu, J. 2013. Substrate-related factors affecting enzymatic saccharification of lig-nocelluloses: Our recent understanding. *Bioenergy Research*, **6**(2), 405–415.

Levy, I., Shoseyov, O. 2002. Cellulose-binding domains: Biotechnological applications. *Biotechnology Advances*, **20**(3–4), 191–213.

Li, N.-W., Zong, M.-H., Wu, H. 2009. Highly efficient transformation of waste oil to bio-diesel by immobilized lipase from *Penicillium expansum*. *Process Biochemistry*, **44**(6), 685–688.

Lynd, L.R., Weimer, P.J., Van Zyl, W.H., Pretorius, I.S. 2002. Microbial cellulose utilization: Fundamentals and biotechnology. *Microbiology and Molecular Biology Reviews*, **66**(3), 506–577.

Ma, H., Wang, Q., Zhang, W., Xu, W., Zou, D. 2008. Optimization of the medium and pro-cess parameters for ethanol production from kitchen garbage by *Zymomonas mobilis*. *International Journal of Green Energy*, **5**(6), 480–490.

Maina, S., Kachrimanidou, V., Koutinas, A. 2017. A roadmap towards a circular and sustain-able bioeconomy through waste valorization. *Current Opinion in Green and Sustainable Chemistry*, **8**, 18–23.

Matharu, A.S., de Melo, E.M., Houghton, J.A. 2016. Opportunity for high value-added chemi-cals from food supply chain wastes. *Bioresource Technology*, **215**, 123–130.

Menon, V., Rao, M. 2012. Trends in bioconversion of lignocellulose: Biofuels, platform chemicals & biorefinery concept. *Progress in Energy and Combustion Science*, **38**(4), 522–550.

Miandad, R., Rehan, M., Nizami, A.-S., Barakat, M.A.E.-F., Ismail, I.M. 2016. The energy and value-added products from pyrolysis of waste plastics. in: Karthikeyan, O., Heimann, K., Muthu, S. (eds) *Recycling of solid waste for biofuels and bio-chemicals. Environmental footprints and eco-design of products and processes*, Springer, Singapore, pp. 333–355.

Mirabella, N., Castellani, V., Sala, S. 2014. Current options for the valorization of food manu-facturing waste: A review. *Journal of Cleaner Production*, **65**, 28–41.

Monlau, F., Barakat, A., Steyer, J.-P., Carrère, H. 2012. Comparison of seven types of thermo-chemical pretreatments on the structural features and anaerobic digestion of sunflower stalks. *Bioresource Technology*, **120**, 241–247.

Moscoviz, R., Trably, E., Bernet, N., Carrère, H. 2018. The environmental biorefinery: State-of-the-art on the production of hydrogen and value-added biomolecules in mixed-culture fermentation. *Green Chemistry*, **20**(14), 3159–3179.

Naik, S.N., Goud, V.V., Rout, P.K., Dalai, A.K. 2010. Production of first and second generation biofuels: A comprehensive review. *Renewable and Sustainable Energy Reviews*, **14**(2), 578–597.

Nair, L.G., Agrawal, K., Verma, P., 2022. An overview of sustainable approaches for bioenergy production from agro-industrial wastes. *Energy Nexus*, **6**, 100086.

Ng, H.S., Kee, P.E., Yim, H.S., Chen, P.-T., Wei, Y.-H., Lan, J.C.-W. 2020. Recent advances on the sustainable approaches for conversion and reutilization of food wastes to valuable bioproducts. *Bioresource Technology*, **302**, 122889.

Ng, S.Y., Ong, S.Y., Ng, Y.Y., Liew, A.H., Ng, D.K., Chemmangattuvalappil, N.G. 2017. Optimal design and synthesis of sustainable integrated biorefinery for pharmaceutical products from palm-based biomass. *Process Integration and Optimization for Sustainability*, **1**(2), 135–151.

Nikodinoska, N., Buonocore, E., Paletto, A., Franzese, P.P. 2017. Wood-based bioenergy value chain in mountain urban districts: An integrated environmental accounting framework. *Applied Energy*, **186**, 197–210.

Nizami, A.-S., Mohanakrishna, G., Mishra, U., Pant, D. 2016. Trends and sustainability criteria for the liquid biofuels. In: Singh, R.S., Pandey, A., Gnansounou, E. (eds) *Biofuels: Production and future perspectives*, CRC Press, New York, pp. 59–95.

Pan, J., Zhang, R., El-Mashad, H.M., Sun, H., Ying, Y. 2008. Effect of food to microorganism ratio on biohydrogen production from food waste via anaerobic fermentation. *International Journal of Hydrogen Energy*, **33**(23), 6968–6975.

Pandey, A., Nigam, P., Soccol, C.R., Soccol, V.T., Singh, D., Mohan, R. 2000. Advances in microbial amylases. *Biotechnology and Applied Biochemistry*, **31**(2), 135–152.

Pasupuleti, S.B., Srikanth, S., Mohan, S.V., Pant, D. 2015. Development of exoelectrogenic bioanode and study on feasibility of hydrogen production using abiotic VITO-CoRE™ and VITO-CASE™ electrodes in a single chamber microbial electrolysis cell (MEC) at low current densities. *Bioresource Technology*, **195**, 131–138.

Pedrolli, D.B., Monteiro, A.C., Gomes, E., Carmona, E.C. 2009. Pectin and pectinases: Production, characterization and industrial application of microbial pectinolytic enzymes. *Open Biotechnology Journal*, 9–18.

Poggi-Varaldo, H.M., Munoz-Paez, K.M., Escamilla-Alvarado, C., Robledo-Narváez, P.N., Ponce-Noyola, M.T., Calva-Calva, G., Ríos-Leal, E., Galíndez-Mayer, J., Estrada-Vázquez, C., Ortega-Clemente, A. 2014. Biohydrogen, biomethane and bioelectricity as crucial components of biorefinery of organic wastes: A review. *Waste Management & Research*, **32**(5), 353–365.

Rabemanolontsoa, H., Saka, S. 2016. Various pretreatments of lignocellulosics. *Bioresource Technology*, **199**, 83–91.

Ragazzi, M., Rada, E.C., Schiavon, M. 2020. Municipal solid waste management during the SARS-COV-2 outbreak and lockdown ease: Lessons from Italy. *Science of the Total Environment*, **745**, 141159.

Ramrakhiani, L., Chand, S. 2011. Recent progress on phospholipases: Different sources, assay methods, industrial potential and pathogenicity. *Applied Biochemistry and Biotechnology*, **164**(7), 991–1022.

Rao, M., Singh, S. 2004. Bioenergy conversion studies of organic fraction of MSW: Kinetic studies and gas yield–organic loading relationships for process optimization. *Bioresource Technology*, **95**(2), 173–185.

Rathore, D., Nizami, A.-S., Singh, A., Pant, D. 2016. Key issues in estimating energy and greenhouse gas savings of biofuels: Challenges and perspectives. *Biofuel Research Journal*, **3**(2), 380–393.

Roni, M.S., Thompson, D.N., Hartley, D.S. 2019. Distributed biomass supply chain cost optimization to evaluate multiple feedstocks for a biorefinery. *Applied Energy*, **254**, 113660.

Ruiz, H.A., Rodríguez-Jasso, R.M., Fernandes, B.D., Vicente, A.A., Teixeira, J.A. 2013. Hydrothermal processing, as an alternative for upgrading agriculture residues and marine biomass according to the biorefinery concept: A review. *Renewable and Sustainable Energy Reviews*, **21**, 35–51.

Rusanowska, P., Zieliński, M., Dudek, M., Dębowski, M. 2018. Mechanical pretreatment of lignocellulosic biomass for methane fermentation in innovative reactor with cage mixing system. *Journal of Ecological Engineering*, **19**(5), 219–224.

Russ, W., Schnappinger, M. 2007. Waste related to the food industry: A challenge in material loops. in: *Utilization of by-products and treatment of waste in the food industry*, Springer, New York, pp. 1–13.

Salapa, I., Katsimpouras, C., Topakas, E., Sidiras, D. 2017. Organosolv pretreatment of wheat straw for efficient ethanol production using various solvents. *Biomass and Bioenergy*, **100**, 10–16.

Sambusiti, C., Monlau, F., Ficara, E., Carrère, H., Malpei, F. 2013. A comparison of different pretreatments to increase methane production from two agricultural substrates. *Applied Energy*, **104**, 62–70.

Saxena, R., Davidson, W., Sheoran, A., Giri, B. 2003. Purification and characterization of an alkaline thermostable lipase from *Aspergillus carneus*. *Process Biochemistry*, **39**(2), 239–247.

Shahzad, K., Narodoslawsky, M., Sagir, M., Ali, N., Ali, S., Rashid, M.I., Ismail, I.M.I., Koller, M. 2017. Techno-economic feasibility of waste biorefinery: Using slaughtering waste streams as starting material for biopolyester production. *Waste Management*, **67**, 73–85.

Shannon, M.A., Bohn, P.W., Elimelech, M., Georgiadis, J.G., Marinas, B.J., Mayes, A.M. 2010. Science and technology for water purification in the coming decades. *Nanoscience and Technology: A Collection of Reviews from Nature Journals*, 337–346.

Srirangan, K., Akawi, L., Moo-Young, M., Chou, C.P. 2012. Towards sustainable production of clean energy carriers from biomass resources. *Applied Energy*, **100**, 172–186.

Sundarram, A., Murthy, T.P.K. 2014. α-amylase production and applications: A review. *Journal of Applied & Environmental Microbiology*, **2**(4), 166–175.

Tang, Y.-Q., Koike, Y., Liu, K., An, M.-Z., Morimura, S., Wu, X.-L., Kida, K. 2008. Ethanol production from kitchen waste using the flocculating yeast *Saccharomyces cerevisiae* strain KF-7. *Biomass and Bioenergy*, **32**(11), 1037–1045.

Task, I.B. 2016. Cascading of woody biomass: Definitions, policies and effects on international trade.

Tonini, D., Astrup, T. 2012. Life-cycle assessment of a waste refinery process for enzymatic treatment of municipal solid waste. *Waste Management*, **32**(1), 165–176.

Ubando, A.T., Felix, C.B., Chen, W.-H. 2020. Biorefineries in circular bioeconomy: A comprehensive review. *Bioresource Technology*, **299**, 122585.

Vaseghi, Z., Najafpour, G.D., Mohseni, S., Mahjoub, S. 2013. Production of active lipase by *Rhizopus oryzae* from sugarcane bagasse: Solid state fermentation in a tray bioreactor. *International Journal of Food Science & Technology*, **48**(2), 283–289.

Vavouraki, A.I., Volioti, V., Kornaros, M.E. 2014. Optimization of thermo-chemical pretreatment and enzymatic hydrolysis of kitchen wastes. *Waste Management*, **34**(1), 167–173.

Wang, Q., Wang, X., Wang, X., Ma, H. 2008. Glucoamylase production from food waste by *Aspergillus niger* under submerged fermentation. *Process Biochemistry*, **43**(3), 280–286.

Wang, Y., Li, J., Jin, Y., Luo, J., Cao, Y., Chen, M. 2018. Liquid-liquid extraction in a novel rotor-stator spinning disc extractor. *Separation and Purification Technology*, **207**, 158–165.

Weber, C.T., Trierweiler, L.F., Trierweiler, J.O. 2020. Food waste biorefinery advocating circular economy: Bioethanol and distilled beverage from sweet potato. *Journal of Cleaner Production*, **268**, 121788.

Worldometer. 2021. World Publication Clock: 7.9 Billion People (2021). *Worldometer*. Avaliable from https://www.worldometers.info/world-population

Xie, S., Frost, J., Lawlor, P.G., Wu, G., Zhan, X. 2011. Effects of thermo-chemical pretreatment of grass silage on methane production by anaerobic digestion. *Bioresource Technology*, **102**(19), 8748–8755.

Zhang, M., Xie, L., Yin, Z., Khanal, S.K., Zhou, Q. 2016. Biorefinery approach for cassava-based industrial wastes: Current status and opportunities. *Bioresource Technology*, **215**, 50–62.

Zhang, Y.-H.P., Himmel, M.E., Mielenz, J.R. 2006. Outlook for cellulase improvement: Screening and selection strategies. *Biotechnology Advances*, **24**(5), 452–481.

2 Valorization of Biowaste to Biowealth Using Cellulase Enzyme During Prehydrolysis Simultaneous Saccharification and Fermentation Process

Akansha Bhatia, Ankur Rajpal, Bhaskar Jyoti Deka, and A.A. Kazmi

Indian Institute of Technology Roorkee, Roorkee, India

Vinay Kumar Tyagi

National Institute of Hydrology, Roorkee, India

CONTENTS

DOI: 10.1201/9781003187721-2

2.1   INTRODUCTION

Biorefinery technology has been shown to be applicable in converting high recalcitrant lignocellulosic biomass into bioethanol. Agricultural biomass is an importnt
source that can be reused to produce biofuels. It is a renewable resource and it is
very feasible to collect and transport the agrowastes to the laboratory (Perlack et al.,
2005; Nair et al., 2022; Agrawal & Verma 2020). One deficiency during the processing of agrobiomass is its resistivity, and pretreatment must be undertaken to obtain
higher yields.

Pretreatment is used for the breakdown of recalcitrant structures formed of cellulose, hemicellulose, and lignin. Cellulose is the major ingredient of agro-industrial
residues. Hemicellulose and cellulose are found in the plant cell wall (Bhardwaj
et al., 2021). Hence, it is necessary to search for an inexpensive and effective method
to remove the cellulose and hemicellulose from agrowastes. Pretreatment processes
(physical, biological, and chemical) can be utilized to treat agrowastes. Physical and
chemical pretreatment methods include milling of feedstock, dilute acid treatment,
hydrothermal pretreatment, and steam explosion. Hydrothermal pretreatment is cost-
effective and ecofriendly, with very less few inhibitors.

SSF processes are inexpensive processes, at about 20% of the cost of ethanol
production from biomass (Hinman et al., 1992; Nguyen & Saddler, 1991). The main
concern for the process is the temperature optima that may be used for saccharification (50°C), and fermentation temperature ranges between 30 and 37°C (Abdel-
Banat et al., 2010; Krishna et al., 2001; Kumar et al., 2020). *Saccharomyces* strains
are used as microorganisms to produce ethanol from biomass. Generally, the
S.cerevisiae strains are resistant to lower pH values, high sugar content, and high titer
of ethanol in comparison to other bacterial or fungal strains, which reduces the contamination rate in the fermentation process (Nevoigt, 2008; Lan et al., 2013). The
metabolism and growth of the strains depend upon the various environmental conditions, like temperature, toxicants formed during the hydrolysis process, varied range
of pH, and availability of nutrients. Ethanol tolerance capacity and resistivity to
higher temperature range also need consideration (Banat et al., 1992). *S.cerevisiae*
strains are well known for high ethanol productivity at a temperature of more than
35°C (Banat et al., 1998).

OD can be used to estimate the cell physiological state of growth. An estimation
is made for the various dilutions of the suspension medium. Different parameters
include cell size and density, temperature, and media composition. OD values are
correlated to cell density. A linear relationship between viable cells and OD values
iss used to quantify the cell biomass of *Mycobacterium tuberculosis* (Bharti et al.,
2021) Biomass is a basic parameter to characterize the growth patterns of microbes.
Measurement of biomass in the SSF process is related to the problems of microbial
biomass separation from the substrate (Raimbault, 1998). The high density of the
recycled yeast can increase the ethanol yield during the SSF processes at high solids
loading. Various physicochemical parameters are responsible for the better efficiency
of the yeasts during fermentation.

Temperature is the basic parameter to be measured during the PSSF process. The
growth pattern and metabolism of the fermenting yeasts at different temperatures

depends upon the composition of the culture medium and related growth parameters. With regard to high-temperature stress, yeast cells can be damaged; this is attributed to the breakdown of hydrogen bonds and denaturation of nucleic acids (Walker, 1998). All living yeast cells respond to heat shock stress. In yeast, *Saccharomyces cerevisiae*, alteration in temperature generates changes in the protein synthesis (Miller et al., 1979). The upper-temperature range of growth for most of the strains of *S.cerevisiae* is 41–42°C. However, temporal sensitivity is observed at 35–38°C. Correspondence to heat shock and other related stress responses (e.g., tolerance to a high titer of self-produced ethanol, acetic acid, and inhibitors) need consideration.

Ethanol, a product during yeast fermentation, can act as a growth-inhibiting parameter for yeast cells. Toxicity due to self-produced ethanol and its tolerance among living fermentative yeast cells have been reviewed by Ingram and Buttke (1985), Casey and Ingledew (1986), Jones and Greenfield (1987), and Jones and Woods (1989). Intracellular ethanol concentration increases as fermentation progresses (D'Amore, T 1992), it is generally accepted that because ethanol diffuses very rapidly across the cell membrane it does not accumulate in the yeast cells (Guijarro and Lagunas, 1984). However, the fact that external ethanol is much less toxic than fermentatively derived ethanol needs to be considered (Jones, 1988). The interrelationship between high temperature and ethanol in affecting yeast physiology has been discussed at length by Slapack et al. (1987).

Nutrient availability is the other type of stress that yeast has to survive within the inhibiting environment during fermentation. *Saccharomyces* species can surpass the long-time shortage of nutrients. The fermenting yeast undergoes many ways to survive on fewer available nutrients (Werner-Washburne et al., 1993).

Besides the aforementioned stresses, the pretreatment process of lignocellulosic biomass generated inhibiting compounds retards the yeast cell's metabolism. Dehydration products inhibit cell growth and ethanol production, denature DNA and protein synthesis, and are major obstacles during the ethanol production process. (Allen et al., 2010; Banerjee et al., 1981; Modig et al., 2002). The properties and concentrations of the final inhibitors are influenced by pretreatment conditions. However, the challenges of using agrowaste-derived feedstock for producing bioethanol are the inconsistent physicochemical qualities of the feedstock, which can result in higher concentrations of degradation byproducts after pretreatment. The inhibition of yeast fermentation by these byproducts can result in low biomass of yeasts, reducing the efficiency of the biorefinery.

In the present work, the effects of the different stress parameters, that is:heat shock, osmotic stress, and nutrient starvation during the fermentation process were optimized. Culture media (YPD) fermentation at elevated culturing temperatures was carried out to develop the interrelationship between the yeast growth parameters such as optical density (OD), total cell biomass, and cell viability. These correlations are significant in estimating cell biomass in lignocellulosic biomass SSF processes. Ethanol tolerance assay was performed at various temperatures to characterize the tested strain's physiological restoration towards osmotic stress. The prehydrolysis simultaneous saccharification and fermentation (PSSF) process was studied.

2.2 MATERIALS AND METHODS

2.2.1 MATERIALS

Agricultural biowastes such as rice husk, wheat husk, and corn stover were collected from the nearby agricultural fields. Commercial cellulose enzyme CTec2 was generously provided by Himedia. The cellulase activity was 55–60 FPU/mL. Buffer (sodium acetate), H_2SO_4, sodium sulfite, hydroxymethylfurfural (HMF), yeast extract peptone dextrose (YPD) broth, and YPD agar was used as purchased from Sigma-Aldrich (St. Louis, MO) and Himedia.

2.2.2 PRETREATMENT

For the pretreatment, 150 g biomass was mixed with 1.5 L of water and 1% w/w H_2SO_4 aqueous solution and pretreated at the designated temperature of 200°C, respectively) for seven minutes. The digester was heated to 200 ± 1°C at a rate of 5°C per minute and maintained at the temperature for seven minutes. Separation of slurry and substrate was conducted after pretreatment. The solid substrate and liquor were tested for their characteristics.

2.2.3 YEAST CULTURE

S. cerevisiae strains were procured from the American Type Culture Collection, that is, ATCC 9763 (designated as A), ATCC 4098 (designated as B). Yeast cells were cultured at 28°C for 48 hours using the YPD agar plates (10 g/L yeast extract, 20 g/L peptone, and 20 g/L agar). Further, the yeast colony was inoculated in YPD broth to the culture at 35°C and 90 rpm for 48 hours. The biomass concentration was measured at 600 nm (OD_{600}) using a UV-Vis spectrophotometer. The cultured medium with an average optical density of 30 at 600 nm was used to inoculate during the SSF experiment.

2.2.4 PSSF EXPERIMENT

Prehydrolysis was conducted in a shaking incubator at 50°C and 200 rpm for altered periods. The yeast was that was activated was inoculated into the shaking flask after 36 hours of prehydrolysis. The temperature range was 28–35°C and the shaking speed was decreased to 130 rpm. Feedstock pH was maintained at pH 5.5–6.2 by using calcium hydroxide. The enzyme cellulase load used was 15 FPU/g-substrate.

Yeast extract peptone dextrose (YPD) media temperature was kept at 35°C for 16–24 h. A colony from the YPD agar plates for the yeast strains was transferred by using a loop into the hydrolysed broth (50 mL) for 72 hours at four different temperatures (30, 35, 37, and 40°C) The shaking speed of the incubator was 90 rpm.

The interrelationship was determined between three growth parameters, that is, OD, cell viability, and biomass. Statistically calculated equations were used to estimate the cell biomass based on cell viability measurements (colony forming units) during the PSSF experiment.

2.2.5 ETHANOL TOLERANCE ASSAY

Physiological characterization of the two strains was tested in terms of their ethanol tolerance efficiency. Various ethanol concentrations, that is, 2, 4, 6, 8, and 10 % were used to illustrate the assay using the modified agar well diffusion method. The measured amount of ethyl alcohol was mixed with a molten YPD agar medium. The culture agar plates were prepared using the aforementioned media. Wells for each agar plate were prepared using the sterile glass rod. 100μL of the respective yeast culture was inoculated in the wells. The plates were finally incubated at 30, 35, 37, and 40°C for 72 hours. The ethanol tolerance activity was calculated by measuring the diameter of the zone of tolerance in millimeters around the tested organism in a petriplate. Methylene blue staining technique was used to enhance the visibility of the zone of tolerance.

2.2.6 ANALYTICAL METHODS

The chemical composition of the pretreated lignocellulose and the liquor were analyzed using the NREL procedure (Dowe and McMillan, 2001). The lignocellulose sample was dried, ground, and screened to 20 mesh (0.9 mm) and hydrolyzed in two stages using sulfuric acid of 72% (v/v) at 30°C for 1 hour and 3.6% (v/v) at 121°C for 1 hour. Monosaccharides (glucose, mannose, xylose, galactose) were analyzed by high-performance anion-exchange chromatography (Shimadzu UFCL) with an RI detector using Bio-Rad Aminex HPX-87P column operated at 85°C (Table 2.1). Double distilled water at a flow rate of 0.6 mL/min was used as an eluent. 20 μL of sample volume was injected. Inhibitor such as HMF and concentration of ethanol were estimated by using the HPLC. Bio-Rad Aminex HPX-87H column was used. A diluted sulphuric acid solution of 5mM was used. 50 μL sample volume was injected. Klason lignin was measured gravimetrically.

Total cell biomass was estimated by measuring the dried weight of the pellet after the centrifugation of YPD broth samples at 3000 rpm. The weight was calculated in g/L.

TABLE 2.1
Chemical Composition of Pretreated Sample

Chemical Composition	Wood (%)	Solid Substrate (%)	Pretreated Liquor (mg/mL)
Glucan	42	48.4	15.2
Xylan	3.7	1.59	7.9
Galactan	4.6	0.58	5.5
Mannan	12	5.05	34.9
Acid soluble lignin	6	19	N.A.
K lignin	23	29	N.A.

Cell viability was determined using the serial dilution pour plate technique and counts were enumerated using a colony counter (Bhatia et al., 2013). Reductions in cell viability values were determined using the formula,

$$\text{Cell viability reduction}(\%) = \frac{CFUmax - CFUobs}{CFUmax} \times 100$$

2.2.7 Simultaneous Saccharification and Fermentation (SSF)

Each solid and liquid mixture contained 50 mL of pretreated biomass slurry diluted to 15% solids. The liquor loading was in the same ratio for the pretreated biomass prior to the separation of solids and liquid. pH was adjusted to 5.5 using solid calcium hydroxide and the pH was controlled by adding a sodium acetate buffer (0.1 M) of pH 5.5. The enzymatic hydrolysis was conducted for the biomass slurry using a commercial enzyme (CTec2) at 20 filter paper units (FPU) per gram glucan of the substrate. Liquefaction of solid substrate was observed within approximately five hours at 50°C and 200 rpm. The samples were then cooled to 35°C and the shaker speed was reduced to 90 rpm and inoculated with yeast culture. Samples of the fermentation broth were taken periodically for glucose, inhibitors ethanol analysis, and cell viability. No additional nutrients were added during the fermentation. The formulation of the fermentation broth is listed in Table 2.2.

Ethanol yield from cellulose ($Y_{Et/Cel}$) was calculated according to the equation:

$$Y_{Et/Cel} = \frac{CET\left(g/g\right)}{Ccelmax} \times 100$$

where C_{Et} is the concentration of ethanol produced in the process (g/L), and C_{cel} is the initial concentration of cellulose in the cultivation medium (g/L) (Paulová et al., 2014).

TABLE 2.2

The Formulation for the Pretreated Substrate Used for Simultaneous Saccharification and Fermentation (SSF)

Formulation	Units	Volume
Solids load	wt%	15
Substrate solids content	wt%	30
Liquid solid content	wt%	7
Cellulase dosage	FPU/g of glucan	15
Liquid load	wt%	10
Yeast volume	%	5

2.3   RESULTS AND DISCUSSION

2.3.1   Culture Media Fermentation

Yeast extract peptone dextrose (YPD) fermentation at different temperatures for the two strains was carried out to study their growth characteristics in terms of optical density (OD), biomass, cell viability, and ethanol yield. *Saccharomyces* species grow best at a temperature range of 25 to 35°C (Watson, 1987). Metabolic activity and growth of the yeast cells first increased with increased temperature and reduced with increasing temperature beyond the specific range (Thevelein, 1984). The selected elevated temporal ranges during the experiment significantly influenced both the biomass and ethanol concentrations. Yields began to drop with the time of fermentation attributing to a longer lag phase for the two strains. However, high resistivity and inhibiting effects were observed at 40°C, characterized by a significant reduction in the production of total cell biomass. It was observed that the optimum temperature for yeast growth is very narrow and monitoring of the metabolic activities of *S. cerevisiae* stress to overheat leads to decreased cell viability, reducing the production of total cell biomass (Figure 2.1).

Optical density (OD) is routinely used to quantify the total number of microbes in suspension. This procedure would be more useful if the number of viable cells can also be quantified. To overcome the limitation, the interrelationship between the cell biomass, OD, and cell viability was obtained. The assumption was that after removing the dissolved part of the medium, remained OD corresponds to the live OD by the presence of living cells only. Based on greater biomass yield, ATCC 9763 was chosen to study the growth characteristics at 35°C to estimate the relationship (Figure 2.1a). However, a linear relationship was found between the live cells OD and biomass or cell viability measured in CFU/mL (Figure 2.1b). Statistically, equations 2.1 and 2.2 were calculated.

$$OD = 0.0085 \ (\text{cell biomass}) + 0.3004 \qquad (2.1)$$

$$OD = 7E - 05 \ (\text{cell viability}) - 0.1326 \qquad (2.2)$$

These equations reveal that measured OD values can be used to estimate the cell biomass and viability during the fermentation process. But, considering the lignocellulosic biomass fermentation, OD measurements cannot be used due to the dark color and high concentration of solid particles. Therefore, it is proposed to use cell viability measurements to calculate the cell biomass during high solids SSF using the equation:

$$\text{Cell viability} \ (\text{CFU/L}) = 1.12 \times \text{biomass} \ (\text{mg/L}) + 12.247 \qquad (2.3)$$

This approach can be very useful for the continuous high solids fermentation processes for understanding the cell biomass growth characteristics in relation to the varied inhibitors. It can help increase ethanol production, especially in industries eliminating the major challenge of the fermentation process.

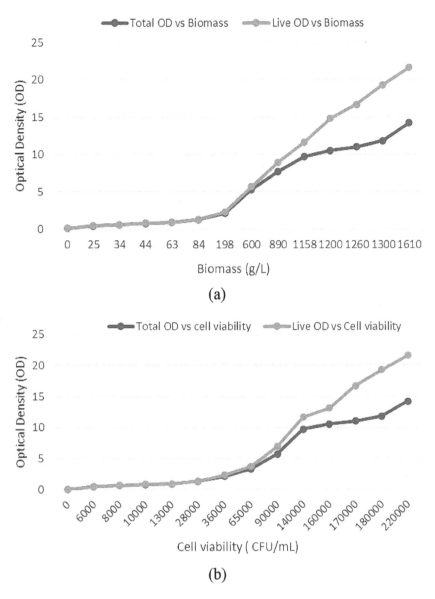

FIGURE 2.1 Relationships between OD (a) biomass and (b) cell viability during YPD fermentation at 35°C.

2.3.2 ETHANOL TOLERANCE ASSAY

Figure 2.2 shows the assay of ethanol tolerance for two strains and the reductions in their cell viability at different temperatures. At 30°C and 35°C, ATCC strains exhibited good growth for the various tested ethanol concentrations. Tolerance limits at these two temperatures can be arranged in the order: 2% > 4% > 6%> 8% > 10% of ethanol, in general, for all strains. However, for the higher temperatures, that is,

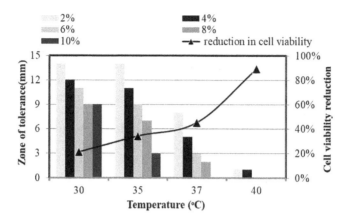

FIGURE 2.2 Ethanol tolerance assay and cell viability reduction using different ethanol concentrations at various temperatures.

37 and 40°C, the diameter for the zone of tolerance was reduced. Reduction in cell viability was also observed for the higher temporal conditions from 20–90% with increased temperatures. Synergistic inhibitory effects developed from the high temperature and ethanol may be attributed to the observance of these results. It was reported that high self-produced ethanol concentration affects the growth factors of yeast cells (Ma & Liu, 2010). Consequently, these factors inhibit the growth and reduce the viability of the yeast cultures. The assay investigation can be helpful to understand the mechanism of heat shock and osmotic stress responses relative to the exposed *Saccharomyces* strains.

2.3.3 SIMULTANEOUS SACCHARIFICATION AND FERMENTATION (SSF)

The liquid fraction of the pretreated feedstock in combination with solids was used for the PSSF process. During the hydrolysis period of 5 hours at 50°C, a 15% solid load of the slurry was analyzed for the release of glucose. The longer period of enzymatic saccharification resulted in the reduction of viscosity of the slurry and enhanced the ethanol yield (Leu and Zhu, 2013; Zhu et al., 2009; Zhou et al., 2013). The observations concur with the elongated prehydrolysis period for 24 hours (Hoyer et al., 2013) On the basis of temperature profile study using YPD medium for the two strains, 35°C was selected as the optimal and elevated fermentative temperature for studying the high solids SSF. Apparently, it was earlier reported that the fermentation at 35°C is a restrictive process for the non–thermophilic *Saccharomyces* strains (Torija et al., 2003). However, the tested yeast strains were forced to adapt at elevated temperatures to optimize their rate of metabolism. Each of the adapted two yeasts has different time-dependent sugar consumptions that can verify dissimilar titers of ethanol produced. ATCC 9763 produced the highest ethanol concentration of 20 g/L in 48 hours at the average glucose consumption rate of 0.27 g/L/h. Other strains produced an optimal amount of ethanol concentrations, that is, 15–18 g/L but the glucose consumption rate was lower. Comparison of the ethanol titers for the two strains, revealed that they produced higher ethanol concentrations of about 10–15% in the initial stages of the SSF process. Therefore, it was found that these two yeasts strains have the ability to adapt to the greater temperatures for the higher ethanol yields. The high titer of ethanol yield attributed to enhanced cell viability during 48–72 hours. Cell viability enumerations increased initially for all the strains and dropped after 72 hours (Figure 2.3), showing the least fall in the number of viable cells for the strain ATCC 9763 (Leu and Zhu, 2013).

The decrease in cell viability after 72 hours is due to the high concentration of self-produced ethanol (Nagodawithana & Steinkraus, 1976). The concentration of

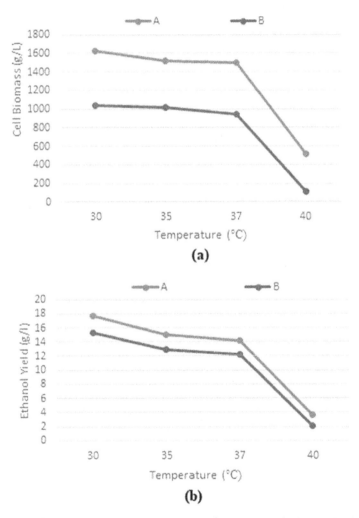

(a)

(b)

* A denotes ATCC 9763, B denotes ATCC 4098

FIGURE 2.3 (a) Biomass yield; (b) Ethanol titer for two strains during YPD fermentation at various temperatures for 72 h.

HMF added in the process using the pretreatment liquid became negligible within 48–72 hours showing the good ability of the strains to metabolize inhibitors. Microbial cells respond negatively to the pretreatment-produced inhibitors (Klinke et al., 2004) (This indicates that the biological conversion and detoxification of inhibitors are key aspects for adapting the cells in the toxic fermentation broth. Compared to the SSF fermentation efficiency, on the basis of maximal ethanol concentrations, ATCC 9763 was found to be the most efficient and ubiquitous, obtaining better capability to adapt to the growth-inhibiting environment.

2.4 CONCLUSIONS

This chapter focuses on the economic application of the agricultural waste-to-bioethanol biorefinery processes. Understanding the metabolic responses of yeast, that is, growth, fermentation, and decay) at elevated temperature is an essential task for process control to increase the productivity of biorefinery. With the increase of tolerance to ethanol at higher temperatures, the selected *S. cerevisiae sp.* can enhance the productivity of ethanol at a high titer (20 g/L), within a relatively short retention time (48 hours). In addition, the relationships among OD value, biomass, and CFU can provide additional information to evaluate the continuity of the SSF processes, in terms of estimating the cell biomass and decay rate.

REFERENCES

Abdel-Banat B M, Hoshida H, Ano A, Nonklang S, Akada R (2010) High-temperature fermentation: how can processes for ethanol production at high temperatures become superior to the traditional process using mesophilic yeast? *Applied Microbiology and Biotechnology* 85: 861–867.

Agrawal K, Verma P (2020). Laccase-mediated synthesis of bio-material using agro-residues. In: *Biotechnological Applications in Human Health*, Springer, Singapore, pp. 87–93.

Allen S A, Clark W, McCaffery J M, Cai Z, Lanctot A, Slininger P J, Liu Z L, Gorsich SW (2010). Furfural induces reactive oxygen species accumulation and cellular damage in *Saccharomyces cerevisiae*. *Biotechnology for Biofuels* 3: 21–24.

Banat I, Nigam P, Singh D, Marchant R, McHale A (1998). Review: Ethanol production at elevated temperatures and alcohol concentrations: Part I–Yeasts in general. *World Journal of Microbiology and Biotechnology* 14(6): 809–821.

Banat I M, Nigam P, Marchant R (1992) Isolation of thermotolerant, fermentative yeasts growing at 52 C and producing ethanol at 45°C and 50°C. *World Journal of Microbiology and Biotechnology* 8(3): 259–263.

Banerjee N, Bhatnagar R, Viswanathan L (1981). Inhibition of glycolysis by furfural in *Sacharomyces cerevisiae*. *European Journal of Applied Microbiology* 11: 226–228.

Bhardwaj N, Kumar B, Agrawal K,Verma P (2021). Current perspective on production and applications of microbial cellulases: a review. *Bioresources and Bioprocessing* 8(1): 1–34.

Bhatia A, Madan S, Sahoo J, Ali M, Pathania R, Kazmi AA (2013). Diversity of bacterial isolates during full scale rotary drum composting. *Waste Management* 33(7): 1595–1601.

Bharti S, Maurya R K, Venugopal U, Singh R, Akhtar M S, Krishnan M Y (2021). Rv1717 is a cell wall-associated β-galactosidase of *Mycobacterium tuberculosis* that is involved in biofilm dispersion. *Frontiers in Microbiology* 11: 611122.

Casey G P, Ingledew W M (1986). Ethanol tolerance in yeasts. *CRC Critical Reviews in Microbiology* 13(3): 219–280.

D'Amore T (1992) Cambridge prize lecture improving yeast fermentation performance. *Journal of the Institute of Brewing* 98(5): 375–382.

Dowe N, McMillan J (2001). SSF experimental protocols: Lignocellulosic biomass hydrolysis and fermentation. *National Renewable Energy Laboratory (NREL) Analytical Procedures.*

Guijarro J M, Lagunas R O (1984) Saccharomyces cerevisiae does not accumulate ethanol against a concentration gradient. *Journal of Bacteriology*, Dec 160(3): 874–878.

Hinman N, Schell D, Riley J, Bergeron P, Walter P (1992). Preliminary estimate of the cost of ethanol production for SSF technology. *Applied Biochemistry and Biotechnology* 34(1): 639–649.

Hoyer K, Galbe M, Zacchi G (2013). Influence of fiber degradation and concentration of fermentable sugars on simultaneous saccharification and fermentation of high-solids spruce slurry to ethanol. *Biotechnology for Biofuels* 6(1): 1–9.

Ingram L O N, Buttke T M (1985). Effects of alcohols on micro-organisms. *Advances in Microbial Physiology* 25: 253–300.

Jones D T, Woods, D R (1989). Solvent production. In: *Clostridia*. Springer, Boston, MA, pp. 105–144.

Jones R P (1988) Intracellular ethanol – accumulation and exit from yeast and other cells. *FEMS Microbiology Letters* 54(3): 239–258.

Jones R P, Greenfield P F (1987). Ethanol and the fluidity of the yeast plasma membrane. *Yeast* 3(4): 223–232.

Klinke H B, Thomsen A B, Ahring B K (2004). Inhibition of ethanol-producing yeast and bacteria by degradation products produced during pre-treatment of biomass. *Applied Microbiology and Biotechnology* 66(1): 10–26.

Krishna S H, Reddy T J, Chowdary G (2001). Simultaneous saccharification and fermentation of lignocellulosic wastes to ethanol using a thermotolerant yeast. *Bioresource Technology* 77(2), 193–196.

Kumar B, Bhardwaj N, Verma, P (2020). Microwave assisted transition metal salt and orthophosphoric acid pretreatment systems: Generation of bioethanol and xylo-oligosaccharides. *Renewable Energy* 158: 574–584.

Lan T Q, Gleisner R, Zhu J, Dien B S, Hector RE (2013). High titer ethanol production from SPORL-pretreated lodgepole pine by simultaneous enzymatic saccharification and combined fermentation. *Bioresource Technology* 127: 291–297.

Leu S Y, Zhu J (2013). Substrate-related factors affecting enzymatic saccharification of ligno-celluloses: Our recent understanding. *Bioenergy Research* 6(2): 405–415.

Ma M, Liu Z L (2010). Mechanisms of ethanol tolerance in *Saccharomyces cerevisiae*. *Applied Microbiology and Biotechnology* 87(3): 829–845.

Miller M J, Xuong N H, Geiduschek E P (1979). A response of protein synthesis to temperature shift in the yeast *Saccharomyces cerevisiae*. *Proceedings of the National Academy of Sciences of the United States of America* 76: 5222–5225.

Modig T, Liden G, Taherzadeh M J (2002). Inhibition effects of furfural on alcohol dehydrogenase, aldehyde dehydrogenase and pyruvate dehydrogenase. *Journal of Biochemistry* 363: 769–776.

Nagodawithana T W, Steinkraus K H (1976). Influence of the rate of ethanol production and accumulation on the viability of *Saccharomyces cerevisiae* in rapid fermentation. *Applied and Environmental Microbiology* 31(2): 158–162.

Nair L G, Agrawal K, Verma P (2022). An overview of sustainable approaches for bioenergy production from agro-industrial wastes. *Energy Nexus* 6(2022): 100086.

Nevoigt E (2008). Progress in metabolic engineering of Saccharomyces cerevisiae. *Microbiology and Molecular Biology Reviews* 72(3): 379–412.

Nguyen Q, Saddler J (1991). An integrated model for the technical and economic evaluation of an enzymatic biomass conversion process. *Bioresource Technology* 35(3): 275–282.

Paulová L, Patáková P, Rychtera M, Melzoch K (2014). High solid fed-batch SSF with delayed inoculation for improved production of bioethanol from wheat straw. *Fuel* 122: 294–300.

Perlack R D, Wright L L, Turhollow A F, Graham R L, Stokes B J, Erbach D C (2005. Biomass as feedstock for a bioenergy and bioproducts industry: The technical feasibility of a billion-ton annual supply. DTIC Document.

Raimbault, Maurice. (1998). General and microbiological aspects of solid substrate fermentation. *Electronic Journal of Biotechnology* 3: 26–27.

Slapack G E, Russell I,Stewart, G G. (1987) *Thermophilic microbes in ethanol production.* United States: N. p., Web.

Thevelein J M (1984) Regulation of trehalose mobilisation in fungi. *Microbiological Reviews* 48(1): 42–59.

Torija M J, Rozes N, Poblet M, Guillamón J M, Mas A (2003). Effects of fermentation temperature on the strain population of *Saccharomyces cerevisiae*. *International Journal of Food Microbiology* 80(1): 47–53.

Walker G M (1998). *Yeast Physiology and Biotechnology*. John Wiley & Sons. New Jersey, USA.

Watson K (1987) Temperature relations. In: *The Yeasts* (eds. A H Rose and J S Harrison), Academic Press, London. 2, pp. 41–72.

Werner-Washburne M, Braun E, Johnston G C, Singer R A (1993). Stationary phase in the yeast *Saccharomyces cerevisiae*. *Microbiological Reviews* 57(2): 383–401.

Zhou H, Zhu J, Luo X, Leu S Y, Wu X, Gleisner R, Dien B S, Hector R E, Yang D, Qiu X. (2013). Bioconversion of beetle-killed lodgepole pine using SPORL: Process scale-up design, lignin coproduct, and high solids fermentation without detoxification. *Industrial & Engineering Chemistry Research* 52(45): 16057–16065.

Zhu J, Pan X, Wang G, Gleisner R. 2009. Sulfite pretreatment (SPORL) for robust enzymatic saccharification of spruce and red pine. *Bioresource Technology* 100(8): 2411–2418.

3 Enzyme-associated Bioconversion of Agro-waste Materials via Macrofungi Cultivation for Sustainable Next-gen Ecosystems

Bhaskar Kalita and Sanjukta Patra

Indian Institute of Technology Guwahati, Guwahati, India

CONTENTS

3.1 INTRODUCTION

Microorganisms are fundamental for the survival of life on Earth. They assist in converting organic material from one form to another, thus sustaining and perpetuating the biogeochemical cycle that is most advantageous to people (Dhar & Shrivastava, 2012). Macrofungal diversity, especially community diversity, which makes up a large

DOI: 10.1201/9781003187721-3

portion of fungal diversity, is essential to world diversity. In ecology, macrofungi are categorized into three groups: saprophytes, parasites, and symbiotic (mycorrhizal) species (Kinge et al., 2017; Agrawal & Verma, 2021; Alam et al., 2021).

Humans began to adapt nature to fit their needs as early as the primitive age as agriculture evolved in various sites around the world. With the passage of time, humans have become more adept at creating crops and livestock tat best meet our needs. In many nation-states, modernization of farming operations, cutting-edge irrigation projects, and the use of agrochemicals have all made substantial contributions to agricultural production. The extension of soils for agricultural use, the technological contribution of the green revolution, which has influenced productivity, and the accelerated rise in world population have all contributed to an increase in agricultural production which has more than trebled in the last 50 years. Agricultural development is typically accompanied by wastes resulting from the indiscriminate use of intensive farming practices and the exploitation of chemicals used in cultivation, both of which substantially influence communities' ecosystems and the global environment in general.

Agricultural waste or agro-waste substrates are recalcitrant molecules containing abundant amounts of lignin, hemicellulose complex, cellulose, extractives, and non-extractive silica, carbonates, and oxalates. The robust and active extracellular enzymatic matrix of macrofungi may quickly decompose this agro-waste. As a futuristic sustainable method, agro-waste can be biotransformed into industrially relevant bioproducts via macrofungal cultivation, resulting in more significant socio-economic benefits. The ability to perform alchemy by converting agricultural and other organic wastes into nourishing and commercial products is a substantial advantage of macrofungal production (Dhar & Shrivastava, 2012; Agrawal & Verma, 2020a; Nair et al., 2022).

Enzymes spontaneously evolve over time, resulting in novel biocatalysts with more attributes that are extraordinary and a broader spectrum of biological applications. Due to a variety of factors such as: depletion of natural resources; less harm to the environment; greater efficiency; economic viability; and the production of high-quality products, enzymes have numerous applications in various industries including baking, and production of beverages, detergent, food, feed, leather, pharmaceuticals, and textiles. This chapter comprehensively portrays the agronomic and environmental benefits of organic waste recycling. Additionally, the importance of biorefinery employing agro-waste via macrofungal cultivations is highlighted in this study.

3.2 AGRO-WASTE AS A KEY TO SUSTAINABILITY

Globally, agriculture is considered as the economic foundation of rural societies. Rural communities all over the world mainly derive their livelihood from agriculture and its allied sectors. Most rural families in developing countries like India, Brazil, Argentina, Indonesia, Myanmar, and Bangladesh are directly or indirectly involved in agricultural crop production. With the growth in the rural economy, agrarian production has increased. The agricultural sector in these countries has experienced a significant leap in crop residue output as a result of considerable changes in cropping patterns, agrochemicals and fertilizers, modernization of farming operations, larger sharing of fertile land, new irrigation projects, availability of labor, and capital. All these components have contributed to an exponential increase in agricultural productivity and thus massive amounts of agro-waste generated throughout the years that is disposed of in inefficient and unscientific ways. According to the Food and Agriculture Organization Corporate Statistical Database (FAOSTAT) emissions database, more than a hundred million tonnes of leftover material are generated by 90.6% of rice produced in Asia itself. Asia is the largest producer of agro-waste contributing 47%, followed by the United States (29%), Europe (16%), Africa (6%), and Oceania (2%). Additionally, agro-waste is also responsible for the production of more than half of the nitrous oxide through decomposition on cultivated soils. It is also one of the major contributors of greenhouse gas and a driver of global warming (FAO 2021).

In India, 657 MT of gross agro-residues are produced, of which 85–100 MT are burned in the field (Devi et al., 2017; MNRE 2009; Sukumaran et al., 2017). With the rise in agriculture production, by 2030 India will reach crop residue production of about 868 MT (Pavlenko and Searle 2019). Figure 3.1 depicts the entire amount of unutilized residue accessible in India. Cereals, oilseeds, and sugarcane account for the majority of crop residue, with gross residue output totaling 566 MT. Fiber crops generate 68 MT of residues, followed by pulses, which generate 20 MT of gross residues. Cereals account for 359 MT (55%) of the crop group, followed by sugarcane (145 MT) (22%). For instance, in India, around 6 MT of straw is created for every 4 MT of wheat and rice grain harvested (Thakur 2003). Overall India produced 155.7 MT of rice and 86.87 MT of wheat in 2011, resulting in about 233.55 MT of rice straw and 130.3 MT of wheat straw as agricultural residue.

Agro-waste management has become a significant challenge, especially in emerging developing countries with rapidly growing populations, production rates, and economies. However, burning agricultural waste (cereal straw, wheat/paddy straw, maize stems, barley grass, and hay) releases various poisonous and hazardous oxides

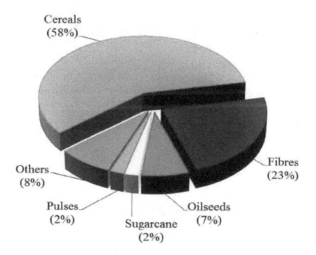

FIGURE 3.1   Surplus crop residue available in India (MNRE 2009).

FIGURE 3.2 Types of agro-waste: (a) Field or primary waste, (b) Processing or secondary waste generated in the agricultural field.

and hydro carbonates into the atmosphere, increasing air pollution and creating respiratory issues (Hyde et al., 2019).

Generally, agro-waste is generated from agriculture crops after harvesting, processing, or both processes. Figure 3.2 illustrates various types of agro residues generated from the agricultural field across the globe. Agro-wastes are chiefly classified into two categories: Field or primary waste and processing or secondary waste. Field or primary waste is those wastes which collect in the field after harvesting the crops: major agro products such as stems, roots, stalks, straw, leaves, seedpods, and so on Processing or secondary waste is produced during the subsequent processing of the crops, for example, husks, cobs, peel, pulp, and shells.

Macrofungi are some of the most efficient decomposers on the planet; thus, they provide a good chance of degrading agro-waste more successfully. Large amounts of degraded agro-waste can now be used as substrates for the large-scale production of scientifically, therapeutically, and pharmacologically active macrofungi, which could be a key strategy for sustainable ecosystems. Agricultural waste has traditionally been used for packing, animal feed, thatching for rural homes, biomass, domestic/industrial fuel, and composting. Agro-waste can also be biotransformed into mushroom-growing substrate, food and animal feed, next-generation product development, biologically active metabolites, enzymes, food flavouring compounds, and other inert substances, among other things. As a result, there is a pressing need to harness the potential of macrofungal species and find novel technologies to prevent their detrimental effects on ecosystems (Gowda & Manvi 2020; Satyendra et al., 2013).

3.3   MACROFUNGI: AN INCREDIBLE ENZYME FACTORY

Macrofungi are increasingly diversified organisms, and an indispensable part of ecosystems that can be found almost anywhere in the world, commonly growing on pasture lands, on forest soils, on tree trunks dead and alive, in association with the roots of forest trees, on dead organic matter, and on insect bodies. From an estimated 1.5–3 million species of fungi, about 150,000 species are macrofungi (mushrooms) taxonomically placed in two phyla, the *Basidiomycota* (class *agaricomycetes*) and *Ascomycota* (class *pezizomycetes*) in the subkingdom *Dikarya*. Approximately 10% (14,000–16,000) of them are scientifically well known, with about 7,000 edible species and 500 dangerous species among them. Ecologically, macrofungi can be classified as pathogens, saprobes, and mutualists (Molina 1993; Sadler 2004; Dhar & Shrivastava, 2012). Ectomycorrhizae are important for forest ecosystem food webs and spore dispersal as they produce reproductive fruiting bodies that emerge above or below ground. The ectomycorrhizal association aids both the fungus and its host, allowing them to withstand environmental stresses such as poor nutrition, drought, disease, extreme temperatures, and heavy metal toxicity. Ectomycorrhizae are macrofungi dominated by members of the *Basidiomycota*, some *Ascomycota*, and, rarely, the *Mucoromycota* species. The *Basidiomycetes* species are a very interesting group of fungi because of their exceptional ability to adjust to adverse environmental conditions where they continue to act as natural lignocellulose destroyers. They include a wide range of environmental groups, including white rot, brown rot, and leaf litter fungi (Hibbett & Taylor 2013; Rinaldi et al., 2016; Taylor & Alexander 2005; Tedersoo et al. 2010).

Macrofungi can be epigeous or hypogeous and flourish amid decaying organic materials, contributing to their distinctive fruiting bodies. Macrofungi have a two-part life cycle in respect of morphology: mycelium (a vegetative or colonisation phase) and fruiting body (a producing phase) (a reproductive phase that bears the spores). Mycelia secrete robust extracellular enzymes that could break down various organic substrates like cellulose and lignin during their vegetative growth phase. Microfungi (mushrooms) are the most important actors in lignocellulose degradation because they produce hydrolytic and oxidative enzymes. The breakdown of lignocellulosic materials is accomplished by integrating hydrolytic and oxidative enzymes. The hydrolytic enzyme system (cellulases and hemicellulases) is responsible for the breakdown of cellulose and hemicellulose, whereas the oxidative enzymes system (ligninases) is involved in the lignin modification and degradation. The ligninolytic system is an extracellular enzymatic complex that produces extracellular hydrogen peroxide through peroxidases, laccases, and oxidases (Bucher et al., 2004; Kumla et al., 2020; Stephen et al., 2005; Wood & Smith 1987; Agrawal et al., 2020; Agrawal et al., 2020a).

3.4 BIOCONVERSION: A FUNDAMENTAL CONCEPT OF BIOREFINERY SYSTEMS

A biorefinery system is an integrated biobased industry referring to biomass conversion processes that encompass a variety of technologies, large and integrated processing systems, and facilities. Additionally, it also provides a platform for several sectors to manufacture multiple bioproducts, thus increasing the economic value of coproduce. Producing multiple products such as chemicals, biofuels, food, industrial enzymes, feed ingredients, biomaterials (including fibers), heat and energy can take advantage of their various components/intermediates. Therefore, when focusing on maximizing the added value derived from agro-waste, it is important to considera the trilemmas of a sustainable ecosystem, circular economy, and next-generation societal aspirations (Biernat and Grzelak 2015; Demirbras, 2010; Kumar & Verma, 2020a).

3.4.1 DIFFERENT BIOCONVERSION PATHWAYS

There are different pathways for bioconversion that include:

i. Fractionation

Fractionation is the prime pathway toward refinery products which is included in biorefinery. Fractionation segregates the constituent components of wood or plant biomass samples (cellulose, hemicelluloses, and lignin). Specifically, steam explosions, aqueous separation, and hot water systems are all examples of fractionation processes (Demirbras, 2010).

ii. Thermochemical

The thermochemical biomass conversion pathway has multifaceted applications potential toward value-added fuels and chemicals from the primary renewable feedstock. For instance, pyrolysis, gasification, and liquefaction are three types of thermochemical conversion processes (Demirbras, 2010).

iii. Hydrolysis

Biochemical conversion occurs at lower temperatures and reaction rates, resulting in greater product selectivity. Ethanol production is a biochemical conversion method that uses biomass to generate energy. Many highly effective biochemical conversion systems have evolved to break down the molecules that make up biomass in nature. Hydrolysis is a conversion pathway which is frequently used to break down cellulose and hemicelluloses from lignocellulosic feedstocks (Adebayo & Martinez-Carrera 2015) into simple sugars. Certainly, fermentation is commonly employed in industry to convert substrates like glucose to ethanol for use in beverages, fuels, and chemicals, as well as other compounds (Demirbras, 2010).

3.4.2 CONCEPTS OF BIOREFINERIES

i. Lignocellulose Feedstock Biorefinery (LCB)

In LCB, straw and various types of wood are used in lignocellulosic feedstock, which chiefly comprises of low molecular extracts, cellulose (linear polymer of glucose units), hemicelluloses (linear and branched polymers of C5- and C6-sugars), and lignin (a three-dimensional molecule consisting of methoxyphenylpropane units). It is noteworthy that globally, the largest cellulose production occurs through separation technology which produces over 100 million MT of cellulose per year and is based on alkaline sodium sulphide or sulfite solutions, organic acids, organic solvents, and combinations of the above.

Recently, new technologies have involved the use of ionic liquids (ILs), which have the potential to offer various benefits due to their ability to dissolve all elements. The biomass type and the targeted outputs heavily influence the procedure and chemicals employed. However, LCB biorefinery technology is still in its early stages and requires certain technological inputs to overcome its shortcomings. Sustainable biomass supply chains, biomass pretreatment, fractionation, saccharification, and sugar and lignin conversion to fuels and chemicals are some of the pertaining issues that need to be addressed for a more sustainable ecosystem (Kamm & Kamm 2004; Singhvi & Gokhale 2015).

ii. Whole Crop Biorefinery (WCB)

Starch and derivatives are the main substrates of WCP. In this biorefinery dextrins and glucose, as well as starch derivatives that is, esters and ethers, have been employed in the food, textile, and paper industries. Additionally, bran is transformed into useful products such as xylose derivatives and ferulic acid in a cereal-based biorefinery. Cereals such as rye, wheat, triticale, and maize serve as raw materials for the WCP (Kamm & Kamm 2004).

iii. Green Biorefinery (GRB)

A GRB is a facility that converts green biomass into a variety of commercial products and energy. Green biomass includes natural and wet biomass, such as crops (e.g., immature cereal, grass, legumes) and agricultural wastes in the context of a green biorefinery (e.g., sugar beet leaves). These crops act as a natural chemical factory, including a wide range of fundamental components such as carbohydrates, proteins, lignin, and lipids, as well as vitamins, colors, and minerals. GRBs are multiproduct systems that function in accordance with the physiology of the relevant plant material in terms of refinery cuts, fractions, and products, preserving and utilizing the diversity of syntheses (Kamm & Kamm 2004).

iv. Sugar-based biorefinery (SBB)

SBB utilizes either sugar cane or sugar beet press juice as a substrate to the system. It can take advantage of all of the glucose-based production experience, which includes chemical changes such as hydrogenation. The hydrogenation process produces sorbitol, or the entire range of fermentation products, which include ethanol, amino acids, and higher-value compounds and medications. In this approach, the residual biomass (bagasse) and other waste streams are primarily used for energy production rather than as a source of products.

3.4.3 Bioconversion of Agro-wastes

Almost all crops produce a voluminous amount of byproducts in the form of stems and straw. Agricultural waste, such as corncobs, rice husks, rice straws, rice straws, sugarcane bagasse, and wheat straw, has accumulated in 2 billion tonnes worldwide. Forest trash makes up about 0.2 billion m^3; Municipal solid waste (MSW) makes up 1.7 billion MT, and industrial waste is 9.1 billion MT. Altogether, it generates more than 10 billion MT of garbage and residuals in total, which is a massive amount that continues to grow over time. According to the Indian Council of Agricultural Research's published data in 2020, every year, India produces over 350 million MT of agricultural waste, and around 1.3 billion MT of food goods for human service are wasted or lost globally, with residential kitchen garbage accounting for roughly one-third of biodegradable municipal solid waste.

However, a large portion of on-farm agro-waste, which farmers often classify as garbage, is burned in the field. Although this is a low-cost, low-labor-intensive, and simple method of agro-waste disposal, it has a significant detrimental influence on the agro-ecosystem and disrupts the soil's physical, chemical, and biological structure, as well as microbial life forms. Apart from burning, some farmers opt for direct application of agro-waste in the fields which has its drawbacks, including decreased crop yields due to microbial infestation, the production of phytotoxins and allelochemicals, immobilization of nutrients, and increased CH_4 emission, thereby contributing toward global warming. Therefore, the present situation necessitates the effective and judicious utilization of agro-waste to pave the way toward environmental sustainability.

Most agricultural waste is lignocellulosic in nature, rich in polysaccharides such as cellulose and hemicellulose, lignin (an aromatic polymer), and other nutrients like proteins and lipids, pectin, and polyphenols. These wastes can be converted into a variety of high-value products, including biofuels, fine value-added chemicals, and low-cost energy sources for microbial fermentation, using the biorefinery concept. Sustainable industries require a steady supply of cost-effective raw materials, such as agro-waste, to operate efficiently. Linking agro-waste sources from farmland to agro-based industries can help alleviate the waste accumulation problem, and increase waste valorization, and waste-to-wealth, leading to a zero-waste idea and a cleaner environment (Kumla et al., 2020).

3.5 MACROFUNGAL CULTIVATION: AN ENZYME-ASSOCIATED BIOREFINERY PROCESS

Macrofungi have been used for decomposition for millennia throughout the world. The organoleptic features of macrofungus, together with a desire for well-balanced meals, has resulted in increased macrofungus consumption. Macrofungi, which include members of the fungi kingdom's phylum *Basidiomycota* and phylum *Ascomycota*, exhibit morphologically different epigeous or hypogeous fruiting bodies and are collectively referred to as mushrooms (Chang and Wasser 2012; Dimitrijevic et al., 2018; Govorushko et al., 2019; Lu et al., 2020a; Lu et al., 2020b). In general, macrofungus production is acknowledged as a practical, effective, and ecologically sustainable means of utilizing agro-waste that provides small farmers with lucrative agribusiness options. Bioconversion of agro-waste is used as a cost-effective and long-term substrate for commercial macrofungi all over the world (Dhar & Shrivastava, 2012).

In Southeast Asia, for optimal use of agro-waste, many edible fungi have been domesticated and are in production, and the most commonly cultivated are Oyster mushrooms (*Pleurotus*), Ear mushrooms (*Auricularia*), Straw mushrooms (*Volvariella*), Shiitake (*Lentinula edodes*), Lentinus (e.g., *Lentinus squarrosulus*), *Ganoderma*, *Macrolepiota*, and *Agrocybe* (i.e., *Cyclocybe* spp) [29]. Likewise in 1926, an American farmer, Lewis Downing, found the first white *Agaricus bisporus* variant. Furthermore, the basic *A. bisporus* cultivation method was devised and developed in Europe by Sinden and Hauser (Sinden & Hauser, 1950), where fungi can be grown in normal compost, containing wheat straw, horse manure, and other agricultural waste. Among the different varieties of mushrooms, the oyster mushroom is a primary decomposer that may be grown on decumbens grass, sugarcane straw, brizantha grass, wheat straw, sugarcane bagasse, rubber sawdust, and rubber sawdust mixed with various supplements such as rice straw, rice husks, or corncobs in a bag-growing system. Mycelium, the fungi's vegetative component, may produce extracellular enzymes and acids capable of breaking down a wide range of organic substrates, including lignin and cellulose. (Bucher et al., 2004). Macrofungi break down organic compounds through several mechanisms, including hydrocarbon-assisted catabolism, and oxidative and reductive processes. Intracellular oxidation and oxygen integration are the first steps in the breakdown process, which are aided by the enzyme catalysts peroxidases, and glutathione peroxidases (Al-Hawash et al., 2018).

Macrofungi are a matter of interest since they can grow on low-cost materials and produce large amounts of enzymes in the culture medium, making downstream processing easier. Amylases, cellulases, lipases, phytases, proteases, and xylanases are among the commercially available fungal enzymes. Table 3.1 shows examples of major fungal enzymes and enzyme sources used in a variety of applications, however, only a few fungal strains reach the commercial production criteria. Enzyme biorefineries are frequently used to produce industrial enzymes, which are then refined in downstream processes. Heterologous expression, recombinant DNA technology, and gene cloning are some of the innovative and next-generation methods which can be utilized to increase the number and activity of enzymes without the difficulties of large-scale manufacture.

To support mycelial growth, macrofungi produce certain enzymes to break down the lignocellulosic materials and detoxify them (Zhang et al., 2019; Lakhtar et al., 2010; Adebayo & Martinez-Carrera, 2015). Cellulases, xylanases, and ligninases are only a few of the lignocellulolytic degrading enzymes produced by macrofungi. The key enzymes mainly involved in lignocellulose conversion are cellulases required in enzyme-based biorefinery. The hydrolysis of cellulose is mediated by the synergistic action of endo-β-(1,4)-glucanases (EGs), exo-β-(1,4)-glucanases, and β-D-glucosidases. Endo-β-(1,4)-glucanases or β-(1,4)-glucan-4-glucohydrolases (EC 3.2.1.4) hydrolyze soluble or insoluble β-(1,4)-glucan substrates. Exo-β-(1,4)-D-glucanases include both the β-(1,4)-D-glucan glucohydrolases (EC 3.2.1.74) and β-(1,4)-D-glucan cellobiohydrolases (CBHs) (EC 3.2.1.91) that release D-glucose and D-cellobiose from β-(1,4)-D-glucans respectively. The β-D-glucosidases or β-D-glucoside glucohydrolases (EC 3.2.1.21) liberate D-glucose from cellobiose (Himmel et al. 2007; Kiran et al., 2014; Bhardwaj et al., 2021a; 2021b).

It is worth noting that some edible and therapeutic macrofungi produce a full complement of enzymes capable of degrading native cellulose and lignin efficiently. Thus, in the waste recycling process, screening of such superior macrofungi lignocellulose-degrading strains has become crucial. Similarly, some hydrolytic enzymes can also be assisted in cellulose degradation and can be categorized as endoglucanases, cellobiohydrolases (exoglucanases), and glucosidases. In the process of degradation, oxidative enzymes such as lignin peroxidase, manganese-dependent peroxidase, and laccase primarily mediate lignin breakdown (Abu Yazid et al., 2017; Fen et al., 2014; Agrawal et al., 2020b).

TABLE 3.1

Major Fungal Enzymes and Enzyme Agro-waste Sources Used in a Variety of Next-generation Applications

Enzyme	Agro-waste Used	Macrofungal Species	Next-generation Technology & Applications	References
Cellulases	Wheat straw	*Lentinula edodes* *Pleurotus dryinus* *Pleurotus ostreatus* *Pleurotus tuber-regium* *Fomitopsis sp.*	Starch processing industry; food processing industry; plastic degradation; next-generation biofuel; textile, and laundry detergent industries;	Deswal et al. (2011); Elisashvili et al. (2003)
	Sorghum straw	*Pycnosporus sanguineus* *Pleurotus ostreatus* *Pleurotus eryngii* *Phanerochaete chrysosporium* *Trametes versicolor*	brewery and wine production	Montoya et al. (2012)
Endoglucanase	Rice straw	*Pleurotus ostreatus* *Lentinus sajor-caju*		Pandit and Maheshwari (2012)
Xylanase	Wheat straw, Tree leaves (*Fagus sylvatica*)	*Lentinula edodes* *Pleurotus dryinus* *Pleurotus ostreatus Pleurotus tuber-regium*	Agricultural silage and grain feed; food and beverage industries; feedstock improvement; Kraft pulp bleaching in poultry industry; de-inking of newspapers; agricultural waste degradation	Elisashvili et al. (2003; Molina (1993)
Laccase	Tree leaves (*Fagus sylvatica*), Wheat straw	*Lentinula edodes* *Pleurotus dryinus* *Pleurotus ostreatus Pleurotus tuber-regium*	Lignin removal; food and feed industries	Elisashvili et al. (2003)
Manganese peroxidase	Tree leaves (*Fagus sylvatica*), Wheat straw	*Lentinula edodes* *Pleurotus dryinus* *Pleurotus ostreatus Pleurotus tuber-regium*	Lignin and organic pollutant degradation systems; dye decolorization, pulp bleaching; biomechanical pulping	Elisashvili et al. (2003)
	Pineapple leaves Rice straw	*Ganoderma lucidum* *Schizophyllum commune*		Moilanen et al. (2015) Usha et al. (2014)

Enzyme	Agro-waste	Species	Application	References
Versatile peroxidase	Banana peel	*Pleurotus eryngii*	Industrial dye effluent treatment	Mehboob et al. (2011)
Amylases	Barley straw, Wheat straw	*Lentinula edodes*, *Hericium erinaceum*	Plastic degradation; next-generation biofuel; starch processing industry; food and dairy industry; textile industry; pulp and paper industry; detergent industry; pharmaceutical industry; animal feed industry	Du et al. (2013); Gaitán-Hernández et al. (2006)
Proteases		*Pleurotus citrinopileatus*	Multifunctional biomedical areas; detergent industry; food processing industry; pharmaceutical industry; leather processing; textile industry	(Inácio et al. 2015)
Pectinases, Glucose dehydrogenase	Oakwood, apple pomace, citrus peel, tomato peel, papaya peel, cucumber peel, and rice husks	*Pleurotus pulmonarius*	Food processing industry (functionality of foods such as prebiotics, low-calorie sweeteners, and rare sugars); Textile industry	(Akoh et al. 2008; Choi et al. 2015)
Ligninases	Coffee pulp, cotton seed hulls, corncobs	*Lentinula edodes*, *Pleurotus sajar-caju*	Lignin biodegradation	(Dudekula et al. 2020)
Endoglucanase, Cellbiohydrolase, β-Glucosidase	Banana leaves	Straw (*Volvariella*)	Food processing industry	(Dudekula et al. 2020)
Laccase, peroxidase, and carboxymethylcellulase	Water hyacinth/water lily, bean straw, cotton straw, cocoa shell waste, coir	*Pleurotus ostreatus*	Food processing industry	(Dudekula et al. 2020)

Prior to the conversion process, proper waste pretreatment technologies will significantly boost biorefining process efficiency and biomass conversion costs. Economic, biological, manufacturing ease, and market acceptance are all factors that play a role in the successful development of a newly evolved macrofungus. The ideal growing conditions must be selected after determining edibility and cultivability. Finding the right growing media, pH, temperature, C source, N source, and C: N ratio, as well as the optimum spawn, compost, and casing, are all crucial cultivation techniques. Exogenous additives such as surfactants, plant oils, organic solvents, fatty acids, phytohormones, and other precursors and chemical elicitors may stimulate mycelial growth and metabolite production during macrofungus farming (Bellettini et al., 2019; Postemsky et al., 2014; Xu et al., 2015).

3.5.1 AN INTEGRATED BIOREFINERY SYSTEM OF MACROFUNGAL CULTIVATION

In this context, biorefineries are being created as a technological means of moving to this economic approach, allowing waste to have added value through more sustainable production. Sustainable development aims to preserve the biosphere's integrity while maintaining economic, social, and industrial progress. Biorefineries convert biological materials (biomass) from live and dead organisms into food, fuels, energy, chemicals, and materials. The development of a biobased economy depends on the incorporation of sustainability into the architecture of biorefineries. In general, an integrative analysis that considers societal ramifications and goes beyond the automatic application of measurements to predetermined problems is required (Chiu et al., 2000; Dhar & Shrivastava, 2012).In Europe and Brazil, wheat straw is the most often utilized raw material for composting. Hay, corncobs, oats, barley, sugarcane bagasse, rye grass, rice straw, a range of other grasses, and discarded mushroom debris are examples of alternative substrates. Composting is required for the industrial production of some mushrooms, such as the *Agaricus* species. *Agaricus* mushroom cultivation differs from mushroom cultivation in a number of ways. *Pleurotus* is a genus of fast-growing oyster macrofungi (Kwon & Thatithatgoon, 2004) with a complete lignocellulolytic enzyme system that can colonize and degrade a wide range of waste, including paddy straw, rice husks, wheat straw, waste paper, corn cobs, sawdust, wood chips, tomato skins, banana leaves, pine needles, sugarcane, bagasse, cotton waste, cottonseed hull, date-palm leaves, chicken and horse manure, and sugarcane, bagasse, cotton waste, cotton waste, cottonseed hulls, olive cake, chicken and horse manure (Hyde et al., 2019)

When new enzyme and catalyst technology, densification techniques, and metabolic pathways are created, input materials processing will become more efficient and cost effective. In this arena, the development of more efficient and robust enzymes, particularly for the conversion of lignocellulosic material from a range of substrates for the cultivation of macrofungus is a significant phenomenon. The many enzymes available for different material processing conditions, as well as the development of novel enzymes to enhance the functionality and cost-effectiveness of the entire biorefinery conversion process. Prior to the biorefining conversion process, selecting appropriate waste pretreatment technologies will significantly improve the efficiency of the biorefining process and reduce biomass conversion costs.

Depending on the feedstock utilized as raw material, biorefineries are classified as first or second generation. Food crop resources such as maize, sugar, and vegetable oil are used in first generation (1G) biorefineries, whilst nonfood materials such as agricultural waste, wood, and energy crops with high lignocellulose content are processed in second generation (2G) biorefineries. Biorefineries that use algal biomass as a feedstock are known as third generation (3G) biorefineries. The 1G biorefinery is the most well established, whereas the 2G and 3G biorefineries are still in the early stages of development, due to technological and economic obstacles (Goswami et al., 2022, Goswami et al., 2021). Most second generation biorefinery plants will not be ready for large-scale commercial production for several years; therefore, this technology is still in its early stages.

A multiproduct biorefinery approach for macrofungal cultivation was explored in the study in order to fully exploit agro-waste, as represented in Figure 3.3. Macrofungal cultivation is a natural consequence of their ecological involvement in the bioconversion of organic waste into edible constituents that can be used as functional foods or as a source of drugs and pharmaceuticals (Wood & Smith 1987). The dynamism of a macrofungus' (mushroom's) mycelium and its ability to activate physiological systems required for appropriate medium utilization influence its opportunities to grow on a lignocellulosic substrate. In cultivating a mushroom on a solid substrate, the rate of hyphal colonization is the first and most critical stage. Mycelia secrete enzymes to break down substrate components like cellulose and lignin during their vegetative growth (Salmones and Mata 2015).

Nature uses enzymes to build and convert biomass; humans use the same enzymes and create them in massive quantities to get the most out of organic matter in biorefineries. The existing/novel enzyme-based biorefinery concept is vast in scope, as it considers all areas of economic, environmental, and social sustainability. These must also become more integrated, incorporating biomass pretreatments, biogenic

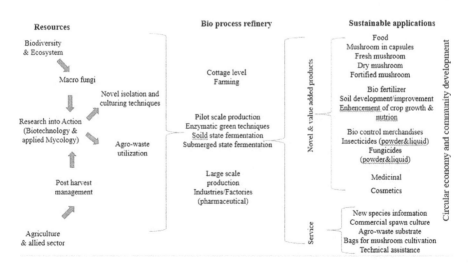

FIGURE 3.3 Integrated system of agro-waste valorization, biorefinery, and sustainability for next-gen ecosystems.

chemical recovery, and bioconversion processes to generate a wide range of products. Enzymes frequently provide a competitive advantage over chemical catalysts as they only require mild conditions and do not produce toxic byproducts. Enzymes are used to catalyze processes in a range of industries, including industrial bioconversion (biocatalyst), environmental bioremediation, agriculture, and biotransformations of various substances. For a few years, proteases, cellulases, hemicellulases, and phytases have been offered as process aids, improving yields, performance, and costs. Bioprospecting for better enzymes, enzyme engineering, and enzyme combination reconstitution are all becoming increasingly appealing (Himmel et al., 2007).

The biological efficiency of farmed macrofungi (defined as fresh macrofungi output divided by the dry weight of the substrate used) ranges from 17% to 250%. As a result, macrofungi cultivation is widely viewed as a possible way to extract elements from agro solid recyclable wastes. Enzymatic hydrolysis is based on the conversion of organic waste (carbohydrate polymers) to monosaccharides. In this reaction, complex sugars are converted to simple sugars, which transform cellulose to glucose, and hemicellulose to pentoses (xylose and arabinose) and hexoses (glucose, galactose, and mannose), respectively. Cellulase and hemicellulase speed up the conversion of cellulose and hemicellulose, respectively, and are also important in cellulose breakdown into fermentable sugars. The hydrolysis of cellulose is aided by endoglucanases, exoglucanases, and glucosidase enzymes (Bellettini et al., 2019; Biernat & Grzelak 2015; Chang & Wasser 2017; Chiu et al., 2000; Dhar & Shrivastava, 2012; Singhvi & Gokhale 2015; Kumar & Verma 2020b).

3.5.2 THE TECHNOLOGICAL EVOLUTION OF THE SYSTEM

The technological evolution of biorefinery began in the 17th century. In 1650, in Paris, the first formal reference to mushroom growing using the composting method was made. However, in 1707, a French botanist Tournefort reported a comprehensive description of the commercial production of mushrooms, utilizing horse manure covered in soil. "This plant is of such very odd growth and temperature," said Abercrombie (1779), "that unless a proper comprehension of its nature and habit is acquired, and the particular means and precautions maintained in the process of its multiplication and development, little success will ensure." (Dhar & Shrivastava, 2012). In 1780, a French gardener was the first to plant mushrooms commercially, and after the American Civil War, mushroom gardening was introduced to North America (Beyer 2003). In Paris, the first cave cultivation of *Agaricus* was accomplished in 1800. At the time, mushrooms were grown in open fields, but in 1810, Chambry (a French gardener) began growing them year-round in underground quarries in Paris. Later, in 1831, Callow demonstrated that indoor mushroom cultivation in heated mushroom houses was possible all year in England. Though the forerunners of current mushroom homes were established in England, it is now widely acknowledged that protected mushroom cultivation was pioneered in caverns in France (Dhar & Shrivastava, 2012).

Europe, North America, Australia, Southeast Asia, and South Asia now produce most of the world's mushrooms. In the early 20th century, when Duggar reported pure mushroom production in 1906, he made one of the most significant contributions to mushroom science in recent times. Sinden and Hauser's simple method for making mushroom compost from agricultural leftovers was also noteworthy (1950 and 1953).

In 1983, the Indian Council of Agricultural Research established the National Research Centre on Mushrooms (NRCM) (currently the Directorate of Mushroom Research), which accelerated mushroom research and development. In 1995–96, three *A. bisporus* varieties (two single spore isolates (SSIs) and one hybrid) and two *A. bitorquis* varieties were released for use by farmers and commercial growers, following intensive trials at NRCM and Coordinated Centres. NRCM played a larger role in mushroom cultivation training for farmers and businesses and offered consulting services to the industry. Fritsche's (1983) contributions to creating two new white button mushroom *A. bisporus* U-1 and U-3 strains changed commercial mushroom growth around the world. It was possible to collect more and more mushrooms per unit area/unit weight of compost as the button mushroom farming method was refined on a continuous scale (Fritsche, 1983; Kerrigan 2005; Dhar & Shrivastava, 2012).

To bridge the gap between rural entrepreneurs and the biotech sector, we need to create a futuristic new biorefinery idea that converts locally available agro-waste into high-value products using enzyme-assisted macrofungal (mushroom) culture. Fungi's distinct characteristics bode well for their use in biotechnology and industry. Furthermore, fungi are relatively easy to grow, allowing for large-scale production. In terms of identifying species with unique commercial applications that will lead to novel goods, the search for fungal biodiversity and the establishment of a live fungi collection have immense economic potential. Consumers all around the world increasingly prefer natural compounds to synthetic chemicals, and commodity chemical makers are increasingly interested in creating sustainable biotechnological procedures to create new genuine goods that will eventually replace old synthetics. Macrofungi have a significant advantage over other biological sources, particularly plants, in that they can be grown in large biorefineries on an industrial scale, and suitable processes for cost-effective fermentation have been available for decades, for example, for the production of certain organic acids, enzymes, and antibiotics (Hyde et al., 2019). Understanding the molecular principles of their biosynthesis could lead to more precise regulation of the manufacturing process in the future, allowing for the generation of innovative natural derivatives. Even the vast majority of known fungal species remain virtually untapped in terms of future uses because they have never been grown and examined for their growth characteristics and physiology. To achieve this aim, new techniques and protocols must be established, which indicates the need for extensive fundamental and futuristic, innovative, revolutionary, next-gen research to understand natural processes. Among macrofungi, *Basidiomycota*, in particular, is a mostly untapped but very promising source of anti-infective, immune suppressants, and other medicines (Badalyan et al., 2019; Sandargo et al., 2019).

Macrofungi cultivation's long-term success depends on research and development to provide new ideas and methods for better cultivation. Agro-waste can be used as a macrofungi culture substrate, resulting in lower costs and better waste management. We can now explore the biovariety of macrofungus much more clearly than ever before in biorefining and product transformation thanks to the rise of novel approaches in genomes, transcriptomics, bioinformatics, analytical chemistry, and biotechnological process development. More public support is needed, however, to ensure that the superior experience gathered over many decades is preserved and that the next generation of researchers can continue to study unique, previously unknown fungal families rather than just model species. The discovery of biological and pharmacological active chemicals in medicinal mushrooms has the potential to alter the trajectory of future technological development and open up new market opportunities in developed and developing countries alike (Fava et al., 2013).

The following list is based on a recent broad analysis of bottlenecks in important research domains as well as integrating enzyme-based biorefineries:

i. Agricultural Wastes
    The availability of agricultural waste determines the sustainability and feasibility of enzyme-based biorefineries. Agro-waste is accumulated at different times of the year. New sorting and collection procedures (automation) and sustainable techniques for integrated characterization of collected biowaste and subsequent decentralized technologies for stabilization, fractionation, and transportation within new logistic value chains are the need of the hour.

ii. Biorefinery Processes Development
    As biorefineries will utilize a variety of feedstocks, a wide range of conversion methods will be required, from pretreatment to refinery processes and finally to the finished product. Depending on the type of waste and desired bio-based products, technical advancements have to affect both physicochemical or thermochemical processes as well as enzymatic processes. Then, in continuous production, these complementary/sequential processes must be optimized. Knowledge exchange and process modeling will also make it easier to make decisions on new biorefinery techniques.

iii. Integration
    Biorefinery entails the collaboration and competitiveness of new industries and sectors with existing enterprises. Further, it results in new linkages between old and new that must be enhanced and developed. Integration of biotechnological and chemical processes will be crucial to design the optimal and most sustainable approach. From harvesting, byproducts, waste, and effluent production to standardized endproducts, logistics connects distributed (typically small-scale) primary production to endproducts, taking into account the entire biomass value chain.

iv. Sustainability Assessment
    An evaluation framework is required to ensure long-term, societally desirable development. The evaluation of sustainability will look at both local and global implications, as well as people and the environment, including nutrient cycles, water management, and food-feed-fuel competition. The acceptability of the enzyme-based biorefining industry can be increased by improving communication channels between different stakeholders, particularly between the farming, forestry, and food processing industries and their downstream sectors, and contact with the general public. Three biorefinery systems are currently being researched and developed. The first is "Whole Crop Biorefinery," which uses cereals like maize as a raw material. "Green Biorefinery" uses "nature-wet" biomasses like green grass, lucerne,

clover, or immature grain. Third, "Lignocellulose Feedstock Biorefinery" uses cellulose-containing biomass as a "nature-dry" raw material (Kamm & Kamm 2004).

In the last decade, 360 changes have been observed in mushroom cultivation technology with more advancement and automation. Not only have strains been enhanced, but the entire mushroom-producing process has been automated. The automated procedures include substrate mixing, bag filling, liquid spawn growth, bag inoculation, bag movement, temperature and humidity control, and packing. Enzymes and whole-cell biomaterials have become a viable technology in the chemical and pharmaceutical businesses for creating a wide range of environmental and physical substances because enzyme-based methods typically lead to a reduction in processing times, number of reaction mechanisms, and waste. Brief descriptions of some technological evolutions associated with macrofungal cultivation are as follows:

(a)   Compost Technology

Agro-wastes generated by agricultural activities are used in a variety of ways by a number of countries. They are used in various agricultural operations, including macrofungi cultivation, in either processed or unprocessed form (Bhuvaneshwari et al., 2019). As a result, the conditions for substrate preparation, inoculation, incubation, and production differ significantly. Composting is a method of preparing the substrate. Mechanical mixing or fermentation are both options for composting. Ribbon, concrete, and auger mixers are examples of mechanical mixing machines. Once the substrate has achieved the required composition, it is delivered to the bottling unit, where it is automatically put into nonreactive polypropylene bottles. The combined substrate is now compost, and is ready to be sterilized. Fermentation is a method of increasing the nutritional content of a substrate by utilizing the natural breakdown process. This method requires the construction of substrate mounds in which worms, insects, and other fungi decompose the substrate and absorb valuable nutrients.

(b)   Bag/Tray Technology

This approach is the most common among small and medium-sized mushroom producers, who employ unoccupied suitable spaces (shafts, vegetable shops, poultry yards, pigsties, and so on) and rooms specifically designed for mushroom production. On farms, where steam treatment of spent compost in a chamber for mushroom culture is impossible, it is better to cultivate mushrooms in bags. In comparison to shelf and container systems, using bags requires less capital. The ability to control pests and diseases more effectively is one of the main benefits of the bag system. It is easier to detect bugs and diseases when bags were utilized as a medium for cultivation (an infected bag can be closed and taken out at any moment). Simultaneously, employing multistands under the bags and a thick layer of compost in the bags improves the efficacy of room space utilization for cultivation. The downsides include a significant amount of manual labor when packing compost into bags (especially when no specific equipment is available), hauling and arranging the bags in place, and applying the casing layer to compost that has become overrun with mycelium. It should also be noted that using additives on this compost is difficult. The amount of compost in the bags, the right filling method, and the bag dimensions are among the technological aspects of employing a bag system.

(c)   Field-grown Technology

With the rise in consumer awareness and demand for farmed mushrooms, there is currently a desire for alternative, cost-effective mushroom cultivation techniques. Inoculation of logs, cultivation in forest understoreys, and usage of managed forests, as well as field-grown mushrooms, are all viable alternatives to more typical high-volume production processes. Field-grown mushrooms make efficient use of space by allowing mushrooms to be grown alongside crops or in between cropping cycles in agricultural fields. This approach has become popular with rural development programs since it provides monetary benefits to the farmers. Furthermore, due to enhanced nutrient cycling rates and organic matter availability in the soil, field cultivation allows for improved soil systems in agricultural fields. This production technique has become popular in many parts of Asia (such as Cambodia), where the climate is conducive, aimed at improving livelihoods and supporting smallholder farmers requiring additional revenue.

(d)   Outdoor Cultivation Technology

Outdoor macrofungi can be cultivated in a variety of ways, with the selection of appropriate species. For instance, in the cooler autumn and winter months, *Coprinus comatus* or *Stropharia rugosoannulata* can be grown, but *Volvariella volvacea* should be grown in the hotter summer months because macrofungi which are produced outside, are-influenced by edaphic factors. In addition to environmental change, outdoor cultivation technology exposes macrofungi to nonsterile conditions, making it easier for slime molds, predatory fungi, and insects to irradiate them.

(e)   Automated Industrial Technology

The global demand for mushrooms has risen fast, resulting in the establishment of large-scale growers, with year-round marketing dominating commercial mushroom production. From 30.2 million MTin 2010 to 48 million MT in 2017(FAO), global production of mushrooms has increased so far. The profitability, efficiency, and competitiveness of the commercial mushroom industry are based on productivity, efficiency, and competitiveness across the entire market value chain. Most recently domesticated species can recycle waste substrates, making them ideal for small-scale farmers in agroecosystems. In developed countries, the installation of sophisticated mushroom facilities with well-equipped enclosed cooling and dis-infected structures is becoming increasingly popular. Temperature, humidity, uniform ventilation, substrate moisture levels, and or light to stimulate the growth of fruit bodies are the most appropriate and regulated conditions for mushroom cultivation. However, maintaining and operating such growing units demands vast amounts of electricity and water, making the process costly and having a significant influence on greenhouse gas emissions. The usage of renewable energy sources provides a new edge to the existing technology. New technologies, such as photovoltaic heating and cooling, artificial intelligence, Sensor-based technology for controlling the environment, and even light formula, will extend the growing season, improve mushroom yield and quality, and lower energy costs.

Substrate bag filling, inoculation, growing, scanning, and picking up contaminated bags, and robots for packing and delivery are all some examples of artificial intelligence or automation in mushroom cultivation. Both small and large-scale manufacturing lines should accept integrated systems, such as the utilization of recycled materials like agricultural waste or a sustainable supply of woody substrates from managed plantations, as well as the integration of renewable energies. The computer control of the cropping room environment has made it possible to harvest mushroom yields of 30–35 kg from 100 kg compost within a cropping period of four weeks in 3–4 flushes (Dhar & Shrivastava, 2012).

(f)   Mycoremediation Technology

Mycoremediation, or bioremediation with fungus, is a novel biotechnological technique that employs live fungi to clean up contamination either total mineralization with enzymes or total elimination of the contaminant via absorption (Stamets 2005). Because their extracellular lignin modifying enzymes have limited substrate specificity, several species of basidiomycete white-rot fungi (Lang et al., 1995) are most suited for mycoremediation . Enzymes like lignin peroxidases (LiP), manganese peroxidases (MnP), and laccases (LAC) secreted by macrofungi (*Lentinus subnudus*, *Phlebia acanthocystis*, and *Pleurotus ostreatus*) are excellent at degrading organo-pollutants such as DDT, lindane, chlorodane, and PCBs.

The absence of a more thorough understanding of the degradation mechanisms and enzymes involved in pollutant degradation by different fungus species makes it difficult to use them in a more targeted manner in mycoremediation. Transgenic fungi with enhanced mycoremediation capacities can be developed by analyzing genes associated to enzyme degradation, mineralization, absorbance, and tolerance to harsh environmental conditions. However, because the remediation ability of naturally occurring fungi is still little understood, suitable fungi should be isolated and screened from their native habitats in order to reclaim damaged sites and preserve a better environment for future generations (Hibbett & Taylor 2013; Lu et al., 2020b)

3.5.3   Solid-state and Submerged Fermentation Systems for Utilization of Agro-waste

The global market for macrofungi, especially medicinal and edible mushrooms has expanded dramatically, owing to increased demand for enzymes and other useful biomolecules, particularly bioactive compounds for human health. Solid-state fermentation (SsF) is a promising technology for waste valorization through the bioconversion of organic wastes used as either substrate or inert support, while macrofungi will play a role in the breakdown of organic wastes into constituents to convert them into high-value-added products. Solid-state fermentation is an ideal fermentation process for macrofungal cultivation using various agro-waste and significant technological advancements that have occurred in recent years, allowing direct and indirect identification of new species. It has been demonstrated that SsF has sustainable properties in the bioconversion of solid wastes, such as high efficiency in terms of product yields and productivities, low energy consumption, and the ability to solve disposal concerns (Abu Yazid et al., 2017).

Solid-state fermentation has ushered in a new era of bioconversion of organic solid wastes by producing physiologically active metabolites on a laboratory and industrial scale. It has also has been used to make a variety of bioproducts, including enzymes, organic acids, biofertilizers, biopesticides, biosurfactants, bioethanol, fragrance compounds, animal feed, pigments, vitamins, antibiotics, and more. Extrinsic factors such as the nature of the substrate phase, the composition of the culture media, temperature, particle size of the support and substrate, moisture, water activity, pH, O_2 and CO_2 concentrations in the media, and the type and design of the bioreactor can all affect mushroom growth (activities related to the secretion of hydrolytic enzymes) in SsF (Bellettini et al., 2019).

Submerged fermentation (SmF) is a cost-effective approach for producing biomass and exquisite bioactive metabolites while reducing contamination risks and manufacturing time. In aqueous media, submerged fermentations are normally carried out with either dissolved or suspended substrates. Different substrates induce different enzyme activities: Simple, all-purpose substrates provide modest enzyme activity, but more difficult substrates can produce abnormally high enzyme activity (Dudekula et al., 2020).

3.5.4   Advantages of Using Macrofungi and Agro-waste

The major components of agro-waste are cellulose, hemicellulose, and lignin, which are all classified as lignocellulosic compounds. This waste can be used as a substrate in a mushroom-based solid-state fermentation process. Macrofungi synthesize lignocellulosic enzymes that degrade lignocellulosic substrates and use the degraded products to make their fruiting bodies. As a result, mushroom cultivation can be regarded as a significant biotechnological approach for reducing and valorizing agro-waste. This type of waste is produced as a result of the environmentally responsible conversion of low-value byproducts into fresh resources that may be used to make value-added products (Kumla et al., 2020).

Besides this, a lot of attention in recent decades on the utilization of agro-waste via macrofungal cultivation has been observed. Some of the advantages are listed below:

1. Agro-waste is rich in nutrients as well as inert/physical support as substrate (and source of carbon).
2. Macrofungal culture on these agricultural wastes is the most environmentally benign means of reducing nutrient levels to a safe use as manure.
3. Agro-wastes can also provide a high output of mushrooms at a low cost.
4. Macrofungal cultivation via agro-waste provides the opportunity for cost-effective farming.
5. Lower energy requirements.
6. Higher product yields.
7. Production of extracellular lignocellulosic enzymes which have industrial significance.
8. The bioconversion of agro-waste into other high-value-added products has huge promise for diverse markets.
9. Utilization of all agro-waste, resulting in zero effluents and ensuring the approach's environmental sustainability
10. Ensuring a biorefinery's overall economic sustainability.

3.6 CONCLUSION

Integrated enzyme-based biorefineries appear to be a promising solution for converting biomass into value-added products. As a result, methodical ways to screen and design an integrated enzyme-based biorefinery should be adopted. Converting agro-waste (organic waste) into different types of chemical compounds or biomaterials and energy to fully exploit and minimize the number of naturally produced chemicals or waste is one strategy to reduce the negative effects on local ecosystems. Furthermore, public pressure for green technology as a result of environmental concerns has stimulated more biocatalytic processes, which are cleaner, safer, and more environmentally friendly, to replace chemical processes. In several domains, enzyme biocatalysis has quickly replaced traditional chemical processes, and the emergence of new technologies in enzyme engineering is likely to expedite this process even further.

Agro-waste can be employed as a mushroom-growing substrate,, resulting in reduced costs and improved waste management. In terms of new antibiotics and other pharmacological compounds, macrofungi are underexplored. It is well worth putting extraordinary effort into this field of research, emphasizing previously overlooked species from underresearched places and ecosystems. Identifying biological and pharmacologically active compounds in medicinal macrofungi has the potential to drive future technology innovation and provide new commercial opportunities in both developed and developing countries.

The accumulation of successful enzyme development anecdotes will give adequate decision strategies. In improving methods of enzyme design, fundamental knowledge of the structure-function and dynamical function interactions should be a new focus of attention to encourage the rational design of next-generation enzymes with desired properties. Customizable next-generation enzymes will be easier to make and apply in the industrial setting, in addition to the advances in technology. In this chapter, agro-waste has been assessed for its prospective usage in mushrooms for lignocellulosic enzyme production. The heterogeneity in waste composition and mushroom species, on the other hand, have an impact on enzyme synthesis. As a result, more research is needed to identify the best circumstances (substrates, mushroom species, and fermentation procedure) for effective lignocellulosic enzyme production both in a pilot study and on a large scale.

REFERENCES

Abu Yazid N, Barrena R, Komilis D, Sánchez A. Solid-state fermentation as a novel paradigm for organic waste valorization: a review. *Sustainability.* 2017 Feb;9(2):224.

Abercrombie J. *The garden mushroom, its nature and cultivation.* Lockyer Davis, London, 1779; 54(16):1–40.

Adebayo EA, Martinez-Carrera D. Oyster mushrooms (*Pleurotus*) are useful for utilizing lignocellulosic biomass. *African Journal of Biotechnology.* 2015;14(1):52–67.

Agrawal K, Shankar J, Verma P. Multicopper oxidase (MCO) laccase from *Stropharia* sp. ITCC-8422: an apparent authentication using integrated experimental and in silico analysis. *3 Biotech.* 2020a; 10(9): 1–18.

Agrawal K, Verma P Laccase-Mediated Synthesis of Bio-material Using Agro-residues. In *Biotechnological Applications in Human Health* (pp. 87–93). Springer, Singapore, 2020a.

Agrawal K, Shankar J, Kumar R, Verma P Insight into multicopper oxidase laccase from *Myrothecium verrucaria* ITCC-8447: a case study using in silico and experimental analysis. *Journal of Environmental Science and Health, Part B.* 2020b. 55(12), pp. 1048–1060.

Agrawal K, Verma P Fungal Metabolites: A Recent Trend and its Potential Biotechnological Applications. In *New and Future Developments in Microbial Biotechnology and Bioengineering* (pp. 1–4), Elsevier, 2021.

Akoh CC, Chang SW, Lee GC, Shaw JF. Biocatalysis for the production of industrial products and functional foods from rice and other agricultural produce. *Journal of Agricultural and Food Chemistry.* 2008 Nov 26;56(22):10445–51.

Alam A., Agrawal K., Verma P. Fungi and Its By-Products in Food Industry: An Unexplored Area. In *Microbial Products for Health, Environment and Agriculture* (pp. 103–120). Springer, Singapore, 2021.

Al-Hawash AB, Alkooranee JT, Abbood HA, Zhang J, Sun J, Zhang X, Ma F. Isolation and characterization of two crude oil-degrading fungi strains from Rumaila oil field, Iraq. *Biotechnology Reports.* 2018 Mar 1;17:104–9.

Badalyan SM, Barkhudaryan A, Rapior S. Recent progress in research on the pharmacological potential of mushrooms and prospects for their clinical application. *Medicinal Mushrooms.* 2019:1–70.

Bellettini MB, Fiorda FA, Maieves HA, Teixeira GL, Ávila S, Hornung PS, Júnior AM, Ribani RH. Factors affecting mushroom *Pleurotus* spp. *Saudi Journal of Biological Sciences.* 2019 May 1;26(4):633–46.

Beyer DM. Basic procedures for Agaricus Mushroom. Publ. Distrib. Center, Pennsylvania State Univ. BioGreenhouse COST Action FA. 2003; 1105.

Bhardwaj N, Kumar B, Agrawal K, Verma P. Current perspective on production and applications of microbial cellulases: a review. *Bioresources and Bioprocessing,* 2021a; 8(1):1–34.

Bhardwaj N, Agrawal K, Verma P. Xylanases: An Overview of Its Diverse Function in the Field of Biorefinery. In *Bioenergy Research: Commercial Opportunities & Challenges* (pp. 295–317). Springer, Singapore, 2021b.

Bhuvaneshwari S, Hettiarachchi H, Meegoda JN. Crop residue burning in India: policy challenges and potential solutions. *International Journal of Environmental Research and Public Health.* 2019 Jan;16(5):832.

Biernat K, Grzelak PL. Biorefinery systems as an element of sustainable development. *Biofuels–Status and Perspective.* 2015 Sep 30 (1): 427–444.

Bucher VV, Pointing SB, Hyde KD, Reddy CA. Production of wood decay enzymes, loss of mass, and lignin solubilization in wood by diverse tropical freshwater fungi. *Microbial Ecology.* 2004 Nov 1;48(3): 331–337.

Chang ST, Wasser SP. The Cultivation and Environmental Impact of Mushrooms. In *Oxford research encyclopedia of environmental science* 2017 Mar 29.

Chang ST, Wasser SP. The role of culinary-medicinal mushrooms on human welfare with a pyramid model for human health. *International Journal of Medicinal Mushrooms.* 2012;14(2): 95–134.

Chiu SW, Law SC, Ching ML, Cheung KW, Chen MJ. Themes for mushroom exploitation in the 21st century: sustainability, waste management, and conservation. *The Journal of General and Applied Microbiology.* 2000;46(6):269–82.

Choi JM, Han SS, Kim HS. Industrial applications of enzyme biocatalysis: current status and future aspects. *Biotechnology Advances.* 2015 Nov 15;33(7):1443–54.

Demirbras A. Biorefinery technologies for biomass upgrading. *Energy Sources, Part A: Recovery, Utilization, and Environmental Effects.* 2010 Jun 23;32(16):1547–58.

Deswal D, Khasa YP, Kuhad RC. Optimization of cellulase production by a brown rot fungus *Fomitopsis* sp. RCK2010 under solid state fermentation. *Bioresource Technology.* 2011 May 1;102(10):6065–72.

Devi S, Gupta C, Jat SL, Parmar MS. Crop residue recycling for economic and environmental sustainability: the case of India. *Open Agriculture.* 2017 Sep 26;2(1):486–94.

Dhar BL, Shrivastava N (2012). Mushrooms and Environmental Sustainability. ResearchGate.

Dimitrijevic MV, Mitic VD, Jovanovic OP, Stankov Jovanovic VP, Nikolic JS, Petrovic GM, Stojanovic GS. Comparative study of fatty acids profile in eleven wild mushrooms of Boletacea and Russulaceae families. *Chemistry & Biodiversity.* 2018 Jan;15(1):e1700434.

Du F, Wang HX, Ng TB. An amylase from fresh fruiting bodies of the monkey head mushroom *Hericium erinaceum.* *Applied Biochemistry and Microbiology.* 2013 Jan;49(1):23–7.

Dudekula UT, Doriya K, Devarai SK. A critical review on submerged production of mushroom and their bioactive metabolites. *3 Biotech.* 2020 Aug;10(8):1–2.

Elisashvili VI, Chichua D, Kachlishvili E, Tsiklauri N, Khardziani T. Lignocellulolytic enzyme activity during growth and fruiting of the edible and medicinal mushroom *Pleurotus ostreatus* (Jacq.: Fr.) Kumm. (Agaricomycetideae). *International Journal of Medicinal Mushrooms.* 2003;5(2):193–198.

FAO, 2021, http://www.fao.org/faostat/en/?#data/GA

Fava F, Zanaroli G, Vannini L, Guerzoni E, Bordoni A, Viaggi D, Robertson J, Waldron K, Bald C, Esturo A, Talens C. New advances in the integrated management of food processing by-products in Europe: sustainable exploitation of fruit and cereal processing by-products with the production of new food products (NAMASTE EU). *New Biotechnology.* 2013 Sep 25;30(6):647–55.

Fen L, Xuwei Z, Nanyi L, Puyu Z, Shuang Z, Xue Z, Pengju L, Qichao Z, Haiping L. Screening of lignocellulose-degrading superior mushroom strains and determination of their CMCase and laccase activity. *The Scientific World Journal.* 2014 Jan 1: 1–6.

Fritsche G. Breeding *Agaricus bisporus* at the Mushroom Experimental Station, Horst. *Mushroom Journal.* 1983; 122:49–53.

Gaitán-Hernández R, Esqueda M, Gutiérrez A, Sánchez A, Beltrán-García M, Mata G. Bioconversion of agrowastes by *Lentinula edodes*: the high potential of viticulture residues. *Applied Microbiology and Biotechnology*. 2006 Jul;71(4):432–9.

Goswami R K, Agrawal K, Verma P. Multifaceted Role of Microalgae for Municipal Wastewater Treatment: A Futuristic Outlook towards Wastewater Management. *CLEAN–Soil, Air, Water*, 2022, p. 2100286. https://doi.org/10.1002/clen.202100286

Goswami R K, Agrawal K, Verma P. Microalgae-based biofuel-integrated biorefinery approach as sustainable feedstock for resolving energy crisis. In N. Srivastava, M. Srivastava, R. Singh (Eds.), *Bioenergy Research: Commercial Opportunities & Challenges* (pp. 267–293). Springer, Singapore, 2021.

Govorushko S, Rezaee R, Dumanov J, Tsatsakis A. Poisoning associated with the use of mushrooms: a review of the global pattern and main characteristics. *Food and Chemical Toxicology*. 2019 Jun 1;128:267–79.

Gowda N.N. and Manvi D. Agriculture crop residues disinfection methods and their effects on mushroom growth. In *Proc Indian Natn Sci Acad* 2020 Sep (Vol. 86, No. 3, pp. 1177–90).

Hibbett DS, Taylor JW. Fungal systematics: is a new age of enlightenment at hand? *Nature Reviews Microbiology*. 2013 Feb;11(2):129–33.

Himmel ME, Ding SY, Johnson DK, Adney WS, Nimlos MR, Brady JW, Foust TD. Biomass recalcitrance: engineering plants and enzymes for biofuels production. *Science*. 2007 Feb 9;315(5813):804–7.

Hyde KD, Xu J, Rapior S, Jeewon R, Lumyong S, Niego AG, Abeywickrama PD, Aluthmuhandiram JV, Brahamanage RS, Brooks S, Chaiyasen A. The amazing potential of fungi: 50 ways we can exploit fungi industrially. *Fungal Diversity*. 2019 Jul;97(1):1–36.

Inácio FD, Ferreira RO, Araujo CA, Brugnari T, Castoldi R, Peralta RM, Souza CG. Proteases of wood rot fungi with emphasis on the genus *Pleurotus*. *BioMed Research International*. 2015 Jun; 9:2015.

Kamm B, Kamm M. Biorefinery-systems. *Chemical and Biochemical Engineering Quarterly*. 2004 Jan;18(1):1–7.

Kerrigan RW. Agaricus subrufescens, a cultivated edible and medicinal mushroom, and its synonyms. *Mycologia*. 2005 Mar 1; 97(1):12–24.

Kinge TR, Apalah NA, Nji TM, Acha AN, Mih AM. Species richness and traditional knowledge of macrofungi (mushrooms) in the awing forest reserve and communities, Northwest region, Cameroon. *Advances in Artificial Neural Systems*. 2017 Oct;2017:2809239.

Kiran EU, Trzcinski AP, Ng WJ, Liu Y. Enzyme production from food wastes using a biorefinery concept. *Waste and Biomass Valorization*. 2014 Dec 1;5(6):903–17.

Kumar B, Verma P, Application of Hydrolytic Enzymes in Biorefinery and its Future Prospects. In *Microbial Strategies for Techno-economic Biofuel Production* (pp. 59–83). Springer, Singapore, 2020a.

Kumar B, Verma P. 2020b. Enzyme mediated multi-product process: a concept of bio-based refinery. *Industrial Crops and Products*. 154, p. 112607.

Kumla J, Suwannarach N, Sujarit K, Penkhrue W, Kakumyan P, Jatuwong K, Vadthanarat S, Lumyong S. Cultivation of mushrooms and their lignocellulolytic enzyme production through the utilization of agro-industrial waste. *Molecules*. 2020 Jan;25(12):2811.

Kwon H, Thatithatgoon S, Mushroom Growing for a Living Worldwide: Mushroom Growing in Northern Thailand. In Gush R., eds., *Mushroom Growers' Handbook1:Oyster Mushroom Cultivation*, Mush World-Heineart Inc, Seoul, Korea, 2004.

Lakhtar H, Ismaili-Alaoui M, Philippoussis A, Perraud-Gaime I, Roussos S. Screening of strains of *Lentinula edodes* grown on model olive mill wastewater in solid and liquid state culture for polyphenol biodegradation. *International Biodeterioration & Biodegradation*. 2010 Jun 1;64(3):167–72.

Lang E, Eller I, Kleeberg R, Martens R Zadrazil F. (1995). Interaction of white-rot fungi and micro-organisms leading to biodegradation of soil pollutants. In *Proceedings of the 5th International FZK/ TNO Conference on Contaminated Soil*. 30th Oct–5 Nov 1995, Maustrient. The Netherlands by Van de Brink WJ, Bosman R and Arend F., 95: 1277–78.

Lu H, Lou H, Hu J, Liu Z, Chen Q. Macrofungi: a review of cultivation strategies, bioactivity, and application of mushrooms. *Comprehensive Reviews in Food Science and Food Safety*. 2020b Sep;19(5):2333–56.

Lu JS, Chang Y, Poon CS, Lee DJ. Slow pyrolysis of municipal solid waste (MSW): a review. *Bioresource Technology*. 2020a Sep 1;312:123615.

Mehboob N, Asad MJ, Imran M, Gulfraz M, Wattoo FH, Hadri SH, Asghar M. Production of lignin peroxidase by *Ganoderma leucidum* using solid state fermentation. *African Journal of Biotechnology*. 2011;10(48):9880–7.

MNRE (2009) Ministry of new and renewable energy resources Govt. of India New Delhi. www.mnre.gov.in/ biomassrsources

Moilanen U, Winquist E, Mattila T, Hatakka A, Eerikäinen T. Production of manganese peroxidase and laccase in a solid-state bioreactor and modeling of enzyme production kinetics. *Bioprocess and Biosystems Engineering*. 2015 Jan;38(1):57–68.

Molina R. Biology, ecology, and social aspects of wild edible mushrooms in the forests of the Pacific Northwest: a preface to managing commercial harvest. US Department of Agriculture, Forest Service, Pacific Northwest Research Station; 1993.

Montoya S, Orrego CE, Levin L. Growth, fruiting and lignocellulolytic enzyme production by the edible mushroom *Grifola frondosa* (maitake). *World Journal of Microbiology and Biotechnology*. 2012 Apr;28(4):1533–41.

Nair LG, Agrawal K, Verma P An overview of sustainable approaches for bioenergy production from agro-industrial wastes. *Energy Nexus*, 2022; 6: 100086.

Pandit NP, Maheshwari SK. Optimization of cellulase enzyme production from sugarcane pressmud using oyster mushroom-*Pleurotus sajor-caju* by solid state fermentation. *Journal of Bioremediation and Biodegradation*. 2012;3(3):140.

Pavlenko N, Searle S. The potential for advanced biofuels in India: assessing the availability of feed stocks and deployable technologies, *The International Council on Clean Transportation*. 2019 Dec.

Postemsky PD, Delmastro SE, Curvetto NR. Effect of edible oils and Cu (II) on the biodegradation of rice by-products by *Ganoderma lucidum* mushroom. *International Biodeterioration & Biodegradation*. 2014 Sep 1;93:25–32.

Rinaldi R, Jastrzebski R, Clough MT, Ralph J, Kennema M, Bruijnincx PC, Weckhuysen BM. Paving the way for lignin valorisation: recent advances in bioengineering, biorefining and catalysis. *Angewandte Chemie International Edition*. 2016 Jul 11;55(29):8164–215.

Sadler MJ. Meat alternatives—market developments and health benefits. *Trends in Food Science & Technology*. 2004 May 1;15(5):250–60.

Salmones D, Mata G. Laccase production by *Pleurotus djamor* in agar media and during cultivation on wheat straw. *Revista mexicana de micología*. 2015;42:17–23.

Sandargo B, Chepkirui C, Cheng T, Chaverra-Muñoz L, Thongbai B, Stadler M, Hüttel S. Biological and chemical diversity go hand in hand: basidiomycota as source of new pharmaceuticals and agrochemicals. *Biotechnology Advances*. 2019 Nov 1;37(6):107344.

Satyendra T, Singh RN, and Shaishav S. Emissions from crop/biomass residue burning risk to atmospheric quality. *International Research Journal of Earth Sciences*. 2013;1(1):1–5.

Sinden JW, Hauser E. The short method of composting. *Mushroom Science*. 1950;1:52–9.

Singhvi MS, Gokhale DV. Biomass exploitation –a challenge finding its way to reality. *Current Science*. 2015;108(9):1593–4.

Stamets P. *Mycelium Running: How Mushrooms Can Help Save the World*. Random House Digital, Inc.; 2005.

Stephen B, Pelling AL, Smith GJ, Reddy CA. Screening of basidiomycetes and xylariaceous fungi for lignin peroxidase and laccase gene-specific sequences. *Mycological Research.* 2005 Jan;109(1):115–24.

Sukumaran RK, Mathew AK, Kumar MK, Abraham A, Chistopher M, Sankar M. First-and second-generation ethanol in India: a comprehensive overview on feedstock availability, composition, and potential conversion yields. *Sustainable Biofuels Development in India.* 2017:223–26.

Taylor AF, Alexander IA. The ectomycorrhizal symbiosis: life in the real world. *Mycologist.* 2005 Aug;19(3):102–12.

Tedersoo L, May TW, Smith ME. Ectomycorrhizal lifestyle in fungi: global diversity, distribution, and evolution of phylogenetic lineages. *Mycorrhiza.* 2010 Apr;20(4):217–63.

Thakur TC. Crop residues as animal feed. Addressing resource conservation issues in rice-wheat systems of South Asia. 2003.

Usha KY, Praveen K, Reddy BR. Enhanced production of ligninolytic enzymes by a mushroom Stereum ostrea. *Biotechnology Research International.* 2014;(2014): 1–10.

Wood DA, Smith JF. The cultivation of mushrooms. *Essays in Agricultural and Food Microbiology (UK).* 1987.

Xu X, Quan L, Shen M. Effect of chemicals on production, composition and antioxidant activity of polysaccharides of *Inonotus obliquus. International Journal of Biological Macromolecules.* 2015 Jun 1;77:143–50.

Zhang K, Lu X, Li Y, Jiang X, Liu L, Wang H. New technologies provide more metabolic engineering strategies for bioethanol production in *Zymomonas mobilis. Applied Microbiology and Biotechnology.* 2019 Mar;103(5):2087–99.

4 Extremophilic Bacteria and Archaea in the Valorization of Metalloids
Arsenic, Selenium, and Tellurium

Devika Nagar and Judith M. Braganca
Birla Institute of Technology and Science Pilani, Zuarinagar, India

Irene J. Furtado
Goa University, Taleigão, India

CONTENTS

4.1 MEANS AND INPUTS OF METALLOIDS INTO THE ENVIRONMENT

Heavy metal/metalloid pollution has serious and long-term environmental problems due to heavy metal/metalloid toxicity, nondegradability, bioaccumulation in the food chain, and long-term harmful impact on ecology, environment, and human health. They show dual effects; at lower concentrations they are essential trace elements, while at higher concentrations they are toxic (Popescu and Dumitru 2009; Voica et al. 2016; Goswami et al. 2021). Heavy metals/metalloids are usually found in nature and although released due to natural processes, the major causes for their increased release and accumulation in the environment are anthropogenic activities like industrialization, use of chemical fertilizers and pesticides, mining and refining mineral ores, and so on(Raffa et al. 2021; Kumar et al. 2021). This chapter focuses on the response of extremophilic microorganisms to metalloids such as selenium, tellurium, and arsenic and their potential as bioagents in the field of environmental bioremediation.

Selenium, with atomic number 34, belongs to the chacolegen family that is, group 16 of the periodic table. Selenium shares similar properties with sulphur and tellurium. The proportion of selenium in the earth's crust is around 10^{-5} to 10^{-6} percent. In the atmosphere, selenium is present as volatile methyl derivatives, like dimethylselenide (DMSe, $[CH_3]_2Se$), dimethyldiselenide (DMDSe, $[CH_3]_2Se_2$), dimethyl selenone ($[CH_3]_2SeO_2$], methane selenol (CH_3SeH) and dimethyl selenyl sulphide (DMSeS, $[CH_3]_2SeS$). It also exists as an organic non-volatile amino acid containing selenium derivatives like selenocysteine, selenocystine, and selenome-thionine (Frankenberger and Arshad 2001). Selenium in the unbound state is rarely found; it appears in various allotropic forms like red crystalline form, a crystalline gray-black metal called metallic selenium, and red amorphous powdered form. In nature, selenium is released due to both natural and anthropogenic activities. Naturally, selenium can be released into the environment as a result of several physical, chemical, and biological processes like the mobilization of selenium from soil, rocks, and sediment into water due to weathering, erosion, or volcanic eruptions, and other natural processes, through which it enters the food chain. Anthropogenic processes such as oil refineries, coal combustion, mining, petroleum industries, electronics industries, metal refineries and processing industries, fertilizer usage, agricultural run-off, and so forth, give rise to selenium pollution which causes damage to the environment, growth abnormalities, tissue and other bodily damage in living organisms, and also other long term after-effects (Sinharoy and Lens 2020; Staicu et al. 2015a). Selenium is a biologically essential element but with an increase in concentration, it is a potent toxin, with a very narrow range of essentiality and toxicity (Staicu et al. 2015b). Selenium in the environment attains oxidation states of –2; +4; +6 and 0. Elemental selenium (0) is comparatively less toxic than other forms. Selenite (+4) and selenates (+6) are the most soluble and thus highly toxic forms of selenium; these exist in seleniferous soil and agricultural drainage water (Di Gregorio et al. 2005; Fresneda et al. 2018). These forms are considered as environmental hazards because selenium oxyanions released in soil and water mobilize and enter the food chain due to bioaccumulation, raising ecological concerns and futuristic impact (Staicu et al. 2015a; Tan et al. 2016). Globally, selenium in its various forms has a range of applications such as use in photovoltaic cells, semiconductors, light sensors, optoelectronic devices (Macaskie et al. 2010), photoelectric cells, xenograpy equipments, among others (Kapoor et al. 1995). Other than this, selenium is also used as an essential trace element in dietary sources in prescribed low concentrations, and is also found in many different cosmetic products, soaps, and shampoos. (Tan et al. 2016). In spite of having many applications in industry and essential elements for living organisms, a small increase in its concentration can be toxic and this can be an issue of concern as there is an only slight change in its concentration, that is, 40 to 400 µg/day from essentiality to toxicity. This problem is prevalent specifically where there is already an excessive amount of selenium released through natural processes, thus leading to increased uptake and mobilization of selenium by plants and animals (El-Ramady et al. 2015; Tan et al. 2016; Kashyap et al. 2021).

Tellurium on the other hand is a rare metalloid and is least abundant in the earth's crust. It is less abundant (10^{-2} to 10^{-8} ppm) globally and is not uniformly distributed on the earth (Maltman and Yurkov 2019). Similar to selenium, it belongs to group 16 of the periodic table with atomic number 52 (Sasani et al. 2020). Like selenium, tellurium also exists in various oxidation states: telluride (–2); elemental tellurium (0); tellurite (+4); tellurate (+6), the oxidized forms with the exception of elemental tellurium (0) are soluble, mobile and are thus toxic in nature. Oxyanion tellurite (+4) is the most toxic and is a strong oxidizing agent. The toxicity of tellurite corresponds to its oxidizing property and is assumed to release reactive oxygen species (ROS) (Ba et al. 2010). It has also been suggested that for tellurite oxyanion toxicity, thiol biochemistry and metabolism play a central role in formation of toxic compounds like telenotrisulphide, however, the molecular mechanism is not fully understood (Tremaroli et al. 2009). Tellurite (+4) is known for its toxicity even at lower concentrations and is found at higher concentrations near waste discharge andwastewater, thus causing environmental concern (Amoozegar et al. 2012; Belzile and Chen 2015). Tellurium is employed in various industrial applications such as rubber manufacturing, copper refining, solar panels, metallurgy, electronics, and many chemical industries. The major toxicity of tellurium is due to its reaction with thiol (SH) group and it forms a toxic telenotrisulphide compound (Safhi et al. 2018). It is reported to be toxic even at 100-fold lower concentrations than other metals and metalloids (Nguyen et al. 2019). Tellurium has many commercial applications;in small amounts, it is used in copper alloys and stainless steel for easier machining and milling to increase strength, while sulphuric acid resistant tellurium is added to lead; used as a colouring agent for glass and ceramic; for the vulcanization of rubber; used in the development of materials like photovoltaic products, fluorescent cadmium tellurium (CdTe) probes; and also used in solar panels and other nanotechnological applications (Belzile & Chen, 2015). Therefore, accumulation of tellurium oxyanions in waste discharge has become a critical pollutant, contaminating the environment.

Arsenic is a naturally occurring nonessential trace element and highly poisonous metalloid. The atomic number of Arsenic is 33 and it is the 20th most abundant element in the earth's crust. (Sanyal et al. 2020). It is released into the environment due to both natural and anthropogenic activities, is ubiquitously distributed in water, soil, and air, and is extremely toxic even at moderate concentrations, therefore is most widely studied for its remediation (Mandal and Suzuki 2002). It can exist in both organic and inorganic states with inorganic forms that are more toxic. Oxidation states in which arsenic occurs are arsines (–3), elemental arsenic (0), arsenite (+3), and pentavalent arsenate (+5). Arsenite (+3) and arsenate (+5) are highly toxic and specifically trivalent arsenite is an even more potent toxin, causing oxidative stress (Ghosh and Sil 2015; Jang et al. 2016). Other than oxidation states, methylation, redox potential, charge at physiological pH, electrostatic attraction and repulsion at the active site on macromolecules, and others are crucial for arsenic toxicity (Ghosh and Sil 2015; Mandal and Suzuki 2002). Arsenic is released into the environment via both natural and human activities. Natural processes like weathering, volcanic activities, seismic activities, and leaching are responsible for the release of arsenic, while human involvement has increased the mobilization of arsenic in environments. Anthropogenic processes like the emission of arsenic fumes, industrial wastes, agricultural applications like insecticides and pesticides, cotton and wool processing, mining, and smelting are responsible for mobilizing arsenic via wind and water into the environment (Sanyal et al. 2020). These anthropogenic activities of utilizing natural resources are leading to the release and circulation of arsenic in the atmosphere, biosphere, lithosphere, and hydrosphere ultimately affecting living organisms and the environment (Jebelli et al. 2018; Mandal and Suzuki 2002). The global increase in arsenic pollution is an issue of concern because this metalloid does not undergo degradation and mobilizes into the food chain leading to bioaccumulation, which is a serious threat to living organisms and the environment (Jang et al. 2016).

4.2 BIOREMEDIATION PROCESSES USING EXTREMOPHILIC MICROORGANISMS; BENEFITS OVER PHYSICOCHEMICAL TECHNOLOGIES

Due to rapid industrial growth, there is excessive release of toxic elements.For the remediation of these toxic metalloids, rather than physical and chemical techniques, green methods like microbial bioremediation of metalloids are gaining interest among researchers (Nguyen et al. 2019). Heavy metal/metalloid contamination has become a worldwide problem. As, although these substances are found naturally in the environment, industrialization and other human activities have resulted in their excessive release into the environment, far above acceptable limits (Raffa et al. 2021). Therefore, environmental clean-up or treatment of such toxic elements has become challenging and has gained attention. Microbial bioremediation of metal(loid)s like selenium, arsenic, and tellurium is feasible due to its cost-effectiveness, lesser energy consumption, and environmental friendliness when compared with physical and chemical-heavy metal remediation processes (Medfu Tarekegn et al. 2020; Mosa et al. 2016; Goswami et al. 2022a; 2022b).

Extremophilic bacteria and archaea have attracted recent attention because these microorganisms can thrive in several extreme conditions, gaining resistance/tolerance for high salt, extreme pH range, wide temperature ranges, heavy metals, and so forth. (Marques 2018). Their resistance/tolerance towards heavy-metal/metalloid, allows them to utilize toxic elements to biotransform them into nontoxic, or beneficial products for cellular metabolism, thus proving to be a potential source for bioremediation or removal of toxic contaminants from the environment (Jeong and Choi 2020). Extremophile assistance could also be utilized for the production of beneficial byproducts like nanoparticles, which alternativelyhave potential applications in various fields (Purcarea et al. 2019). Strategies acquired by extremophiles, as mentioned in Figure 4.1 for overcoming heavy metal/metalloid toxicity are

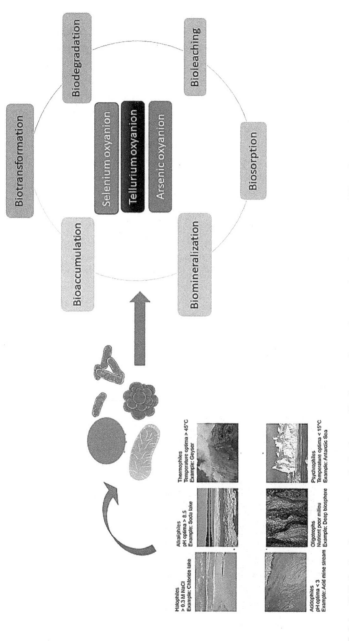

FIGURE 4.1 Cellular level processing of selenium, tellurium, and arsenic oxyanions by extremophilic microorganisms.

bioreduction, volatilization, biosorption, biotransformation, and bioaccumulation (Giovanella et al. 2020; Mosa et al. 2016).

Biotransformation or conversion of toxic selenium oxyanions to less toxic forms like elemental or biogenic selenium, or volatile selenium forms with the help of microorganisms is being more widely studied (Stefaniak et al. 2018). Oxidation, methylation, demethylation, and reduction (assimilatory and dissimilatory) reaction mechanisms are being employed by microorganisms and thus play a critical role in the biotransformation of selenium oxyanions (Eswayah et al. 2016; Paul and Saha 2019). These metabolic reactions, oxidation, reduction, methylation, and demethylation are performed in nature by extremophilic microorganisms known to thrive in the seleniferous environment and to regulate the selenium oxyanion concentrations (Dungan and Frankenberger 1999). The use of such microorganisms living in an extreme condition known for bioremediation has been widely studied and improvised further to be employed at a full scale for the same purpose. Likewise, although tellurium has a toxic effect on prokaryotic and eukaryotic cells, there are only a few bacteria, specifically extremophiles, which are resistant to tellurium and reduce the toxicity by biotransforming the tellurium oxyanions to a less toxic elemental tellurium (Sasani et al. 2020). Extremophilic microorganisms encounter many extreme conditions including metalloid contamination, they thrive in a high ionic environment and therefore are potential candidates for bioremediation and biotransformation of toxic heavy metal(loid)s (Jeong and Choi 2020). The applicability of arsenic in various fields such as medicine, agriculture, electronics, metallurgy, and several other industrial sectors, has led to profound use of this metalloid. However, recognizing its several health hazards even at low levels, measures are been made for its replacement with nontoxic elements and remediation processes (Mandal and Suzuki 2002; Paul and Saha 2019). For arsenic removal, physicochemical processes such as chemical precipitation, filtration, reverse osmosis, ion exchange, and adsorption have been performed, but these techniques are reported to have limited productivity under natural conditions, are expensive, and release toxic residues. Therefore, bioremediation seems to be a better alternative because it is cost-effective, toxin-free, and feasible in natural conditions for the removal of arsenic (Irshad et al. 2021; Rahman and Singh 2020). Microbial bioremediation has an added advantage as microorganisms tolerate a wide range of environmental conditions, especially extremophiles and various extreme heavy metal contaminated conditions, are capable of removing arsenic and another metalloids (Marques 2018). Arsenic is naturally available and, with an increase in their mobilization into the environment, extremophilic microorganisms have evolved several mechanisms for remediation of arsenic either via utilizing arsenic as a metabolite or by modifying the toxic forms of arsenic to nontoxic and nonsoluble forms via arsenic sorption, oxidation reaction, reduction reaction, and so on. (Cavalca et al. 2013). Thus, microbial bioremediation is gaining attention and employment of such organisms for the detoxification and removal of toxic metalloids/heavy metals from the contaminated region is encouraged. Extremophilic microorganisms with the capability to resist toxic metalloid stress, like selenium, tellurium, and arsenic, are thus a green alternative for metalloid removal by bioremediation.

4.3 EXTREMOPHILIC MICROORGANISMS AND THEIR VERSATILITY IN THE REMOVAL OF METALLOIDS

Extremophiles that can thrive under various extreme conditions have also recently been investigated for their resistance and tolerance towards heavy metal/metalloid contaminated environments and their capability of biotransformation of toxic pollutants to sustainable or into any beneficial less toxic compounds (Jeong and Choi 2020) Table. 4.1 refers to such extremophilic microorganisms and their minimum inhibitory concentrations (MICs) for metalloids like selenium, tellurium, and arsenic. Extremophiles inhabit extreme econiches, which makes the prospect of survival for most organisms difficult. Extremophiles also thrive in polluted regions resembling extreme environments (Jeong & Choi 2020).

The extremophilic microorganisms that show resistance/tolerance against selenium can be utilized for the biotransformation of some beneficial forms of selenium. Halophiles which are salt-loving microorganisms are studied for their capability in selenium oxyanion resistance and removal. Recently, halophilic archaeon *Halogeometricum* sp. and halophilic bacteria *Bacillus* sp. were reported to show selenite resistance and reduction of toxic selenite compounds to elemental selenium nanoparticles (Abdollahnia et al. 2020). In another report, *Halococcus salifodinae* BK18 reduces selenite to selenium nanoparticles having antiproliferative properties against HeLa cell lines, thus exploiting the selenium nanoparticles as chemotherapeutic agents (Srivastava et al. 2014). *Halorubrum xinjiangense* strain 106, isolated from Tuz (salt) lake showed high resistance to selenite and exported the reduced elemental selenium out of the cell (Güven et al. 2013). Halophilic bacterium *Halomonas* sp. strain MAM, isolated from hypersaline soil of Iran, was reported to exhibit tolerance towards selenium oxyanions, and its application in bioremediation of contaminated soils and waste discharge sites has been evaluated (Kabiri et al. 2009). Amoozegar et al. have isolated ten moderately halophilic spore-forming *Bacilli* sp. with intrinsic resistance towards chromate, tellurite, arsenate, selenate, selenite among others. They have determined that these halophilic bacteria showed resistance to selenoxyanions with MIC between 10 to 40 mM (Amoozegar et al. 2005). Recently, moderately halotolerant *Shewanella* sp. strain 9a, a selenate reducing bacteriumwas isolated and characterized by Soda et al. (Soda et al. 2018). They reported that *Shewanella* sp. 9a anaerobically reduces both selenite and selenate into elemental selenium nanomaterial, specifically suggesting this to be dissimilatory reduction enabling 40 to 70% removal of 1mM selenate and selenite. In another report, *Bacillus* sp. strain QW90 which can grow at 3% (w/v) NaCl exhibited maximum MIC for selenite that is, 550 mM, and thus is resistant to high selenite concentrations with reduction of selenite to red elemental selenium (Khalilian et al. 2014). Two strains of moderately halotolerant *Bacillus megaterium*, BSB6 and BSB12 have been identified to resist and reduce selenite to elemental selenium, thus these strains share potential in bioremediation and synthesis of selenium nanoparticles (Mishra et al. 2011). Haloalkaliphiles are a group of microorganisms that are able to grow at both extreme conditions of high salt and alkaline pH. Such haloalkaliphile *Bacillus beveridgei* strain MLTeJB isolated from Mono Lake, California was reported to respire on oxyanions of selenium, tellurium, and arsenic (Baesman et al. 2009).

TABLE 4.1

Resistance/Tolerance of Extremophiles to Selenium, Tellurium, and Arsenic Oxyanions

Microorganism	Microbial Culture	MIC	References
Selenium			
Haloarchaea	Halogeometricum sp.	5 mM selenite	(Abdollahnia et al. 2020)
	Halococcus salifodinae BK18	6 mM selenite	(Srivastava et al. 2014)
	Halorubrum xinjiangense	25 mM selenite	(Güven et al. 2013)
Halotolarant Bacteria	Halophilic Bacillus	5 mM selenite	(Abdollahnia et al. 2020)
	Halomonas sp. MAM	25 mM to 200 mM	(Kabiri et al. 2009)
	Bacilli sp. (10 species of Halophilic bacterium)	10 mM selenite and 40 mM biselenite	(Amoozegar et al. 2005)
	Shewanella sp. 9a	1 mM selenate and selenite	(Soda et al. 2018)
	Bacillus sp. QW90	550 mM selenite	(Khalilian et al. 2014)
	Bacillus megaterium BSB6 & BSB12	0.05 mM to 2 mM and complete reduction of selenite at 0.25 mM	(Mishra et al. 2011)
Haloalkaliphiles	Bacillus beveridgei MLTeJB		(Baesman et al. 2009)
Thermophiles	Tepidibacillus infernus MBL–TLP	10mM selenate and 5mM selenite	(Podosokorskaya et al. 2016)
	Bacillus thermoamylovorans SKC1	8mM selenite	(Slobodkina et al. 2007)
Tellurium			
Haloarchaea	Haloferax alexandrines GUSF-1	Grown at 1 mM tellurite	(Alvares and Furtado 2021)
	Halococcus salifodinae BK3	6 mM tellurite	(Srivastava et al. 2015)
	Staphylococcus xylosus QWTm6	26.39 mM tellurite	(Sasani et al. 2020)
	Thermoactinomyces sp. QS-2006	500 mM tellurite	(Amoozegar et al. 2012)
Halotolarant Bacteria	Halomonas sp. MAM	0.5 mM tellurite	(Kabiri et al. 2009)
	Salinicoccus sp. strain QW6	26.39 mM tellurite	(Amoozegar et al. 2008)
	Paenibacillus sp. TeW	50 to 90% tellurite oxyanions removal	(Chien and Han 2009)
	Shewanella sp. Taa	70 to 90% of 0.4 mM tellurite removal	(Soda et al. 2018)
	Psychrobacter glacincola BNF20	2.3 mM tellurite oxyanions	(Muñoz-Villagrán et al. 2018)
Haloalkaliphiles	Bacillus beveridgei MLTeJB		(Baesman et al. 2009)

	Organism	Condition	Reference
Thermophiles	Bacillus thermoamylovorans SKC1	4 mM tellurite	(Slobodkina et al. 2007)
Arsenic			
Haloarchaea	Halococcus hamelinensis		(Gudhka et al. 2015)
	Haloarcula sp. IRU1	90 mg/L arsenate	(Taran et al. 2013)
Halotolarant Bacteria	Halomonas sp. ZM3	9 mM and 700 mM for As(III) and As(V) respectively	(Dziewit et al. 2013)
	Halarsenatibacter silvermanii SLAS-1[T]		(Blum et al. 2009)
	Marinobacter santoriniensis sp.	Growth media contained 5mM of arsenic and 5 mM of arsenate	(Handley et al. 2009)
	Bacillus sp. KM02	51.45% and 53.29% for arsenite and arsenate respectively	(Dey et al. 2016)
	Aneurinibacillus aneurinilyticus BS-1	51.99% and 50.37% for arsenite and arsenate respectively	
Haloalkaliphiles	Desulfohalophilus alkaliarsenatis SLSR-1		(Blum et al. 2012)
	Desulfuribacillus alkaliarsenatis AHT28	40 mM arsenate	(Sorokin et al. 2012)
	Alkalilimnicola ehrlichii MLHE-1		(Richey et al. 2009)
Alkaliphiles	Alkaliphilus oremlandii sp. OhILAs	10 mM arsenate and 1 mM arsenite	(Fisher et al. 2008)
Acidophiles	Acidithiobacillus ferrooxidans BYQ-12	32 mM arsenite and 64 mM arsenate	(Yan et al. 2017)
	Acidithiobacillus ferrooxidans BY3		(Gao et al. 2018)
Thermophiles	Thermus thermophilus HB27	40 mM arsenite and 44 mM arsenate	(Antonucci et al. 2017; Antonucci et al. 2018)
	Tepidibacillus infernus MBL-TLP[T]	10 mM	(Podosokorskaya et al. 2016)
	Geobacillus kaustophilus	15 mM arsenite and 80 mM arsenate	(Cuebas et al. 2011)

Tepidibacillus infernus strain MBL-TLP, a moderate thermophile respired on selenate and arsenate (Podosokorskaya et al. 2016). *Bacillus thermoamylovorans* SKC1 also a moderately thermophilic bacteria was reported to reduce selenite and also other oxyanions like chromate, tellurite and others (Slobodkina et al. 2007).

Several studies have been conducted in order to isolate, characterize, and understand molecular mechanisms of extremophiles so that they could be helpful with the removal or treatment of tellurium toxicity. A study on halophilic archaeon*Haloferax alexandrinus* GUSF-1, showed tellurite resistance by its reduction and conversion to tellurium nanorods (Alvares and Furtado 2021). Similarly, *Halococcus salifodinate* BK3 showed a MIC of 6mM for tellurite with tellurite reduction leading to the synthesis of tellurium nanoparticles (Srivastava et al. 2015). Recently, a halophilic bacterium, *Staphylococcus xylosus* strain QWTm$_6$ was identified and reported to accumulate tellurite salt equal to 26.39 mM concentration and reduce it to less toxic elemental tellurium (0) (Sasani et al. 2020). A halophilic bacterium from the hypersaline environment of Iran, *Thermoactinomyces* sp. strain QS-2006, has been identified to tolerate high tellurite concentrations with growth and removal even at 500mM (Amoozegar et al. 2012). *Halomonas* sp. strain MAM isolated from hypersaline soil (Kabiri et al. 2009) was reported to have resistance to both metalloid selenium and tellurium. Characterization of moderate halophilic bacterium, *Salinicoccus* sp. strain QW6 showed removal of tellurium oxyanions at wide-range variance of pH (5.5 to 10.4), temperature (25 to 45°C), and with different salts in the culture medium, and showed reduction of toxic tellurite to less toxic elemental tellurium (Amoozegar et al. 2008). A halotolerant, *Paenibacillus* sp. strain TeW isolated from heavy-metal contaminated sediment has exhibited high resistance against tellurite, and up to 50 to 90% of tellurite oxyanions were removed by the bacteria from the culture media (Chien and Han 2009). A moderately halotolerant *Shewanella* sp. strain Taa was identified as tellurite-reducing bacteria, which could anaerobically and aerobically remove from 70 to 90% of 0.4mM tellurite from water with 6% (w/v) of NaCl within three days (Soda et al. 2018). Another halotolerant bacterium *Psychrobacter glacincola* BNF20, which was isolated from the Antarctic Peninsula was reported to show high tellurite (2.3mM) resistance (Muñoz-Villagrán et al. 2018). Thermophilic *Bacillus thermoamylovorans* SKC1 was observed to show resistance towards toxic tellurite with MIC of 4mM, and resistance towards the toxic oxyanions was observed to be due to tellurite reduction to a less toxic form (Slobodkina et al. 2007).

In order to overcome arsenic toxicity/contamination extremophiles have efficiently developed innate properties and adopt different mechanisms to survive in the presence of arsenic oxyanion environments. Genome analysis of *Halococcus hamelinensis*, a halophilic stromatolites archaeon revealed the presence of genes encoding for arsenic resistance enzymes like arsenical pump-driving ATPase, arsenate reductase, and arsenical resistance protein (Gudhka et al. 2015). In another report, halophilic archaeon, *Haloarcula* sp. IRU1 showed resistance towards arsenic and efficient arsenic bioaccumulation was obtained by this archaea (Taran et al. 2013). A halophilic bacterium *Halomonas* sp.ZM3 was reported to have tolerance towards high concentrations of inorganic arsenic species with MICs of 9 mM and 700 mM for As(lll) and As(V) respectively (Dziewit et al. 2013). *Marinobacter santoriniensis* strain NKSG1[T] a halophile isolated from hydrothermal sediment was identified as an arsenate-respiring and arsenite-oxidizing bacteria (Handley et al. 2009). *Halarsenatibacter silvermanii* strain SLAS-1[T], a halophile isolated from salt-saturated Searles Lake, California was identified as chemoautotrophic arsenate-respiring microorganismby (Blum et al. 2009). Isolated from the same place was *Desulfohalophilus alkaliarsenatis* strain SLSR-1, a haloalkaliphilic reported to have arrA gene for arsenate respiration and able to grow by arsenate reduction (Blum et al. 2012). Halotolerant bacteria *Bacillus* sp. strain KM02 and *Aneurinibacillus aneurinilyticus* strain BS-1, were reported to remove of about 51.45% and 51.99% of arsenite and 53.29% and 50.37% of arsenate, respectively. These bacteria oxidized arsenite to less toxic arsenate and thus could be applicable in arsenic bioremediation (Dey et al. 2016). *Desulfuribacillus alkaliarsenatis* strain AHT28 was identified as arsenate reducing gram-positive, spore-forming obligate anaerobic haloalkaliphilic bacterium which is capable of respiratory growth using arsenate as electron acceptor (Sorokin et al. 2012). In the case of another haloalkaliphilic bacterium, *Alkalilimnicola ehrlichii* genome analysis highlighted the presence of two operons encoding for respiratory arsenate reductase (Arr) as a bidirectional enzyme that can function as a reductase and oxidase (Richey et al. 2009). *Alkaliphilus oremlandii* sp. OhILAs capable of growth at alkaline pH was reported to reduce arsenate, which is used as a terminal electron acceptor and is also reported to transform both inorganic and organic arsenic (Fisher et al. 2008). Different strains of *Acidithiobacillus ferrooxidans* isolated from Wudalianchi volcanic lake have been reported as being capable of arsenic tolerance and bioleaching (Yan et al. 2017), while strain BY3 of the same acidophilic bacterial species was reported to be capable of bioadsorption and bio-transformation of toxic arsenic (Gao et al. 2018). Arsenic resistance studies were conducted by Antonucci et al. on *Thermus thermophilus* HB27 a thermophilic bacterium. Initially they had reported two adjacent genes TtsmtB, encoding for an ArsR/SmtB transcriptional repressor, and TTC0354, encoding for a Zn^{2+}/Cd^{2+} dependent membrane ATPase responsible for arsenic resistance because a mutation in these genes resulted in arsenic sensitivity (Antonucci et al. 2017). Later, the same group have reported the system composed of TtSmtB and TtArsx genes responsible for both arsenic and cadmium resistance. The genes responsible for arsenic resistance reported were TtArsC: arsenate reductase, TtArsX: arsenic efflux membrane protein, and TtSmtB: transcriptional repressor (Antonucci et al. 2018). *Tepidibacillus infernus* strain MBL-TLP[T], a moderate thermophilic bacterium isolated from a gold mine was identified as arsenate respiring bacteria and used arsenate as an electron acceptor for its growth (Podosokorskaya et al. 2016). Another thermophilic bacterium, *Geobacillus kaustophilus* isolated from geothermal soil was identified as arsenic resistant even at high arsenic concentrations (Cuebas et al. 2011).

4.4   ECONOMICS AND FEASIBILITY FOR BIOREMEDIATION PROCESSES

Any preferred technique for the removal of heavy metal/metalloid remediation from the environment requires several essential factors like cost-efficacy, feasibility, time duration, removal capacity, metalloid concentration, and so forth (Raffa et al. 2021).

The heavy metal/metalloid waste generated and released in the environment has gained economical value by recovering these metals and further using them in other applications. Extremophilic microorganisms are involved in bioremediation and producing value-added by-products, like biosynthesis of nanoparticles, which have applications in industrial, medical, and biosensor fields. There are various examples of the bioconversion of waste by microorganisms into a new material, like utilizing palladium- and platinum-containing wastes for clean electricity generation. The bioconversion of gold-containing wastes into a nano-gold catalyst for glycerol oxidation, bioconversion of selenium oxyanions to optically active material, and various others have been reported with economical application along with bioremediation (Macaskie et al. 2010).

REFERENCES

Abdollahnia, M., A. Makhdoumi, M. Mashreghi and H. Eshghi (2020). "Exploring the potentials of halophilic prokaryotes from a solar saltern for synthesizing nanoparticles: The case of silver and selenium." *PLOS ONE* **15**(3): e0229886.

Alvares, J. J. and Furtado, I. J.. (2021) "Anti-*Pseudomonas aeruginosa* biofilm activity of tellurium nanorods biosynthesized by cell lysate of *Haloferax alexandrinus* GUSF-1(KF796625)." *Biometals* **34**(5): 1007–1016.

Amoozegar, M. A., M. Ashengroph, F. Malekzadeh, M. Reza Razavi, S. Naddaf and M. Kabiri (2008). "Isolation and initial characterization of the tellurite reducing moderately halophilic bacterium, *Salinicoccus* sp. strain QW6." *Microbiological Research* **163**(4): 456–465.

Amoozegar, M. A., J. Hamedi, M. Dadashipour and S. Shariatpanahi (2005). "Effect of salinity on the tolerance to toxic metals and oxyanions in native moderately halophilic spore-forming bacilli." *World Journal of Microbiology and Biotechnology* **21**(6): 1237–1243.

Amoozegar, M. A., M. Khoshnoodi, M. Didari, J. Hamedi, A. Ventosa and S. A. Baldwin (2012). "Tellurite removal by a tellurium-tolerant halophilic bacterial strain, *Thermoactinomyces* sp. QS-2006." *Annals of Microbiology* **62**(3): 1031–1037.

Antonucci, I., G. Gallo, D. Limauro, P. Contursi, A. L. Ribeiro, A. Blesa, J. Berenguer, S. Bartolucci and G. Fiorentino (2017). "An ArsR/SmtB family member regulates arsenic resistance genes unusually arranged in *Thermus thermophilus* HB27." *Microbial Biotechnology* **10**(6): 1690–1701.

Antonucci, I., G. Gallo, D. Limauro, P. Contursi, A. L. Ribeiro, A. Blesa, J. Berenguer, S. Bartolucci and G. Fiorentino (2018). "Characterization of a promiscuous cadmium and arsenic resistance mechanism in *Thermus thermophilus* HB27 and potential application of a novel bioreporter system." *Microbial Cell Factories* **17**(1): 78.

Ba, L. A., M. Döring, V. Jamier and C. Jacob (2010). "Tellurium: An element with great biological potency and potential." *Organic & Biomolecular Chemistry* **8**(19): 4203–4216.

Baesman, S. M., J. F. Stolz, T. R. Kulp and R. S. Oremland (2009). "Enrichment and isolation of *Bacillus beveridgei* sp. nov., a facultative anaerobic haloalkaliphile from Mono Lake, California, that respires oxyanions of tellurium, selenium, and arsenic." *Extremophiles* **13**(4): 695–705.

Belzile, N. and Y.-W. Chen (2015). "Tellurium in the environment: A critical review focused on natural waters, soils, sediments and airborne particles." *Applied Geochemistry* **63**: 83–92.

Blum, J. S., S. Han, B. Lanoil, C. Saltikov, B. Witte, F. R. Tabita, S. Langley, T. J. Beveridge, L. Jahnke and R. S. Oremland (2009). "Ecophysiology of "*Halarsenatibacter silvermanii*" strain SLAS-1T, gen. nov., sp. nov., a facultative chemoautotrophic arsenate respirer from salt-saturated Searles Lake, California." *Applied and Environmental Microbiology* **75**(7): 1950–1960.

Blum, J. S., T. R. Kulp, S. Han, B. Lanoil, C. W. Saltikov, J. F. Stolz, L. G. Miller and R. S. Oremland (2012). "*Desulfohalophilus alkaliarsenatis* gen. nov., sp. nov., an extremely halophilic sulfate- and arsenate-respiring bacterium from Searles Lake, California." *Extremophiles* **16**(5): 727–742.

Cavalca, L., A. Corsini, P. Zaccheo, V. Andreoni and G. Muyzer (2013). "Microbial transformations of arsenic: Perspectives for biological removal of arsenic from water." *Future Microbiol* **8**(6): 753–768.

Chien, C. C. and C. T. Han (2009). "Tellurite resistance and reduction by a *Paenibacillus* sp. isolated from heavy metal-contaminated sediment." *Environ Toxicol Chem* **28**(8): 1627–1632.

Cuebas, M., D. Sannino and E. Bini (2011). "Isolation and characterization of arsenic resistant *Geobacillus kaustophilus* strain from geothermal soils." *Journal of Basic Microbiology* **51**(4): 364–371.

Dey, U., S. Chatterjee and N. K. Mondal (2016). "Isolation and characterization of arsenic-resistant bacteria and possible application in bioremediation." *Biotechnology Reports* **10**: 1–7.

Di Gregorio, S., S. Lampis and G. Vallini (2005). "Selenite precipitation by a rhizospheric strain of *Stenotrophomonas* sp. isolated from the root system of *Astragalus bisulcatus*: A biotechnological perspective." *Environment International* **31**(2): 233–241.

Dungan, R. S. and W. T. Frankenberger (1999). "Microbial transformations of selenium and the bioremediation of seleniferous environments." *Bioremediation Journal* **3**(3): 171–188.

Dziewit, L., A. Pyzik, R. Matlakowska, J. Baj, M. Szuplewska and D. Bartosik (2013). "Characterization of *Halomonas* sp. ZM3 isolated from the Zelazny Most post-flotation waste reservoir, with a special focus on its mobile DNA." *BMC Microbiol* **13**: 59.

El-Ramady, H., N. Abdalla, T. Alshaal, É. Domokos-Szabolcsy, N. Elhawat, J. Prokisch, A. Sztrik, M. Fári, S. El-Marsafawy and M. S. Shams (2015). "Selenium in soils under climate change, implication for human health." *Environmental Chemistry Letters* **13**(1): 1–19.

Eswayah, A. S., T. J. Smith and P. H. E. Gardiner (2016). "Microbial transformations of selenium species of relevance to bioremediation." *Applied and Environmental Microbiology* **82**(16): 4848–4859.

Fisher, E., A. M. Dawson, G. Polshyna, J. Lisak, B. Crable, E. Perera, M. Ranganathan, M. Thangavelu, P. Basu and J. F. Stolz (2008). "Transformation of inorganic and organic arsenic by *Alkaliphilus oremlandii* sp. nov. strain OhILAs." *Annals of the New York Academy of Sciences* **1125**: 230–241.

Frankenberger, W. T., Jr. and M. Arshad (2001). "Bioremediation of selenium-contaminated sediments and water." *Biofactors* **14**(1–4): 241–254.

Gao, Q., D. Tang, P. Song, J. Zhou and H. Li (2018). "Bio-adsorption and Bio-transformation of Arsenic by *Acidithiobacillus ferrooxidans* BY3." *International Microbiology* **21**(4): 207–214.

Ghosh, J. and P. C. Sil (2015). Mechanism for Arsenic-induced Toxic Effects. *Handbook of Arsenic Toxicology*. S. J. S. Flora. Oxford, Academic Press: 203–231.

Goswami R. K., K. Agrawal, M. P. Shah and P. Verma (2021). "Bioremediation of heavy metals from wastewater: A current perspective on microalgae-based future." *Letters in Applied Microbiology* **2021**: 13564. https://doi.org/10.1111/lam.13564

Goswami, R. K., K. Agrawal and P. Verma (2022b). "An exploration of natural synergy using microalgae for the remediation of pharmaceuticals and xenobiotics in wastewater." *Algal Research*, **64**: 102703.

Goswami, R. K., S. Mehariya, O. P. Karthikeyan, V. K. Gupta and P. Verma (2022a). "Multifaceted application of microalgal biomass integrated with carbon dioxide reduction and wastewater remediation: A flexible concept for sustainable environment." *Journal of Cleaner Production*, 130654.

Giovanella, P., G. A. L. Vieira, I. V. Ramos Otero, E. Pais Pellizzer, B. de Jesus Fontes and L. D. Sette (2020). "Metal and organic pollutants bioremediation by extremophile microorganisms." *The Journal of Hazardous Materials* **382**: 121024.

Gudhka, R. K., B. A. Neilan and B. P. Burns (2015). "Adaptation, ecology, and evolution of the halophilic stromatolite archaeon *Halococcus hamelinensis* inferred through genome analyses." *Archaea* **2015**: 11.

Güven, K., M. B. Mutlu, C. Çırpan and H. M. Kutlu (2013). "Isolation and identification of selenite reducing archaea from Tuz (salt) Lake In Turkey." *Journal of Basic Microbiology* **53**(5): 397–401.

Handley, K. M., M. Héry and J. R. Lloyd (2009). "*Marinobacter santoriniensis* sp. nov., an arsenate-respiring and arsenite-oxidizing bacterium isolated from hydrothermal sediment." *International Journal of Systematic and Evolutionary Microbiology* **59**(4): 886–892.

Irshad, S., Z. Xie, S. Mehmood, A. Nawaz, A. Ditta and Q. Mahmood (2021). "Insights into conventional and recent technologies for arsenic bioremediation: A systematic review." *Environmental Science and Pollution Research* **28**(15): 18870–18892.

Jang, Y., Y. Somanna and H. Kim (2016). "Source, distribution, toxicity and remediation of arsenic in the environment–a review." *International Journal of Applied Environmental Sciences* **11**(2): 559–581.

Jebelli, M. A., A. Maleki, M. A. Amoozegar, E. Kalantar, F. Gharibi, N. Darvish and H. Tashayoe (2018). "Isolation and identification of the native population bacteria for bioremediation of high levels of arsenic from water resources." *The Journal of Environmental Management* **212**: 39–45.

Jeong, S.-W. and Y. J. Choi (2020). "Extremophilic microorganisms for the treatment of toxic pollutants in the environment." *Molecules* **25**(21): 4916.

Kabiri, M., M. A. Amoozegar, M. Tabebordbar, K. Gilany and G. H. Salekdeh (2009). "Effects of selenite and tellurite on growth, physiology, and proteome of a moderately halophilic bacterium." *Journal of Proteome Research* **8**(6): 3098–3108.

Kapoor, A., S. Tanjore and T. Viraraghavan (1995). "Removal of selenium from water and wastewater." *International Journal of Environmental Studies* **49**(2): 137–147.

Kashyap, S., R. Chandra, B. Kumar, P. Verma (2021). "Biosorption efficiency of nickel by various endophytic bacterial strains for removal of nickel from electroplating industry effluents: An operational study." *Ecotoxicology* **31**(4): 565–580.

Khalilian, M., M. Zolfaghari, M. Soleimani and M. R. Monfared (2014). "*Bacillus* sp. strain QW90, a bacterial strain with a high potential application in bioremediation of selenite." *Report of Health Care* **1**: 6–10.

Kumar, B., K. Agrawal and P. Verma (2021). "Current perspective and advances of microbe assisted electrochemical system as a sustainable approach for mitigating toxic dyes and heavy metals from wastewater." *ASCE's Journal of Hazardous, Toxic, and Radioactive Waste* **25**(2): 04020082.

Macaskie, L. E., I. P. Mikheenko, P. Yong, K. Deplanche, A. J. Murray, M. Paterson-Beedle, V. S. Coker, C. I. Pearce, R. Cutting, R. A. D. Pattrick, D. Vaughan, G. van der Laan and J. R. Lloyd (2010). "Today's wastes, tomorrow's materials for environmental protection." *Hydrometallurgy* **104**(3): 483–487.

Maltman, C. and V. Yurkov (2019). "Extreme environments and high-level bacterial tellurite resistance." *Microorganisms* **7**(12): 601.

Mandal, B. K. and K. T. Suzuki (2002). "Arsenic round the world: A review." *Talanta* **58**(1): 201–235.

Marques, C. R. (2018). "Extremophilic microfactories: Applications in metal and radionuclide bioremediation." *Frontiers in Microbiology* **9**(1191).

Medfu Tarekegn, M., F. Zewdu Salilih and A. I. Ishetu (2020). "Microbes used as a tool for bioremediation of heavy metal from the environment." *Cogent Food & Agriculture* **6**(1): 1783174.

Mishra, R. R., S. Prajapati, J. Das, T. K. Dangar, N. Das and H. Thatoi (2011). "Reduction of selenite to red elemental selenium by moderately halotolerant *Bacillus megaterium* strains isolated from Bhitarkanika mangrove soil and characterization of reduced product." *Chemosphere* **84**(9): 1231–1237.

Mosa, K. A., I. Saadoun, K. Kumar, M. Helmy and O. P. Dhankher (2016). "Potential biotechnological strategies for the cleanup of heavy metals and metalloids." *Frontiers in Plant Science* **7**: 303–303.

Muñoz-Villagrán, C. M., K. N. Mendez, F. Cornejo, M. Figueroa, A. Undabarrena, E. H. Morales, M. Arenas-Salinas, F. A. Arenas, E. Castro-Nallar and C. C. Vásquez (2018). "Comparative genomic analysis of a new tellurite-resistant *Psychrobacter* strain isolated from the Antarctic Peninsula." *PeerJ* **6**: e4402.

Nguyen, V. K., W. Choi, Y. Ha, Y. Gu, C. Lee, J. Park, G. Jang, C. Shin and S. Cho (2019). "Microbial tellurite reduction and production of elemental tellurium nanoparticles by novel bacteria isolated from wastewater." *Journal of Industrial and Engineering Chemistry* **78**: 246–256.

Paul, T. and N. C. Saha (2019). "Environmental arsenic and selenium contamination and approaches towards its bioremediation through the exploration of microbial adaptations: A review." *Pedosphere* **29**(5): 554–568.

Podosokorskaya, O. A., A. Y. Merkel, S. N. Gavrilov, I. Fedoseev, E. V. Heerden, E. D. Cason, A. A. Novikov, T. V. Kolganova, A. A. Korzhenkov, E. A. Bonch-Osmolovskaya and I. V. Kublanov (2016). "*Tepidibacillus infernus* sp. nov., a moderately thermophilic, selenate- and arsenate-respiring hydrolytic bacterium isolated from a gold mine, and emended description of the genus *Tepidibacillus*." *International Journal of Systematic and Evolutionary Microbiology* **66**(8): 3189–3194.

Popescu, G. and L. Dumitru (2009). "Biosorption of some heavy metals from media with high salt concentrations by halophilic archaea." *Biotechnology & Biotechnological Equipment* **23**(sup1): 791–795.

Purcarea, C., G. Necula-Petrareanu and A. Vasilescu (2019). Extremophile-assisted nanomaterial production and nanomaterial-based biosensing. *Functional Nanostructured Interfaces for Environmental and Biomedical Applications*. Elsevier: 153–180.

Raffa, C. M., F. Chiampo and S. Shanthakumar (2021). "Remediation of metal/metalloid polluted soils: A short review." *Applied Sciences* **11**(9): 4134.

Rahman, Z. and V. P. Singh (2020). "Bioremediation of toxic heavy metals (THMs) contaminated sites: Concepts, applications and challenges." *Environmental Science and Pollution Research* **27**(22): 27563–27581.

Richey, C., P. Chovanec, S. E. Hoeft, R. S. Oremland, P. Basu and J. F. Stolz (2009). "Respiratory arsenate reductase as a bidirectional enzyme." *Biochemical and Biophysical Research Communications* **382**(2): 298–302.

Ruiz Fresneda, M. A., J. Delgado Martín, J. Gómez Bolívar, M. V. Fernández Cantos, G. Bosch-Estévez, M. F. Martínez Moreno and M. L. Merroun (2018). "Green synthesis and biotransformation of amorphous Se nanospheres to trigonal 1D Se nanostructures: Impact on Se mobility within the concept of radioactive waste disposal." *Environmental Science: Nano* **5**(9): 2103–2116.

Safhi, M., D. M. Alam, G. Khuwaja, M. Ashafaq, A. Khan, F. Islam, T. Anwer, G. Khan, S. Sivagurunathan Moni and F. Islam (2018). "Selenium in combination with tellurium protects the toxicity of tellurium in the liver mitochondria of rats." *Bulletin of Environment, Pharmacology and Life Sciences* **7**: 90–95.

Sanyal, T., P. Bhattacharjee, S. Paul and P. Bhattacharjee (2020). "Recent advances in arsenic research: Significance of differential susceptibility and sustainable strategies for mitigation." *Frontiers in Public Health* **8**(464).

Sasani, M., S. Heidarzadeh, M. R. Zolfaghari, M. Soleimani and S. Serajian (2020). "High potential of tellurite bioremediation by moderately halophilic *Staphylococcus xylosus*." *SN Applied Sciences* **2**(8): 1338.

Sinharoy, A. and P. N. L. Lens (2020). "Biological removal of selenate and selenite from wastewater: Options for selenium recovery as nanoparticles." *Current Pollution Reports* **6**(3): 230–249.

Slobodkina, G. B., E. A. Bonch-Osmolovskaya and A. I. Slobodkin (2007). "Reduction of chromate, selenite, tellurite, and iron (III) by the moderately thermophilic bacterium *Bacillus thermoamylovorans* SKC1." *Microbiology* **76**(5): 530–534.

Soda, S., W. Ma, M. Kuroda, H. Nishikawa, Y. Zhang and M. Ike (2018). "Characterization of moderately halotolerant selenate- and tellurite-reducing bacteria isolated from brackish areas in Osaka." *Biosci Biotechnol Biochem* **82**(1): 173–181.

Sorokin, D. Y., T. P. Tourova, M. V. Sukhacheva and G. Muyzer (2012). "*Desulfuribacillus alkaliarsenatis* gen. nov. sp. nov., a deep-lineage, obligately anaerobic, dissimilatory sulfur and arsenate-reducing, haloalkaliphilic representative of the order Bacillales from soda lakes." *Extremophiles* **16**(4): 597–605.

Srivastava, P., J. M. Braganca and M. Kowshik (2014). "In vivo synthesis of selenium nanoparticles by *Halococcus salifodinae* BK18 and their anti-proliferative properties against HeLa cell line." *Biotechnology Progress* **30**(6): 1480–1487.

Srivastava, P., E. V. Nikhil, J. M. Bragança and M. Kowshik (2015). "Anti-bacterial TeNPs biosynthesized by haloarcheaon *Halococcus salifodinae* BK3." *Extremophiles* **19**(4): 875–884.

Staicu, L. C., E. D. van Hullebusch and P. N. L. Lens (2015a). "Production, recovery and reuse of biogenic elemental selenium." *Environmental Chemistry Letters* **13**(1): 89–96.

Staicu, L. C., E. D. van Hullebusch, M. A. Oturan, C. J. Ackerson and P. N. L. Lens (2015b). "Removal of colloidal biogenic selenium from wastewater." *Chemosphere* **125**: 130–138.

Stefaniak, J., A. Dutta, B. Verbinnen, M. Shakya and E. R. Rene (2018). "Selenium removal from mining and process wastewater: A systematic review of available technologies." *Journal of Water Supply: Research and Technology-Aqua* **67**(8): 903–918.

Tan, L. C., Y. V. Nancharaiah, E. D. van Hullebusch and P. N. L. Lens (2016). "Selenium: Environmental significance, pollution, and biological treatment technologies." *Biotechnology Advances* **34**(5): 886–907.

Taran, M., M. Safari, A. Monaza, J. Z. Reza and S. Bakhtiyari (2013). "Optimal conditions for the biological removal of arsenic by a novel halophilic archaea in different conditions and its process optimization." *Polish Journal of Chemical Technology* **15**(2): 7–9.

Tremaroli, V., M. L. Workentine, A. M. Weljie, H. J. Vogel, H. Ceri, C. Viti, E. Tatti, P. Zhang, A. P. Hynes, R. J. Turner and D. Zannoni (2009). "Metabolomic investigation of the bacterial response to a metal challenge." *Applied and Environmental Microbiology* **75**(3): 719–728.

Voica, D. M., L. Bartha, H. L. Banciu and A. Oren (2016). "Heavy metal resistance in halophilic Bacteria and Archaea." *FEMS Microbiol Lett* **363**(14) 146.

Yan, L., H. Hu, S. Zhang, P. Chen, W. Wang and H. Li (2017). "Arsenic tolerance and bioleaching from realgar based on response surface methodology by *Acidithiobacillus ferrooxidans* isolated from Wudalianchi volcanic lake, northeast China." *Electronic Journal of Biotechnology* **25**: 50–57.

5 Enzyme Purification Strategies

Dixita Chettri, Manswama Boro, and Anil Kumar Verma

Sikkim University, Sikkim, India

CONTENTS

5.1 INTRODUCTION

Enzymes are important biomolecules that have a vital role in carrying out all bio-chemical reactions in living organisms. Their specificity and ability to enhance the rate of reaction in a biochemical process are essential for the functioning of all life forms. With the development of technology, these enzymes are finding further appli-cations in the commercial sector in areas such as food, agriculture, biomedical, and so on. (van Beilen and Li 2002) (Figure 5.1).

DOI: 10.1201/9781003187721-5

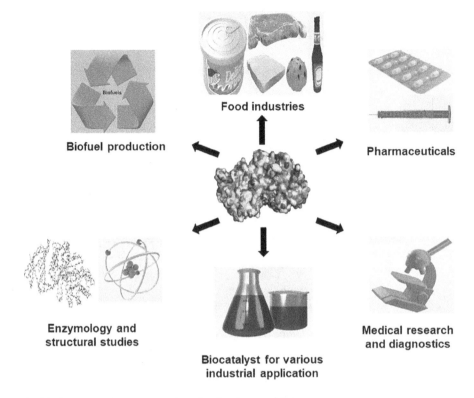

FIGURE 5.1 Different fields of application for a purified enzyme.

However, utilizing these enzymes for a different purpose requires them to be present in purified form and free of any contaminants which could be cell debris or the growth media component. Depending on the application, different levels of enzyme purity are required, with molecular biology, food, and biochemical sectors requiring a higher grade of enzyme purity. Purification of enzymes is a highly expensive step and technological advancement is required to make the overall process of using enzymes for different processes economical (Abid et al. 2018; Agrawal et al., 2019; Bhardwaj & Verma, 2021).

For ensuring higher economical yield a purification scheme needs to be designed considering all available data regarding the enzyme of interest along with different purification techniques known at the present time. The different strategies currently being followed for enzyme purification are described in the given chapter (Figure 5.2).

5.2 PREREQUISITES FOR ENZYME PURIFICATION

For designing an enzyme purification strategy, some information about the enzyme is required beforehand, such as the source of enzyme, some unique characteristics of the enzyme, and other information such as the biochemical properties of the enzyme and whether the enzyme is tagged for generation of recombinant enzyme.

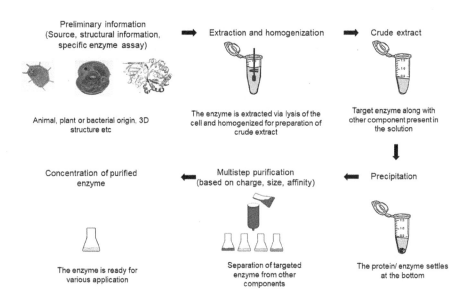

FIGURE 5.2 Schematic designing of an enzyme purification process.

Whether the enzyme is expressed intracellularly or extracellularly, any contaminants present, and the final application of the purified enzyme, are some other factors that need to be known prior to deciding the purification process (Banerjee 2006; Kumar & Verma, 2020; Kumar et al., 2018). Most essential is the knowledge of source, and assay techniques applicable to all samples since all enzyme purification processes require the sample to be prepared.

5.2.1 SOURCE AND ENZYME TYPE

The source selected for enzyme purification should be the one in which the enzyme required is present in sufficient amounts. Further, for recombinant enzymes, a cell source that has a suitable expression system should be used (Ramos and Malcata 2011; Bhardwaj et al., 2020). For enzymes expressed within the cell or that are periplasmic in nature, the purification strategy involves cell disruption to form a homogenate. Further fractionation is required to identify the cell component containing the aimed enzyme, after which differential centrifugation is required to separate this cellular component from other debris (Ramos and Malcata 2011). On the other hand, if the enzymes are extracellular in nature the purification steps become simpler as centrifugation is sufficient to separate the cell and the enzyme-containing supernatant for the purification steps to be performed (Bhatia 2018). Furthermore, based on the cell type, different techniques can be used for cell lysis such as the use of detergents like sodium dodecyl sulfate, enzymatic treatment using lysozyme, mechanical lysis using ultrasonication, electrolysis. (Shehadul Islam, Aryasomayajula, et al. 2017; Verma, 2022).

5.2.2 ASSAY

The enzyme assay is a necessary step to identify the protein or enzyme of interest among a large number of other enzymes and protein present in the source sample, with the target enzyme sometimes present in proportions as low as 1%. Before proceeding with the extensive and laborious step of enzyme purification, we have to ensure that the targeted enzyme is present in the given sample and in sufficient quantity. To establish this, an enzyme assay for the determination of enzyme activity is an essential prerequisite in the process of enzyme purification. Furthermore, assays are essential to confirm the success of every purification step, where the concentration of total protein present in the sample analyzed is measured. Thus, based on the ratio of enzyme activity of the target enzyme and the concentration of total protein present in the sample, the specific activity of the target enzyme is calculated, which is predicted to increase with every successive purification step. Therefore, the selection of a specific enzyme assay is necessary as a prerequisite to first identify the presence of the target enzyme and also to determine the efficiency of subsequent purification steps (Berg, Tymoczko, et al. 2002).

In an assay, some unique property specific for the targeted enzyme of interest is exploited to detect the presence of the enzyme among other proteins and enzymes. Though there are numerous assay techniques, the determination of a specific assay for a particular enzyme is difficult, and finding one makes the overall purification process highly effective. For the enzymes, usually the assays are designed based on the biochemical reaction they catalyze in the cell. For example, the enzyme lactase dehydrogenase catalyzes the conversion of lactate to pyruvate along with the reduction of NAD^+ to NADH in the anaerobic process of energy generation. Among the reaction components, the reduced form of nicotinamide adenine dinucleotide, that is, NADH, possesses a unique characteristic of absorbance of 340nm of light in the UV range. Thus, an enzyme assay where the absorbance at 340nm wavelength of light is measured in a given unit of time is used for determining the occurrence of the enzyme in question, that is, lactase dehydrogenase.

5.3 CONVENTIONAL ENZYME PURIFICATION STRATEGIES

Once the source and assay type for an enzyme has beenestablished, further purification techniques to be used are designed based on the enzyme properties. The classical enzyme/protein purification methods exploit these enzyme characteristics such as solubility, polarity, molecular weight, and binding affinities with more than one property being employed for the process. Thus, enzyme purification is achieved in a multistep process with an assay conducted at each step to ensure the yield of a purified enzyme with a retained activity which is an essential aspect of the process (Kumar & Sharma 2015). Some of the purification strategies are discussed below (Figure 5.3):

5.3.1 BASED ON CHARGE/SOLUBILITY

On the basis of the net surface charge of the enzyme and its interaction with varying pH, salt concentration, hydrophobic interactions, and so on, the enzymes can be separated using techniques like precipitation, ion-exchange chromatography, and electrophoresis.

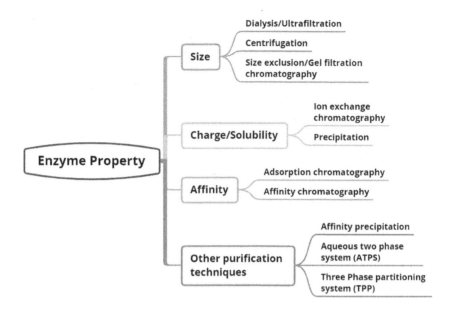

FIGURE 5.3 Different purification techniques developed based on enzyme properties.

5.3.1.1 Precipitation

Different proteins have varying solubility to different components of a solvent such as salt concentration and nature of the solvent. Salting out is a precipitation technique where the particular concentration of salt (for example ammonium sulphate) is used to precipitate the enzyme of interest depending upon its solubility since different enzymes precipitate at a different salt concentration (Burgess 2009). Similarly, the addition of an organic solvent such as acetone, methanol, and ethanol changes the dielectric constant of the solvent (Novák & Havlíček 2016) while trichloroacetic acid disrupts the hydration sphere of protein (Rajalingam, Loftis, et al. 2009) to bring about enzyme precipitation. Water miscible polymers such as cationic polymer polyethyleneimine can also be used to bring about the precipitation of negatively charged proteins (Burgess 2009). Isoelectric point precipitation is another precipitation method where the pH of the solution is brought to a point at which the net charge of the target protein is zero which is also referred to as the pI. At this point, the repulsive force present in an enzyme is balanced out with the dominance of the attractive force causing the enzyme to precipitate (Novák & Havlíček 2016).

5.3.1.2 Ion Exchange Chromatography

Chromatography is a technique developed for the separation of biomolecules based on the principle that when a mobile phase carrying a mixture of biomolecules passes through a stationary phase, these biomolecules separate from each other based on varying interactive properties of the surface of the stationary phase with each biomolecule. The biocompound with higher interaction with the stationary phase moves slowly in the chromatography system while the other molecules exit the system quickly (Coskun 2016).

Ion exchange chromatography is a type of chromatography where the separation of biomolecules is made based on the net charge of the biomolecule. The surface of the stationary phase or the column is coated with polymers bearing a charge which attracts the biomolecule with an opposite charge present in the mobile phase. This technique is commonly being used in the separation of different enzymes and proteins with H^+, Na^+, K^+, Ca^{2+}, Mg^{2+}, SO_4^{2-}, PO_4^{3-}, NR_2H^+ being used as the matrix for anion exchange chromatography for binding with negatively charged enzymes and COO^- OH^-, Cl^- as for cation ion-exchange chromatography (Cummins, Rochfort, et al. 2017).

5.3.1.3 Isoelectric Focusing Electrophoresis

This is a purification method in which an electric field is developed for the movement of proteins and enzymes through a highly polymerized matrix that separates the enzymes based on their molecular weight and charge density. Isoelectric focusing electrophoresis, similarly to isoelectric point precipitation, separates the enzymes on the basis of their charge with the protein molecules moving according to the charge they bear until they reach their pI with each enzyme having a unique pI. Along with SDS-PAGE, a type of electrophoresis method in which the overall charge of all the proteins and enzymes are made constant by treating them with sodium dodecyl sulphate, thus, separation made on the basis of the molecular weight, a two-dimensional (2-D) electrophoresis method has been devised. The proteins are first separated based on their pI by running the sample through a tube gel after which the gel is placed on the top of an SDS-PAGE for further separation of these enzyme molecules on the basis of their molecular weight. This method has been found to be highly effective in the separation of enzymes and proteins with similar properties (Smith 2017).

5.3.2 BASED ON THE SIZE

Since the protein and the enzyme can have a molecular mass ranging from 10 kDa to 1000 kDa, size is one of the factors which is taken into consideration for the purification of enzymes (Smith 2017). Some of the methods used for enzyme separation and purification of enzymes using the size as a property are:

5.3.2.1 Dialysis/Ultrafiltration

The method uses a semipermeable membrane with micropores which allow small-sized molecules to pass through for their separation from larger enzymes and other proteins molecules. A dialysis tube/membrane, with one sealed end and the other end being clamped, is filled with the sample solution which is placed on a buffer solution and continuously stirred to allow the diffusion of the lower molecular weight molecules from the membrane to the solution. Depending upon the molecular size of the enzyme of interest, the partially purified enzyme can be collected either from the tube or the buffer solution. In simple dialysis, the enzymes and protein molecules move passively via the process of osmosis. Ultrafiltration is a type of dialysis that uses extreme pressure using a mechanical pump, centrifugation, or gas pressure for the separation of the enzymes through the semipermeable membrane (Craig 1968).

5.3.2.2 Centrifugation

The principle of the centrifugation technique in enzyme separation is that different proteins having different mass/density will settle down at different rates, with heavier ones settling quickly compared to the lighter ones. In a typical protein/enzyme purification, differential centrifugation is applied where the duration and the centrifugal force are adjusted to separate the soluble enzymes and proteins from other insoluble components which are collected as a pellet. The supernatant containing the enzyme is further purified using other techniques (Lodish, Berk, et al. 2000).

5.3.2.3 Size Exclusion/Gel Filtration Chromatography

This chromatographic technique works on the same basic principle of chromatography; however, the molecules are separated on the basis of their molecular weight, where the sample containing the enzyme of interest is passed through a stationary phase which in this case is made up of polymer beads forming pores of different sizes. The large enzyme molecules with a size larger than the pores will pass through the column without entering the bead whereas the smaller-sized particles will enter the bead, thus taking a longer route to elute out of the column. Thus, the large enzyme molecules are released earlier with the small molecules eluted later. To speed up the process and elution time, external pressure can also be applied (Burgess 2018). Depending upon the protein of interest, different resins can be used for the preparation of pores of different sizes along with the suitability of the polymer. For example, the polymer which can be degraded by the targeted enzyme cannot be used for its purification. Silica (Ousalem, Zhu, et al. 2000), agarose, sephadex, polyacrylamide, polyvinyl pyrrolidone, among others are the stationary phase used in size exclusion chromatography (Hong, Koza, et al. 2012).

5.3.3 BASED ON AFFINITY

This section evaluates the different techniques used for separation and purification enzymes based on their affinity to different molecules.

5.3.3.1 Affinity Chromatography

In this commonly applied chromatographic technique the stationary phase is developed using the property of high affinity of the enzymes to different chemical groups. In this method, the sample is added to the column which is coated with the chemical group to which the protein of interest has high affinity so that only the targeted enzyme is retained in the column with the other molecules being eluted out of the column using a buffer solution. Finally, the desired enzyme is then removed from the column via the addition of a soluble form of the chemical group in high concentration (Mondal, Gupta, et al. 2006b, Hage, Anguizola, et al. 2017). Depending on the specificity of the enzyme, different ligands can be used for the purification process. For most of the enzymes, their substrate can be a suitable ligand (Hoeppner, Schmitt, et al. 2013). Based on the ligand used, different affinity chromatography can be used such as bio-affinity, immunoaffinity, dye-ligand, and immobilized metal ion-based affinity chromatography (Hage, Anguizola, et al. 2017). The technique is highly efficient for purification tagged proteins and enzymes which are developed via the process of recombination with histidine being a common tag and immobilised metal ions being used for the preparation of the solid phase (Mateo, Fernandez-Lorente et al. 2001, Andreescu, Bucur, et al. 2006).

5.3.3.2   Adsorption Chromatography

Similar to affinity chromatography, adsorption chromatography is based on the affinity of the biomolecule to adsorb to a solid adsorbent. The different biomolecules are adsorbed into a particular site (active site) of the adsorption column,based on the effective distribution coefficient, which is the ratio of solute distribution throughout the different phases of the chromatography setup. Eventually, a solvent is used to remove and collect the adsorbed enzyme of interest. Some of the adsorbent material used for the preparation of the column includes diatomaceous earth, silica, alumina, and starch (Hansen, Helboe, et al. 1984). In the case of enzymes and proteins, the commonly used adsorbent is hydroxyapatite which is a complex organic compound used for the purification of acidic and basic proteins (Surovtsev, Borzenkov, et al. 2009).

5.3.3.3   Affinity Precipitation

The technique exploits the ability of bio-affinity ligands to bind and bring about the precipitation of different biomolecules including the enzymes. When a ligand specific for the desired protein is added to a sufficient amount in the crude sample solution, the ligand and the target enzyme form a complex which can then be precipitated by applying an appropriate stimulus. Thus, the desired protein can be recovered from the precipitate with all other undesired molecules being present in the solution. The process is simple, requiring no sophisticated equipment as in the case of affinity column chromatography, and can be scaled up easily (Mondal, Jain, et al. 2006a).

All these steps require enzyme activity assay and protein quantification assay to check for the efficiency of each process. Table 5.1 summarizes the principles and advantages of each of these purification techniques along with the volume of sample that can be used for the purification, with a small volume indicating less than 100mg of protein sample.

5.4   OTHER ENZYME PURIFICATION STRATEGIES

Other than the above-mentioned techniques, different schemes are available for enzyme purification. From crystallization technique, which has long been used, to newly developed two-phase and three-phase partitioning which are emerging and establishing an effortless way for achieving enzymes with high purity.

5.4.1   Aqueous Two-phase System (ATPS)

The aqueous two-phase partitioning system which has been widely acknowledged as a highly resourceful and advanced liquid-liquid fractionation method is an effortless and proficient way for purification and recovery of different biomolecules (Mohammadi & Omidinia 2013; Bhardwaj &Verma 2020). An ATPS consists of a

TABLE 5.1

Principles, Advantages, and Sample Sizes of Different Purification Methods

Principle	Method	Advantage	Scale of Application
Charge/ Solubility	Precipitation	An easy and cost-effective method to concentrate protein and enzyme.	Small
	Ion exchange chromatography	Reliable with high efficiency and selectivity.	Large and small
	Isoelectric focusing	Easy handling and providing higher resolution with high sample capacity for purification of the enzyme with low abundance.	Small
	Electrophoresis	Easy separation with modifications for further separation based on charge and size. Visible confirmation of effectiveness of purification steps.	Small
Size	Dialysis/ ultrafiltration	Inexpensive method of concentrating enzymes by removal of smaller unwanted compounds.	Small
	Centrifugation	Separation of crude enzyme from other cellular components.	Large and small
	Size exclusion/ gel filtration chromatography	Separation of larger molecules from smaller compounds with small elution volume and any solution can be used to run the column without interference to the process.	Small
Affinity	Affinity chromatography	High specificity with target enzyme obtained in high purity and yield.	Small
	Adsorption chromatography	A wide range of mobile phases can be used with the separation of complex mixtures which is not possible by other purification methods.	Small
	Affinity precipitation	Economical and rapid requiring fewer steps for separation of target enzyme compared to chromatographic methods.	Small

liquid-liquid two-phase aqueous system that is prepared either by combining a polymer with a salt or integrating the liquid solution of two different polymers. Based on the source, the ATPS system parameters that is the polymer/salt components of particular molecular weight, their concentration, pH, and volume are selected and finally evaluated for the purification and recovery of the target enzyme (Rosa, Ferreira, et al. 2010). The polymers and salt components used are incompatible with each other which, when added in a concentration above a critical value, are separated into two phases (Nadar, Pawar, et al. 2017). The most commonly used polymer is polyethylene glycol (PEG) which is used in combination with other different polymers such as starch, dextran, and polyvinyl alcohol.

The biphasic system of ATPS provides increased selectivity for protein separation thus providing a very high yield in the preliminary step of enzyme extraction itself. The integration of preliminary down-streaming phases with clarification and concentration of protein and enzymes along with partial purification in one single unit makes the processing an easy and rapid task, which is an advantage of the ATPS method (Rachana and Lyju Jose 2014; Agrawal and Verma, 2020). Further, scaling up of the process for largescale purification of proteins and enzyme using an aqueous two-phase system is relatively simple. The ATPS can also be used as an alternative for the micro- and ultrafiltration steps of protein purification for concentrating the proteins as well as separation of membrane proteins and replacement of chromatography techniques (Kumar and Sharma 2015). Further, the system provides environmental benefits of lower toxicity, lower energy requirements, and biocompatibility other than retaining the structural and catalytic activity of the enzymes (Neelwarne 2012). This type of purification method is particularly desirable for industrial application for the generation of purified enzymes with increasing prominence in the commercial purification of enzymes and proteins (Kumar and Sharma 2015).

However, a major shortcoming of the ATPS is in the lower specificity for purification of the desired biomolecule and so advanced ATPSs are being developed such as an aqueous two-phase affinity partitioning system where the specificity is increased by manipulating the interaction of the targeted biomolecule with an affinity ligand. Similarly in the ionic liquid-based aqueous two-phase partitioning system, the electrolytic property of the ionic liquids (ILs) is used, whereby the competition for water molecule between ILs and the salt, with water molecules moving toward salts with higher affinity, results in separation of the two phases. The thermo-separating polymer-based aqueous two-phase system and the aqueous micellar two-phase system are other advanced ATPSs along with microfluidic technology merging to meet the increasing biotechnological demands (Nadar, Pawar et al. 2017).

5.4.2 Three-phase Partitioning (TPP) System

In recent years a new method of biomolecule separation has developed termed three-phase partitioning (TPP). This system has arisen as a highly simple, economical, and efficient method for rapid purification of enzymes and proteins via integration of the enrichment, purification, and recovery process of different biomolecules including the enzymes (Roy and Gupta 2002). This technique consists of three phases and has been successfully used for the separation of enzymes and proteins with the proteins or the enzyme of interest being trapped in the mid-layer (Kumar and Sharma 2015). Since no polymer is added in this method, the task of removing them is not applicable here. Instead, a salt such as ammonium sulphate, sodium citrate, or potassium sulphateis added sufficiently to the available crude sample followed by an organic solvent (3-butanol, n-propanol, sec-propanol). After proper blending of these components, the solution separates into an aqueous layer followed by a mid-protein/enzyme enriched layer followed by an organic alcohol/solvent layer on the top which accumulates all the nonpolar contaminations in the form of lipids, detergents, and so forth, along with compounds of small molecular weight. The polar impurities like the cell debris, unwanted proteins, and saccharides are collected at the bottom in the aqueous phase of the reaction mixture (Dennison and Lovrien 1997, Kumar and Rapheal 2011, Tan, Wang, et al. 2015).

For cases where even after the separation of the three phases proteins are found mainly in the aqueous layer, a modified two-step TPP has been performed where in the second step the concentration of the salt and the organic solvent is increased

(Jain, Singh, et al. 2004). Similarly, a metal affinity-based step can be integrated with TPP (Roy and Gupta 2002) or in some cases macroligand facilitated three-phase partitioning for purification of different enzymes (Sharma and Gupta 2002).

The advantage of the TPP system includes fewer processing steps and a shorter processing time with approximately five to ten minutes required for the processing of milliliters of the crude protein sample. Unlike chromatography, no dilution is required in the TPP and thus proteins and enzymes concentrated in fold up to 100 times can be achieved with this method compared to other techniques. Since TPP eliminates all the undesirable compounds such as the phenolics, lipids, and detergent present in the crude sample, the process is also called a depigmenting method. Based on the sample

TABLE 5.2

Comparison of ATPS and TPP System of Enzyme Purification

ATPS	TPP	Reference
A simple and fast method of enzyme purification with low energy requirements, biocompatibility and eco-friendly nature making it easy for scaling up for processing in large volumes for commercial application.	The use of ammonium sulphate and butanol as salt and solvent respectively along with the process being carried out at room temperature makes the process economical for industrial use.	(Pike, Dennison et al. 1989, Show, Tan et al. 2012, Avhad, Niphadkar et al. 2014)
Separation of the enzymes take place by separation of the solution into two phases with the protein/ enzyme of interest present in the aqueous phase.	The solution separates into three phases with the enzyme being found to be present in the mid-layer.	(Roy and Gupta 2002, Yan, Wang et al. 2018, Albertsson, Johansson et al. 2020)
The properties of nontoxicity, flammability, and interfacial tension provided by the process makes the condition mild and suitable for maintaining the structural and functional stability of the biomolecules.	Some of the enzymes are inhibited by the solvent used such as t-butanol while others are stabilized with increased activity.	(Becker, Thomas et al. 2009, Kuepethkaew, Sangkharak et al. 2017, Vobecká, Romanov et al. 2018)
Can be used for processing of samples with both small as well as large volumes in a relatively shorter period of time.	Applicable for both small scale lab research as well as large scale industrial use.	(Iqbal, Tao et al. 2016, Chew, Ling et al. 2019)
Since different factors such as temperature, pH, choice of polymer and salt used influence the partitioning, the method is more effective.	The process is also affected by the concentration of the reagents used, temperature and pH.	(Raja, Murty et al. 2011, Yan, Wang et al. 2018)
The process involves combination of either polymer-polymer, polymer-salt or alcohol-salt for preparation of the two-phase system.	A salt is added to an aqueous sample followed by addition of an organic solvent.	(Ooi, Tey et al. 2009, Kuepethkaew, Sangkharak et al. 2017)
Does not require expensive equipment and reagents.	No requirement of heavy or high-priced equipment and reagents.	(Dobreva, Zhekova et al. 2019)

the TPP can be adjusted for selecting a wide range of pH, temperature other than salts, and solvent alcohol used and can be controlled at room temperature. As the most commonly used reagents of TPP that is ammonium sulphate and 3-butanol are inexpensive, the overall process is highly economic (Kumar and Sharma 2015). A comparative analysis of the difference between ATPS and TPP methods of purification is given in Table 5.2.

5.5 CONCLUSION

Enzymes are important biomolecules with wide application and their study as well as commercial application has become an integrated part of the modern world. From the academic to industrial sector a certain level of enzyme purity is required for their application. The development of new biotechnological techniques has reduced the initial upstream steps for enzyme production via the production of recombinant proteins and enzymes shifting the burden to the process of purification and concentration of the protein and enzymes. Thus, rapid development is needed in this area to make the study and commercial application of the enzyme economic. The existing conventional techniques, though efficient, are laborious and expensive. The newly emerging techniques like ATPS, TPP are can have certain advantages on the economic front with multiple steps being integrated. Thus, a proper scheme including both the conventional as well as the newly emerging methods needs to be made for making the process of enzyme purification economically feasible.

ACKNOWLEDGMENTS

The authors would like to thank the Department of Microbiology, Sikkim University for providing the computational infrastructure and central library facilities for procuring references.

REFERENCES

Abid, F., M. Zahid, Z. Abedin, S. Nizami, M. Abid, S. Kazmi, S. Khan, H. Hasan, M. Ali and A. Gul (2018). "Omics technologies and bio-engineering."
Agrawal, K., N. Bhardwaj, B. Kumar, V. Chaturvedi and P. Verma (2019). "Process optimization, purification and characterization of alkaline stable white laccase from *Myrothecium verrucaria* ITCC-8447 and its application in delignification of agroresidues." *International Journal of Biological Macromolecules* 125; 1042–1055.
Agrawal, K. and P. Verma (2020). "Potential removal of hazardous wastes using white laccase purified by ATPS–PEG–salt system: an operational study." *Environmental Technology & Innovation*, 17, 100556.
Albertsson, P.-Å., G. Johansson and F. Tjerneld (2020). "Aqueous two-phase separations." In *Separation Processes in Biotechnology*, CRC Press Boca Raton: 287–328.
Andreescu, S., B. Bucur and J.-L. Marty (2006). "Affinity immobilization of tagged enzymes." In *Immobilization of Enzymes and Cells*, Springer, Humana Totowa, NJ: 97–106.
Avhad, D. N., S. S. Niphadkar and V. K. Rathod (2014). "Ultrasound assisted three phase partitioning of a fibrinolytic enzyme."*Ultrasonics Sonochemistry* 21(2): 628–633.
Banerjee, R. (2006). "Isolation and purification of enzymes." In *Enzyme Technology*, Springer, New York, NY: 515–532.

Bhardwaj, N. and P. Verma (2020). "Extraction of fungal xylanase using ATPS-PEG/sulphate and its application in hydrolysis of agricultural residues." In *Biotechnological Applications in Human Health*, Springer, Singapore: 95–105.

Bhardwaj, N. and P. Verma (2021). "Microbial xylanases: A helping module for the enzyme biorefinery platform." In *Bioenergy Research: Evaluating Strategies for Commercialization and Sustainability*, John Wiley & Sons Ltd. West Sussex, UK 129–152.

Bhardwaj, N., V.K. Verma, V. Chaturvedi and P. Verma (2020). "Cloning, expression and characterization of a thermo-alkali-stable xylanase from *Aspergillus oryzae* LC1 in Escherichia coli BL21 (DE3)." *Protein Expression and Purification* **168**, 105551.

Becker, J., O. Thomas and M. Franzreb (2009). "Protein separation with magnetic adsorbents in micellar aqueous two-phase systems." *Separation and Purification Technology* **65**(1): 46–53.

Berg, J., J. Tymoczko and L. Stryer (2002). "The purification of proteins is an essential first step in understanding their function." *Biochemistry* **5**: 20.

Bhatia, S. (2018). "Technologies and procedures involved in enzyme production." *Introduction to Pharmaceutical Biotechnology* In IOP Sciences, Bristol UK **2**: 2–42.

Burgess, R. R. (2018). "A brief practical review of size exclusion chromatography: rules of thumb, limitations, and troubleshooting." *Protein Expression and Purification* **150**: 81–85.

Burgess, R. R. (2009). "Protein precipitation techniques." *Methods in Enzymology* **463**: 331–342.

Chew, K. W., T. C. Ling and P. L. Show (2019). "Recent developments and applications of three-phase partitioning for the recovery of proteins." *Separation & Purification Reviews* **48**(1): 52–64.

Coskun, O. (2016). "Separation techniques: chromatography." *Northern Clinics of Istanbul* **3**(2): 156.

Craig, L. (1968). "Dialysis and ultrafiltration." *Methods in Immunology and Immunochemistry* **2**: 119–133.

Cummins, P. M., K. D. Rochfort and B. F. O'Connor (2017). "Ion-exchange chromatography: basic principles and application." In: *Protein Chromatography*, Springer, Humana Press, New York: 209–223.

Dennison, C. and R. Lovrien (1997). "Three phase partitioning: concentration and purification of proteins." *Protein Expression and Purification* **11**(2): 149–161.

Dobreva, V., B. Zhekova and G. Dobrev (2019). "Use of aqueous two-phase and three-phase partitioning systems for purification of lipase obtained in solid-state fermentation by *Rhizopus arrhizus*." *The Open Biotechnology Journal* **13**(1).

Hage, D. S., J. A. Anguizola, R. Li, R. Matsuda, E. Papastavros, E. Pfaunmiller, M. Sobansky and X. Zheng (2017). "Affinity chromatography." In *Liquid Chromatography*, Elsevier: 319–341.

Hansen, S. H., P. Helboe and U. Lund (1984). "Adsorption and partition chromatography." In *New Comprehensive Biochemistry*, Elsevier. **8**: 167–204.

Hoeppner, A., L. Schmitt and S. H. Smits (2013). *Proteins and their Ligands: Their Importance and How to Crystallize Them*, InTech Rijeka.

Hong, P., S. Koza and E. S. Bouvier (2012). "A review size-exclusion chromatography for the analysis of protein biotherapeutics and their aggregates." *Journal of Liquid Chromatography & Related Technologies* **35**(20): 2923–2950.

Iqbal, M., Y. Tao, S. Xie, Y. Zhu, D. Chen, X. Wang, L. Huang, D. Peng, A. Sattar and M. A. B. Shabbir et al. (2016). "Aqueous two-phase system (ATPS): an overview and advances in its applications." *Biological Procedures Online* **18**(1): 1–18.

Jain, S., R. Singh and M. N. Gupta (2004). "Purification of recombinant green fluorescent protein by three-phase partitioning." *Journal of Chromatography A* **1035**(1): 83–86.

Kuepethkaew, S., K. Sangkharak, S. Benjakul and S. Klomklao (2017). "Use of TPP and ATPS for partitioning and recovery of lipase from Pacific white shrimp (*Litopenaeus vannamei*) hepatopancreas." *Journal of Food Science and Technology* **54**(12): 3880–3891.

Kumar, B., N. Bhardwaj, A. Alam, K. Agrawal, H. Prasad, and P. Verma (2018). "Production, purification and characterization of an acid/alkali and thermo tolerant cellulase from *Schizophyllum commune* NAIMCC-F-03379 and its application in hydrolysis of lignocellulosic wastes." *AMB Express* **8**(1): 1–16.

Kumar, B. and P. Verma (2020). "Enzyme mediated multi-product process: a concept of bio-based refinery." *Industrial Crops and Products* **154**: 112607.

Kumar, P. and S. M. Sharma (2015). "An overview of purification methods for proteins." *Indian Journal of Applied Research* **1**(12): 450–459.

Kumar, V. V. and V. S. Rapheal (2011). "Induction and purification by three-phase partitioning of aryl alcohol oxidase (AAO) from *Pleurotus ostreatus*." *Applied Biochemistry and Biotechnology* **163**(3): 423–432.

Lodish, H., A. Berk, S. Zipursky, P. Matsudaira, D. Baltimore andJ. Darnell (2000). "Section 3.5, purifying, detecting, and characterizing proteins." In *Molecular Cell Biology*, WH Freeman: New York, NY.

Mateo, C., G. Fernandez-Lorente, B. C. Pessela, A. Vian, A. V. Carrascosa, J. L. Garcia, R. Fernandez-Lafuente and J. M. Guisan (2001). "Affinity chromatography of poly-histidine tagged enzymes: new dextran-coated immobilized metal ion affinity chromatography matrices for prevention of undesired multipoint adsorptions." *Journal of Chromatography A* **915**(1–2): 97–106.

Mohammadi, H. S. and E. Omidinia (2013). "Process integration for the recovery and purification of recombinant *Pseudomonas fluorescens* proline dehydrogenase using aqueous two-phase systems." *Journal of Chromatography B* **929**: 11–17.

Mondal, K., M. N. Gupta and I. Roy (2006b). *Affinity-based Strategies for Protein Purification*, Analytical Chemistry, 78(11): 3499–3504.

Mondal, K., S. Jain, S. Teotia and M. N. Gupta (2006a). "Emerging options in protein bioseparation." *Biotechnology Annual Review* **12**: 1–29.

Nadar, S. S., R. G. Pawar and Rathod (2017). "Recent advances in enzyme extraction strategies: a comprehensive review." *International Journal of Biological Macromolecules* **101**: 931–957.

Neelwarne, B. (2012). *Red Beet Biotechnology: Food and Pharmaceutical Applications*, Springer Science & Business Media.

Novák, P. and V. Havlíček (2016). "Protein extraction and precipitation." In *Proteomic Profiling and Analytical Chemistry*, Elsevier: 51–62.

Ooi, C. W., B. T. Tey, S. L. Hii, S. M. M. Kamal, J. C. W. Lan, A. Ariff and T. C. Ling (2009). "Purification of lipase derived from *Burkholderia pseudomallei* with alcohol/salt-based aqueous two-phase systems." *Process Biochemistry* **44**(10): 1083–1087.

Ousalem, M., X. Zhu and J. Hradil (2000). "Evaluation of the porous structures of new polymer packing materials by inverse size-exclusion chromatography." *Journal of Chromatography A* **903**(1–2): 13–19.

Pike, R. and C. Dennison (1989). "Protein fractionation by three phase partitioning (TPP) in aqueous/t-butanol mixtures." *Biotechnology & Bioengineering* **33**(2): 221–228.

Rachana, C. and V. J. Jose (2014). "Three phase partitioning-a novel protein purification method." *International Journal of ChemTech Research* **6**(7): 3467–3472.

Raja, S., V. R. Murty, V. Thivaharan, V. Rajasekar and V. Ramesh (2011). "Aqueous two phase systems for the recovery of biomolecules–a review." *Science and Technology* **1**(1): 7–16.

Rajalingam, D., C. Loftis, J. J. Xu and T. K. S. Kumar (2009). "Trichloroacetic acid-induced protein precipitation involves the reversible association of a stable partially structured intermediate." *Protein Science* **18**(5): 980–993.

Ramos, O. and F. J. C. B. Malcata (2011). Food-grade enzymes. In *Comprehensive Biotechnology*, 2nd ed, Elsevier: USA 555–569.

Rosa, P., I. Ferreira, A. Azevedo and M. Aires-Barros (2010). "Aqueous two-phase systems: a viable platform in the manufacturing of biopharmaceuticals." *Journal of Chromatography A* **1217**(16): 2296–2305.

Roy, I. and M. N. Gupta (2002). "Three-phase affinity partitioning of proteins." *Analytical Biochemistry* **300**(1): 11–14.

Sharma, A. and M. N. Gupta (2002). "Macroaffinity ligand-facilitated three-phase partitioning (MLFTPP) for purification of xylanase." *Biotechnology & Bioengineering* **80**(2): 228–232.

Shehadul Islam, M., A. Aryasomayajula and P. R. Selvaganapathy (2017). "A review on macroscale and microscale cell lysis methods." *Micromachines* **8**(3): 83.

Show, P. L., C. P. Tan, M. S. Anuar, A. Ariff, Y. A. Yusof, S. K. Chen and T. C. Ling (2012). "Primary recovery of lipase derived from *Burkholderia cenocepacia* strain ST8 and recycling of phase components in an aqueous two-phase system." *Biochemical Engineering Journal* **60**: 74–80.

Smith, D. M. (2017). "Protein separation and characterization procedures." In *Food Analysis*, Springer , Cham: 431–453.

Surovtsev, V., V. Borzenkov and K. J. B. Detushev (2009). "Adsorption chromatography of proteins." *Biochemistry Moscow* **74**(2): 162–164.

Tan, Z.-J., C.-Y. Wang, Y.-J. Yi, H.-Y. Wang, W.-L. Zhou, S.-Y. Tan and F.-F. Li (2015). "Three phase partitioning for simultaneous purification of aloe polysaccharide and protein using a single-step extraction." *Process Biochemistry* **50**(3): 482–486.

van Beilen, J. B. and Z. J. Li (2002). "Enzyme technology: an overview." *Current Opinion in Biotechnology* **13**(4): 338–344.

Verma, P. (2022) *Industrial Microbiology and Biotechnology*, Springer, Singapore.

Vobecká, L., A. Romanov, Z. Slouka, P. Hasal and M. Přibyl (2018). "Optimization of aqueous two-phase systems for the production of 6-aminopenicillanic acid in integrated microfluidic reactors-separators." *New Biotechnology* **47**: 73–79.

Yan, J.-K., Y.-Y. Wang, W.-Y. Qiu, H. Ma, Z.-B. Wang and J.-Y. Wu (2018). "Three-phase partitioning as an elegant and versatile platform applied to nonchromatographic bioseparation processes." *Critical Reviews in Food Science and Nutrition* **58**(14): 2416–2431.

6 Overview of the Enzyme Support System of Immobilization for Enhanced Efficiency and Reuse of Enzymes

Cecil Antony, Praveen Kumar Ghodke, and Saravanakumar Thiyagarajan

Department of Biochemistry and Molecular Biology, Michigan State University, East Lancing, Michigan – 48824, USA

CONTENTS

DOI: 10.1201/9781003187721-6

6.1   INTRODUCTION

Enzymes are molecular biocatalysts that are present universally in all organisms. Enzymes increase the rate of reaction without being altered by the reactions. They also maintain the reactant and product equilibrium (Cooper 2000). Furthermore, enzymes speed up biochemical reactions, and without them reactions in complex life systems take place at a much slower rate (Hernandez & Fernandez-Lafuente 2011). In industrial applications the majority of enzyme usage is hindered by operational stability over a prolonged time, and by technical difficulties with the reuse of enzymes during the recovery process (Krajewska 1991). For the proper functioning of enzymes, their amino acid sequence and interactions between residues in the enzyme active site must naturally be in an optimized condition (Demetrius 2002). Enzymes are extensively used in many industries such as: food (juice, brewery, and dairy), textile, pharmaceutical, chemical, water treatment, paper, and pulp-making (DiCosimo et al. 2013). Apart from industrial applications, enzymes are innovatively utilized in environmental biotechnology sectors like waste management and wastewater treatment systems (Tonini & Astrup 2012; Bhardwaj et al. 2021a). At the beginning of the 21st century, many countries around the globe are focusing on research funding for the development of biofuels by converting waste biomass (Atadashi et al. 2010; Bhardwaj & Verma 2021; Kumar & Verma 2020).

The overall industrial and environmental application of enzymes could be feasible if hese enzymes have robust stability and reusability during the process. The ability to stabilize and reuse an enzyme catalyst through immobilization has been shown to be one of the essential components in making an enzymatic process economically viable in modern biotechnology. (Parmar et al. 2000). The availability of immobilized enzyme catalysts with increased activity and stability is also projected to reduce product costs. (Blanco et al. 2007). Furthermore, enzyme immobilization has recently been revealed as an exceptionally strong tool for enhancing enzyme properties, if meticulously designed with regard to, for example: stability, activity, specificity and selectivity, and inhibition reduction. (Ottolina et al. 1992). However, compared with soluble enzymes, immobilized enzymes produced unpredicted results such as reduction or increase in activity during experiments. The cost of enzyme and enzymatic reactions reduces remarkably with enzyme immobilization (Sheldon 2007b).

6.2 ADVANTAGES OF ENZYME IMMOBILIZATION

There are numerous reasons for producing and using immobilized enzymes such as:

(i) More convenient handling of enzyme preparations.
(ii) Ease of separation of the enzyme from the product.
(iii) Reuse and efficient recovery after the reaction.

The convenience with which the enzyme can be separated from the products simplifies enzyme applications and promotes a reliable and efficient reaction technology. Enzyme reuse, on the other hand, offers cost advantages, which are often a prerequisite for establishing an enzyme-catalyzed process. In immobilized enzyme preparations, the properties of the enzyme and carrier material help to determine their properties.

Of the numerous parameters which must be considered, the most important factors are outlined in Table 6.1. Considering product cost, the immobilization technique and the amount of soluble enzyme utilized are most crucial for the immobilized

---

TABLE 6.1
Characteristics of Immobilized Enzymes

1	Enzyme	Biochemical properties of enzymes: Molecular weight of enzymes, functional groups, and prosthetic groups present on protein surface, purity of enzymes utilized. Kinetic parameters of enzymes: Specific activity of enzyme, optimum pH, temperature and reaction kinetics for activity and inhibition, enzyme stability against variable pH, temperature, solvents concentrations.
2	Carrier support	Chemical features: Chemical composition of support molecules, functional groups on the surface, swelling or shrinkage properties, volume of matrix and porous size in case of entrapment/ encapsulation and chemical stability of carrier support molecules. Mechanical properties: Diameter of particle, compression behavior for each particle, flow resistance, sedimentation velocity.
3	Immobilized enzyme	Immobilization method: Amount of protein binding, yield in immobilized enzyme activity, kinetic parameters of bounded enzymes. Mass transfer effects: Different concentrations of solutes inside and outside the catalyst particles, porous diffusion rate bot externally as well internally. This shows the efficacy in proportion to free enzyme measured under suitable reaction conditions. Stability: Immobilized enzyme operational stability (activity decay and reusage), stability in storage. Performance: Productivity of immobilized enzymes (product per unit of enzyme), consumption of enzyme (e.g., units kg^{-1} product, until half-life)

enzyme activity. The resulting activity may be further decreased by mass transfer effects under process conditions. The yield of enzyme activity after immobilization is affected not only by binding-related losses but also by decreased availability of enzyme molecules within pores or by slowly diffusing substrate molecules.

6.3 FACTORS FOR THE COST OF ENZYME IMMOBILIZATION

To determine the cost advantages of immobilized enzymes, the individual manufacturing steps and their contribution to overall costs must also be considered.

The three main factors are:

1) The expense of biomass derived from plants and animals, as well as biomass derived through microbial fermentations (costs are also determined mainly by the scale of fermentation and the rate of enzyme expression).
2) Type of down-streaming to achieve purity with less loss of activity (Larger fermentation scales may be required to compensate for activity loss and a cost increase).
3) Reduction in the cost of immobilization procedures by using cheap biopolymer and less expensive cross-linking agents (the selection of a stable biopolymer and characterization of it with immobilized enzyme will eventually increase the reuse cycles)

In conclusion, all these steps will lead to dramatic cost reductions.

6.4 ENZYME IMMOBILIZATION METHODOLOGY AND STRATEGIES

It is critical in choosing the right mode of immobilization to avoid losing enzyme function and altering the chemical structure or reactive groups in the enzyme's active site. It is beneficial to have a good knowledge of the nature of the enzyme's active site. The active site can be protected by the attachment of protective groups, which could be removed later without adversely impacting enzyme activity. In a few conditions, a substrate or competitive inhibitor of the enzymes can perform this protective function. Adsorption, covalent coupling, trapping, and cross-linking are the widely prevalent techniques in enzyme immobilization strategy (Brady & Jordaan 2009).

Enzyme immobilization methods based on different physical and chemical principles are summarized in Figure 6.1.

6.4.1 ADSORPTION OF ENZYMES

Enzyme adsorption onto insoluble is a simple technique with a variety of applications. This method involves the physical adherence of the enzyme to the immobilization matrix or surface. Compared to other immobilization methods it has a high enzyme loading capability. Immobilization can be simply achieved by mixing enzymes with the appropriate adsorbent carrier at optimum pH and ionic strength. The enzyme immobilized is directly usable after washing off the unbound enzymes.

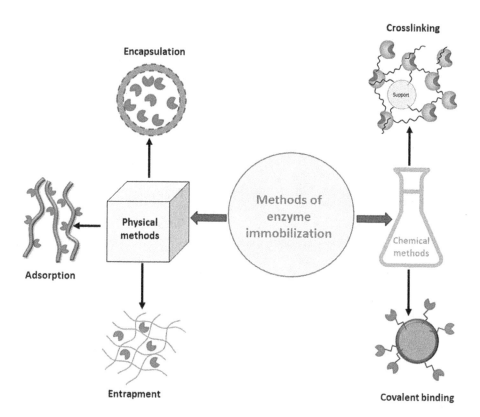

FIGURE 6.1   Enzyme immobilization methods.

Weak forces like Vander Waal forces, hydrogen or ionic bonding, and hydrophobic interactions are primarily involved in the adsorption process. Ion exchangers were the first immobilization matrix or surface used by the adsorption method. A specific charge is imparted on the enzyme surface by the pH and isoelectric point that can be calculated for selecting the compatible immobilization matrix or surface. Ion exchangers have the disadvantage of inhibiting enzyme activity upon pH alterations arising due to the high charges of the ionic matrix Apart from ionic interaction, affinity binding has also been used for the immobilization of the enzymes. Affinity binding works on the mechanism of attraction between two complementary molecules which offers high selectivity. The confirmation of the immobilized enzymes is less affected with greater retention in its activity. It is possible to protect adsorbed enzymes against agglomeration, proteolysis, and interactions with hydrophobic surfaces (Minteer 2014).

To minimize chemical modification and enzyme denaturation, retain existing surface properties of enzymes, and minimize leakage of enzymes used during immobilization, steps must be taken for the choice of adsorbents. The main disadvantage of this method is the changes that occur in the active site microenvironment of enzymes during the binding of a single enzyme molecule with several binding sites in the

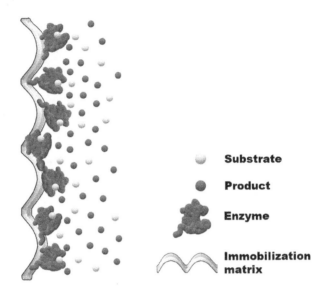

FIGURE 6.2   Immobilization of enzymes by adsorption.

adsorbent surfaces (Figure 6.2). Other factors which affect the adsorption are fluctuations in temperature and changes in ionic concentrations (Johnson et al. 1996).

6.4.2   ENTRAPMENT/ENCAPSULATION OF ENZYMES

This method of enzyme immobilization is found to be irreversible as the immobilization matrix or support entraps the enzymes using covalent or noncovalent. The immobilization matrix or support only allows the substrates and products to pass through them. preventing entry by the enzymes. The porous gel or fiber might be of natural or synthetic origin. This method has been reported to enhance mechanical stability with high retention capacity. The enzyme denaturation is prevented as it cannot chemically react with the immobilization matrix or support. The enzyme entrapment is done by various methods as shown in Figure 6.3, such as gel or fiber microencapsulation (a) or entrapping (b). For reactions with low molecular weight substrates and products, this method is very convenient. It is cost effective with an increase in stability as well as being effective while providing better contact between enzyme and substrates, thus reducing the mass transfer of substrate to enzyme molecules captured inside the carrier support due to its small pore size matrix. Both entrapment and encapsulation are purely caging of molecules that involve covalent binding (Shen et al. 2011). The disadvantages in these types of methods are low enzyme loading capacity, abrasion of carrier support, and deactivation during the immobilization steps, but these can often be circumvented by the addition of cross-linking agents (Mohamad et al. 2015). For entrapment/encapsulation both natural and synthetic polymers are used. Commonly used natural materials are agar, agarose, alginate, and carrageenan (Datta et al. 2013). In the case of synthetic polymers, polyacrylamide, and polyvinyl alcohol are extensively used (Grosová et al. 2007).

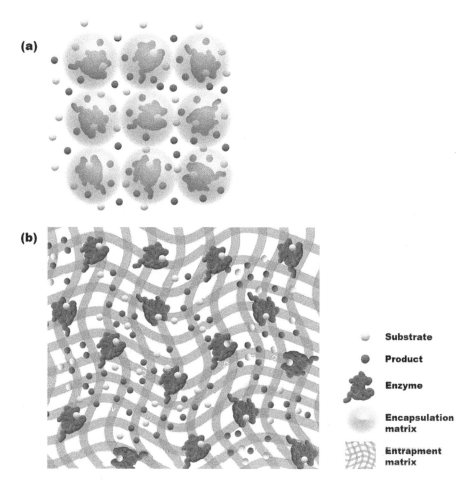

FIGURE 6.3 Immobilization of enzymes by entrapment or encapsulation.

6.4.3 Covalent Immobilization of Enzymes

The covalent bond generation between the functional groups of enzymes that are not involved in the catalytic activity and the support matrix is known as covalent immobilization of enzymes, as shown in Figure 6.4. Usually, the binding reaction is performed under optimum conditions to avoid the loss of enzymatic activity. The side chains of a few amino acids like cysteine, lysine, glutamic, aspartic that contains thio, e-amino, phenolic, carboxylic and imidazole groups that are not required for the enzyme activity have been targeted for covalent formation with the immobilization matrix or support. The enzyme activity of covalently attached enzymes are determined by the material of their chemical constitution and parameters maintained while linking (Fu et al. 2011). In general, the maximum enzyme activity is preserved upon sparing the amino acids present in the active site. However, the increase in the number of covalent bonds between the enzyme and the immobilization matrix or surface

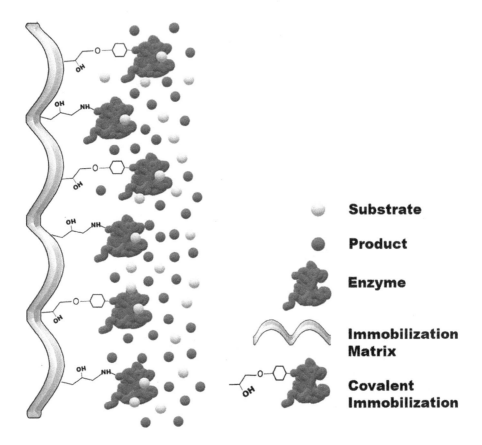

FIGURE 6.4   Immobilization of enzymes by covalent bond.

proportionately increases the enzyme stability. High specific activity and stability with controlled protein orientation could be achieved by using peptide-modified sur- face enzyme linkages. Thermal stability has been conferred to covalently attached enzymes by CNBr-agarose and CNBr-activated-sepharose comprising carbohydrate moiety and glutaraldehyde as a spacer arm. The carrier and enzyme can be linked directly or via an intercalated link of various lengths, known as a spacer. The spacer molecule increases the mobility of the linked biocatalyst, allowing the activity to be increased incomparison to directly coupled biocatalyst (Hartmann & Kostrov 2013).

6.4.4   Cross-linking of Enzymes

Biocatalyst attachment to each other using bifunctional as well as multifunctional ligands is known as the cross-linking of enzyme immobilization as shown in Figure 6.5. It is a type of irreversible method where high molecular weight insoluble aggre- gates are formed. This method does not require the immobilization matrix or sup- port and hence it is termed carrier-free enzyme immobilization (Wang et al. 2008). Glutaraldehyde is the most used cross-linking reagent and is both economically

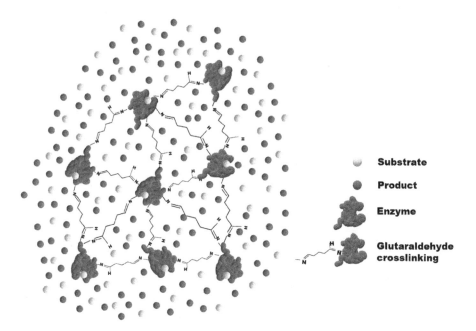

FIGURE 6.5 Immobilization of enzymes by chemical cross-linking.

feasible and easily available in large quantities. The free amino groups present in lysine residues are cross-linked to the surface amino acid of enzyme molecules via aldol condensations. The cross-linking methodology is pH-dependent and Schiff's base formation or Michael reactions of cross-linking are involved (Hanefeld et al. 2009).

Some inert proteins like albumin, collagen, and gelatine are used for cross-linking of enzymes. Due to the reaction of catalytic groups, loss of activity may arise in this type of immobilization and it is not easy to control the cross-reactivity. Cross-linked enzyme aggregates (CLEAs) are more recently developed by precipitation of the enzyme using an aqueous solution, which results in the formation of physical aggregates of protein molecules, or by simply adding salts, water-miscible organic solvents, or non-ionic polymers (Sheldon 2007a). The CLEAs type of enzyme immobilization is illustrated in Figure 6.6.

These physical aggregates of proteins are maintained through non-covalent bonding without disturbing their non-denatured tertiary structure. Then they are linked in a permanently insoluble method still retaining its preorganized superstructure and therefore its enzymatic catalytic activity. This approach leads to a novel family of enzymes in immobilized form namely CLEAs. This sort of immobilized enzyme is a particularly potent biocatalyst since it can be generated economically and efficiently. CLEAs are easily reusable and provide adequate stability and performance for a variety of applications (Homaei et al. 2013). For the preparation CLEAs, the enzymes are aggregated in precipitants like acetone or ammonium sulfate and the addition of cross-linker to the mixture.

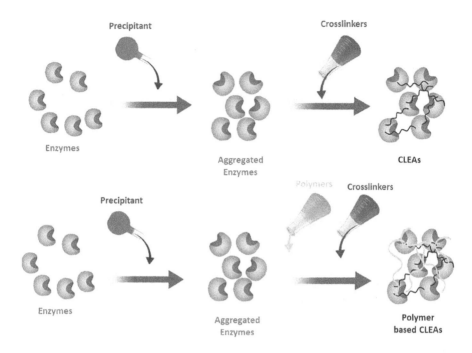

FIGURE 6.6   Cross-linked enzyme aggregates (CLEAs).

6.5   MATERIAL SELECTION AND TYPES FOR IMMOBILIZATION OF ENZYMES

The process of immobilization of enzymes directly or indirectly alters the physio-chemical properties of the enzymes. Basing it on the immobilization support and the procedure followed for chemically attaching the enzyme to the support influences the enzyme stability. The products formed in the reaction can also affect the stability and processivity of the enzymes. The immobilization surface or matrix that harbors the enzyme greatly influences the structure of the enzyme depending on the type of bond (covalent or hydrogen). Therefore, one has to pay more attention to selecting the enzyme immobilization technique, and preferably it should not compromise the substrate binding site on the enzyme.

The movement of the substrates and products should not be restricted by the immobilization surface or matrix. The materials used for immobilization should be non-reactive organic and inorganic materials. The selection of immobilization surface based on the following criteria would lead to the better performance of enzymes as well as cost-effectiveness:

a) Selection of inexpensive material as immobilization matrix
b) Non-reactive and mechanically stable
c) Highly stable and resistant to disintegration
d) Should not inhibit product or enzyme activity
e) Should not allow microbial growth

f) Should have temperature stability

g) Should have pH stability across a range of variable pH

h) Porosity of the material should allow easy transfer of reactants and products

Various organic and inorganic materials are available for enzyme immobilization and have been previously used in multiple enzyme catalyzed reactions.

6.5.1 ORGANIC MATERIALS USED FOR IMMOBILIZATION

6.5.1.1 Natural Polymers

6.5.1.1.1 Alginate, Cellulose, Chitin and Chitosan, Starch, Sepharose

Alginate: Sodium and calcium salts of alginic acids are used for immobilization in both entrapment/encapsulation methods as calcium alginate beads and sodium alginate beads. The utilization of divalent ions and glutaraldehyde for cross-linking of alginates increases the stability of enzymes (Flores-Maltos et al. 2011).

Cellulose: Cellulose is the most abundant natural biopolymer present in the biomass (Bhardwaj et al. 2021b). Modified forms like diethylaminoethyl (DEAE)-modified cellulosic has been shown to increase the storage capacity of enzymes (Al-Adhami et al. 2002). Magnetic nanoparticles coated with cellulose have also been used for the degradation of starch in which the binding of amylase into cellulose di-aldehyde coated magnetite nanoparticles forms the novel starch degrading system (Namdeo & Bajpai 2009). Ionic liquid-based cellulose film used for enzyme immobilization produces better flexibility when activated with glutaraldehyde (Klein et al. 2011).

Chitin and chitosan: Both polymers are used as a support material for the immobilization of enzymes. Chitosan contains an amine group in which protein or carbohydrate residues of enzymes are used for binding (Vaillant et al. 2000). The combination of alginate with chitosan for enzyme coating decreases the leaching effect. Mixing chitosan with clay for enzyme immobilization increases the hydrophilicity and porosity due to the formation of amino and hydroxyl groups (Chang & Juang 2007).

Starch: The basic composition of starch is made up of linear molecules of amylase and highly branched amylopectin. Starch alginate hybrids are used for the immobilization of peroxidase. In industrial purposes, components such as acrylamide and dimethyl aminoethyl methacrylate are grafted onto starch for better yield of products (Matto & Husain 2009).

Pectin: Pectin has been utilized for immobilization for the development of new materials in skin injury treatment, when mixed with 0.2–0.8% of glycerol thus reducing the brittleness of career supports. Due to the formation of highly stable polyelectrolyte complexes pectin coated supports have enhanced the thermal stability and increased the catalytic properties of entrapped enzymes (Gómez et al. 2006).

Sepharose: Sepharose activated with CNBr used to immobilize enzymes plays a major role in maintaining the activity even at extreme pH, elevated temperatures, and at high salt concentrations due to alkyl residues of support and hydrophobic clusters of enzymes (Hosseinkhani et al. 2003).

6.5.2 Some Synthetic Polymers for Immobilization

Synthetic polymers: Acrylic resins, such as Eupergit®-C are composed of macroporous copolymers of glycidyl methacrylate, N, N′-methylene-bi- (methacrylamide), methacrylamide, and allyl glycidyl ether with a pore diameter of about 20–30 nm and with average particle size around 170 mm, which have been most used synthetic polymers for enzyme immobilization. These polymers are stable both physically and chemically, over a wide range of pH from 0–14 and even at drastic pH changes it retains the structure without swelling or shrinkage. It is basically high hydrophilic in nature, but due to its extreme hydrophilic status these resins have limitations in diffusion (Boller et al. 2002). Due to its high binding capacity and easier immobilization procedure without the usage bi-functional reagents or spacer arms like polyketone polymers –[–$COCH_2CH_2$–] n–, as it appear to be a promising support for enzyme immobilization. Polyketone polymers are produced with ethane and CO by copolymerization using palladium as catalyst (Agostinelli et al. 2007).

Smart polymers: The polymers which proceed conformational modifications dramatically in response to small fluctuations in the temperature, pH, and ionic strength are known as smart polymers (stimulus-responsive polymers). The critical solution temperature of a PolyNIPAM aqueous solution (LCST) is around 32°C: below that temperature polymer readily dissolves into water whereas, beyond the LCST, due to the removal of the water molecules from the polymer network, it becomes intractable. The biotransformation may therefore be carried out under conditions in which the enzyme is soluble, thereby reducing diffusional restrictions and loss of activity due to protein conformational changes on the support surface. Increasing the temperature above the LCST causes the immobilized enzyme to precipitate, allowing for easier recovery and reusage. Another advantage of utilizing thermo-selective immobilized enzymes is that such circumstances are avoided since the catalyst precipitates and the reaction shuts off when the reaction temperature exceeds beyond the LCST (Klouda & Mikos, 2008).

6.5.3 Inorganic Supports for Immobilization

Activated carbon and charcoal: Carbon which is modified by HCl is porous containing large sites for enzyme immobilization. This activated carbon has a high surface area of 600–1000 m^2g^{-1} and is suitable for use in the immobilization of proteases and lipases (Daoud et al. 2010). Charcoal has also been used for immobilizing enzymes in food industries, especially amyloglucosidase without any cross-linking for starch hydrolysis, and retained 90 % of catalytic activity. Charcoal is known to be an excellent adsorbent with high adsorption capacity and minimum release in fine particulate matter (Rani et al. 2000).

Zeolites: Zeolites, often known as "molecular sieves," are microporous crystalline solids with well-defined morphologies and shape-selective characteristics, which are frequently employed for molecular adsorption. Because of the existence of higher amount hydroxyl groups, they create strong hydrogen bonds with the enzyme, thus microporous zeolites were shown to provide superior support for a-chymotrypsin immobilization when compared to microporous dealuminized ones. Zeolites' heterogeneous surfaces with numerous adsorption sites are thought to be useful for controlling enzymes and supporting interactions (Serralha et al. 2001).

Ceramics: Ceramic foams are inert supports that comprise both micropores (45 nm) and macropores (77 nm) have been shown to reduce the rate of diffusion and increase the surface area (Huang & Cheng 2008). Tonite is unique ceramic support. Under acidic circumstances, it was hydrothermally produced from kaolin minerals. The average particle size of the support is 150 µm, while the porous size is 60 nm. The TN could be modified with two silane coupling agents to generate TN-M and TN-A supports with methoacryloyloxy and amino functional groups which can be used for various industrial enzyme immobilization processes (Kamori et al. 2000).

Silica and glass beads: Amination of hydroxyl and reactive siloxane groups on the surface of silica, as well as the addition of methyl or polyvinyl alcohol groups, enhance enzyme and immobilization support linkages. They have been utilized because of their nano-sized structures with a large surface area, organized organization, and excellent chemical and mechanical resistance (Pogorilyi et al. 2007; Agrawal & Verma 2020). Peroxidase enzymes like heme-containing lignin peroxidase and horseradish peroxidase are used for immobilization procedures to chlorolignins from eucalyptus kraft effluent (Dezott et al. 1995). Glass is a transparent viscous liquid that has been used to immobilize α-amylase, and phthaloyl chloride-containing amino group functionalized glass beads have been shown to be durable and reusable in this process (Rosa et al. 2002).

Both organic and inorganic materials have been used as immobilization support with different enzymes as shown in Table 6.2.

6.6 SURFACE ANALYSIS TECHNOLOGIES FOR THE CHARACTERIZATION OF IMMOBILIZATION

Enzymes dictate the rate of biochemical reactions by their biocatalytic properties. When enzymes alter the biochemical reactions by increasing their rate, they do not get altered or used up upon completion of the reaction. Although enzymes have the advantage of fastening a biochemical reaction, they have a few drawbacks like susceptibility to getting inhibited, or low sensitivity due to reaction conditions (Guisan 2006). Many industrial applications are functioned in a continuous manner that can affect enzyme stability, recovery, and reuse (Sheldon 2007b). Immobilization is a method that efficiently uses the enzymes by chemically cross-linking specific regions to the support,without compromising their catalytic properties for use in continuous industrial processes. This method not only facilitates the recovery and reuse of the enzymes, but also the easy separation of products from the enzymes, thereby reducing the cost of production and product purification (Saifuddin et al. 2013).

Immobilization also reduces the cost of enzyme production as the immobilized enzymes are reused with higher degrees of activity retention. Yet another advantage is the immediate cessation of the reaction that is possible while using immobilized

TABLE 6.2

Inorganic and Organic Materials Used for Immobilization of Enzymes

Type		Matrix Material	Immobilized Enzymes	References
Inorganic		Charcoal	Amyloglucosidase,	(Rani et al. 2000)
		Silica	Oxidoreductases, Transferases, Hydrolases, Isomerases and Lipases	(Godjevargova et al. 2006; Kramer et al. 2010; Falahati et al. 2011; Zdarta et al. 2015; Kołodziejczak-Radzimska et al. 2018)
		Glass	Alkaline phosphatase	(Sharmin et al. 2007)
		Activated carbon	Pancreatin	(Silva et al. 2005)
		Ceramics (cordierite)	Horseradish peroxidase, Beta-galactosidase	(Ebrahimi et al. 2010, Wang et al. 2012)
		Mineral (bentonite)	Glucose oxidase	(Chrisnasari et al. 2015)
		Gold-nano particles	Trypsin	(Kotal et al. 2011)
		Carbon-nano tubes	Alpha-glucosidase	(Mohiuddin et al. 2016)
		Graphene oxide	Peroxidase	(Zhang et al. 2010)
Organic	Synthetic	Polyvinyl alcohol	Laccase	(Bai et al. 2014)
		Polyaniline	Alpha-amylase	(Ashly et al. 2011)
		Polypropylene	Glucose oxidase	(Vartiainen et al. 2005)
		Polyamide 66	Tyrosinase	(Hamann & Saville 1996)
		Polystyrene	Lipase	(Wang et al. 2015)
		Amberlite	Alpha-amylase	(Kumari & Kayastha 2011)
		Polyurethane	Inulinase	(Silva et al. 2013)
		Sepabeads	Alcohol dehydrogenase	(Alsafadi & Paradisi 2014)
	Natural	Agarose	Beta-1,4-endoxylanase, Alpha-amylase	(Prakash & Jaiswal 2011; Martins de Oliveira et al. 2018)
		Calcium alginate	Lipase	(Betigeri & Neau 2002)
		Chitosan	Nuclease, Glucose isomerase	(Shi et al. 2011, Cahyaningrum et al. 2014)
		Copper alginate	Polyphenol oxidase	(Betigeri & Neau 2002)
		Kappa-carrageenan hydrogels	Lipase	(Tümtürk et al. 2007)

enzymes. The enzyme activity limiting factors such as temperature, pH, and buffer conditions, etc. could be manipulated for improving its activity for use in industrial applications (Xie et al. 2009). Selection of required enzymes and customizing the compatible matrix a well as the type of chemical cross-linking would play a major role in the process of enzyme immobilizations. There are various methods available to immobilize the enzymes such as: affinity attachment, ionic linking, and physical adsorption that are all reversible methods, while attaching enzymes through covalent bonds by amide, thiol-ether, carbamate, and ether bonds turn out to be irreversible methods (Guisan 2006). Apart from attaching the enzymes to the matrix support, the quantification of immobilized enzymes sounds more important as it is directly related to the enzyme performance during biocatalysis.

6.6.1 Analysis Technologies for Immobilized Enzymes

The loss of activity and stability followed by enzyme immobilization are due to the changes in alignment and structural characteristics of the enzyme. Often during the immobilization process, the enzyme active sites are not exposed which results in reduced activity. There are various techniques to determine the orientation and structure of the immobilized enzyme:

a. *Attenuated Total Reflection-Fourier Transform Infrared (ATR-FTIR)*: This functions on a surface-sensitive vibration method which can analyze the secondary structure, interaction, orientation, and protonation of the immobilized enzymes/proteins. The backbone carbonyl bond stretching of the enzyme/proteins results in an active band at amide I IR. The frequency of this band is used to study the secondary structure and is deconvoluted to components corresponding to the secondary structure (Schartner et al. 2013).

b. *Circular Dichroism (CD)*: This technique analyses the secondary structure of the protein sample by following its folding characteristics. (Nygren et al. 2008). It is a nondestructive technique and requires a smaller amount of sample. CD measures the difference in absorbance of left and right circularly polarized light by the sample. The peptide bonds in the proteins are optically active and the absorbance or ellipticity dependent on the native confirmation of the protein sample. Each structural confirmation such as alpha-helices, beta-sheets, beta-turns, and random coils have a characteristic spectra. To determine the secondary structure, the protein samples are analyzed at the far-UV region (180–250 nm). CD provides information on protein-ligand, protein-protein, and structural compositions of proteins which is useful to characterize an immobilized enzyme (Mittal and Singh 2013).

c. *Surface Plasmon Resonance (SPR)*: The binding of immobilized substrates or ligands to their respective enzymes or proteins that leads to change in mass is actively analyzed by this technique. Of late, the changes to the dielectric properties of the immobilized proteins/enzymes by the ligands/substrates binding can be actively detected. The variation in SPR signals corresponds to the secondary structure of the immobilized analyte. The increase in SPR signals is proportional to high alpha-helix content, whereas the decline in the signal relates to the presence of high beta-turn content or other disordered structures that is analogous to CD spectroscopy (Mannen et al. 2001).

d. *Sum frequency generation (SFG)*: This is a label-free method that can analyze different interfaces *in situ* and is recommended for studying secondary structures as well as orientations of enzymes/proteins. It is a second-order non-linear vibrational spectroscopy that is highly sensitive to the bond vibrations based on its orientation. The orientation of the amide bonds to the immobilization matrix or surface can be inferred based on the recording of the spectra at different polarization conditions. Further combining the obtained data with the three-dimensional structure of the immobilized analysis ascertains its orientation with the carrier surface

(Chen et al. 2002). This technique is also analogous to SPR and ATR-FTIR-based label-free methods.

e. *Polarization Modulation Infrared Reflectance Absorption Spectroscopy (PM-IRRAS)*: This is a non-damaging and label-free technique. It can deliver information on the structures that contain hydrogen bonds as well as the orientation of the functional groups bound to the carrier. The strong near-surface absorptions are filtered out by increasing the signal-to-noise ratio in the polarization modular type of IRRAS that distinguishes the oriental and structural changes related to the enzyme/protein-carrier interactions (Kreider et al. 2014).

f. *Surface-enhanced Raman Spectroscopy (SERS)*: This technique exploits the optical characteristics of colloidal Au or Ag metal nanostructures in the range of 10–100 nm to enrich Raman-active vibrations to an extent of 10^{11} fold. Although, it has a high resolution of analyzing even a single molecule the interaction of the nanoparticles with the protein analytes affects the reproducibility of this technique. Later, Ag nanoparticles were modified with iodide to overcome the non-specific interaction with proteins that gives spectra similar to Raman spectra (Xu et al. 2014). Since SERS can be used to analyze protein samples with low concentration it has emerged as a powerful analytical tool to study protein-carrier interactions (Michota-Kaminska et al. 2006, Iosin et al. 2009).

g. *Time-of-flight Secondary Ion Mass Spectrometry (TOF-SIMS)*: These are energies with quite a few keV of ionized molecules or elements (i.e., SF5+ or Xe+) that form the primary ions beam in mass spectrometry. When the primary ion beam strikes the analyte it generates emission of secondary ions from its surface which are further analyzed (Youn et al. 2015). The emitted secondary ions carry only the information about the outer surface of the analyte as the sampling depth is 10Å in TOF-SIMS. The surface chemical attributes of the proteins are analyzed by TOF-SIMS as the proteins have a diameter larger than 10Å with which the orientation, denaturation, and three-dimensional structure can be determined. However, air-dried samples cannot be analyzed by this technique.

6.7 GENERAL REACTORS UTILIZED IN INDUSTRY FOR BIOCATALYSIS REACTIONS

Enzyme immobilization: the expertise to enhance the stability and reusability of an enzyme catalyst through immobilization has been an established technique for some time. The enzyme immobilization process is economically viable and supports many industries such as the chemical, pharmaceutical, and food industries. The catalytic efficiency of enzyme biocatalysts works well in mild reaction conditions. Whereas, enzyme stabilization decreases in the harsh conditions of industrial processes, thus shortening their industrial lifespan. To improve operational stability and enhance enzyme catalytic properties enzyme immobilization technology provides an effective solution. The type of support material and cross-linkers used are critical in the immobilization process. Immobilization of enzymes onto supports is accomplished through a variety of techniques like physical adsorption, covalent bonds, chemical cross-linking, entrapment, and encapsulation. Detailed study of interactions between the enzyme protein structure and support surface plays an important role in successful immobilization methods. For determining the efficacy of immobilization, a thorough knowledge of the above-mentioned techniques is necessary to lead to the advancement of future enzyme immobilization technologies. Initially one has to identify a suitable support material with customized properties in order to produce an effective biocatalytic system for its utilization in different industrial processes.

6.7.1   ENZYME APPLICATION

Currently, immobilized enzymes are routinely used in the chemical, pharmaceutical, and food industries to process a variety of industrial products. Some enzymes, like lipases, have high efficiency naturally and are exploited for producing a variety of compounds in commercial applications (Adrio and Demain 2014). In comparison with chemical methods, the usage of an immobilized enzyme has brought substantial benefits. As in the case of transaminases, protein engineering is required for using them in Active Pharmaceutical Ingredients (APIs) synthesis as the wild-type enzyme would be too harsh for the process (Olempska-Beer et al. 2006). To build cost-effective and long-lasting industrially relevant enzyme immobilization is highly recommended. Enzymes may have tremendous potential in enhancing the overall sustainability of the chemical processing, food processing, and pharmaceutical industries (Agrawal & Verma 2021) A better understanding of support material or immobilization matrix influences the biocatalysts' operation (Moslemi 2021).

6.7.2   IMMOBILIZED ENZYMES IN THE FOOD INDUSTRY

The food industry, unlike the pharmaceutical and a few segments of the chemical industries, must generate a substantial quantity of cost-sensitive items in a short amount of time. (Andler & Goddard 2018). The required biocatalyst (immobilized enzyme) must be inexpensive in this scenario and should have strong operational stability for its continuous operation. However, in the food industry, where goods must be produced, the continuous flow operating arrangement has typically been chosen over batch procedures. Few immobilized enzymes are employed during the scale-up continuous processes for the synthesis of foodstuffs like high cocoa butter analogs, high fructose corn syrup (HFCS), allulose, galacto-oligosaccharides, and others. More are listed in Table 6.3. and discussed in the following section (Jeong et al. 2013).

6.7.3   HFSC PRODUCTION BY D-GLUCOSE/XYLOSE ISOMERASES

For both enzyme quantity and product volume, any use of glucose/xylose isomerase (EC 5.3.1.5) in the manufacturing of high fructose corn syrup is the largest commercial process using an immobilized enzyme (Chanitnun and Pinphanichakarn 2012). About 500 metric tons of D-glucose/xylose isomerase that are immobilized have

TABLE 6.3

Application of Immobilized Enzymes in Food Processing at Industrial Scale (Kilara et al. 1979, Basso and Serban 2019)

Product	Immobilized Enzymes	Type of Immobilization	Reaction Type	Application
Hydrolyzed lactose	β-Galactosidase (GLB1)	Adsorption over ion exchange fiber like modified cellulose	Hydrolysis	Hard cheese, probiotic yogurt, some dairy protein powders, kefir, heavy cream
Omega-3 ethyl ester	Candida antarctica lipase B (CALB)	Adsorption over methacrylate and ion exchange co-polymer	Esterification	Eicosapentaenoic acid (EPA) and Omega-3 fish oil (e.g. docosahexaenoic acid (DHA)
Triglycerides	Rhizomucor miehei lipase (RML)	Adsorption on silica or ionic exchanger polymer	Transesterification	Cocoa-based confectionery products.
Vitamin C ester	Candida antarctica lipase B (CALB)	Adsorption on methacrylate and ion exchange copolymer	Transesterification	Bilirubin, thiols, NADPH and NADH, ubiquinone (coenzyme Q10), uric acid.
Tagatose, allulose, and HFCS	D-glucose/xylose isomerase, 3 -epimerase, β-galactosidase (GLB1)	Adsorption on silica/ ionic exchanger polymer	Isomerization, epimerization, and hydrolysis	Saccharine, sucralose, and aspartame

been consumed each year, allowing the manufacture of almost 10^6 tons of HFCS. Fructose can also be made from HFCS that is utilized as a sweetener in meals and beverages.

Although D-glucose/xylose isomerase's natural substrate is D-xylose, the enzyme has a wide substrate specificity which has been exploited for converting D-glucose to D-fructose in industrial applications (Nagasawa et al. 2016). The position of isomerization equilibrium varies with temperature as higher temperatures favor fructose production. However, the conversion of dextrose yields roughly about 50% of fructose at 60°C in the HFCS commercially whereas the equilibrium mixture contains 56% of fructose at 90°C. Nowadays, HFCS manufacturing is done in fixed bed reactors that are connected in parallel and run continuously. D-glucose syrup generated from corn has been transformed into a mixture known as HFCS-42, which contains around 50% of D-glucose, 42% of D-fructose, 2% of maltotriose, 6% of maltose, and trace amounts of other sugars (Nagasawa et al. 2016). Higher fructose concentrations like 55% HFCS grades used in the majority of soft beverages achieved through enrichment by chromatography of 42% HFCS grade to the final concentration HFCS-90 (i.e., D-fructose concentration of 90%). Later HFCS-90 has to be combined with HFCS-42 to obtain HFCS-55 (i.e., D-fructose concentration of 55%).

If HFCS-42 is the starting material then industrial crystalline D-fructose can be made with a purity above 99% (w/w). After glutaraldehyde cross-linking, the two commercial preparations of D-glucose/xylose isomerase enzyme immobilized have been extensively employed in these productions were immobilized on low-cost inorganic based support or carrier like diatomaceous earth and bentonite clay. The Food and Drug Administration (FDA) has approved the glutaraldehyde cross-linking reagent for immobilization of enzymes under the Code of Federal Regulations CFR 21: 173.357 of the United States. The final composite must be dehydrated and physically squeezed out before being dried in a fluidized bed drier and then transferred to the packed bed reactor where a constant temperature of 60°C is maintained. The immobilized D-glucose/xylose isomerase (DuPont Industrial Biosciences) has exceptional stability with a half-life over a year (Zittan et al. 1975).

6.7.4   EPIMERASE ACTION ON ALLULOSE

Allulose is a sweetener that resembles the properties of dextrose and has nutritional energy of zero calories. Structurally D-fructose and D-allulose are similar, whereas the latter is referred to as the C-3 epimer of the former. Allulose is not digested by humans and hence has no calories. Thus, allulose is an apt choice sweetener since it has characteristics similar to monosaccharides. Food and beverage makers are the major target market for allulose; fructose, substituting HFCS or dextrose in their goods with allulose saves calories while maintaining sugar characteristics qualities, for instance, bulking, browning, sweetness, and texture (Wang et al. 2020).

In general steam-treated coffee, high fructose corn syrup, wheat plant products, processed cane, and beet molasses naturally contain allulose with daily consumption larger than 200 mg per day despite being unmetabolized by the human system. In the United States, allulose has been designated as GRAS (Generally Recognized as Safe) for its safe use and it is now becoming profitable due to its utilization as food in nations including Chile, Costa Rica, Mexico, and Columbia. Even though allulose not an approved food in the European Union, various allulose manufactures have submitted their applications to the European Union for its approval (Jiang et al. 2020).

Ketose-3-epimerases (EC 5.1.3.31) produced by a variety of microbes have the potential to interconvert fructose and allulose according to CJ CheilJedang Corp. The great technological task was to generate an enzyme that could transform fructose to D-psicose through the activity of D-psicose epimerase obtained from *Agrobacterium tumefaciens*. The conversion is accomplished by epimerizing the hydroxy group of fructose in order to obtain D-psicose. The novel enzyme was noted for its high activity and thermostability. Tate & Lyle, on the other hand, took a huge stride toward industrialization in 2014 when they developed a comprehensive industrial process based on enhanced immobilization of enzymes over ion exchange resin, and sweetener compositions that acted as substrates were allulose, fructose, and sucralose. Therefore, the introduction of enzyme immobilization technology has allowed allulose to be produced in continuous packed bed designs, resulting in significant cost savings (Zhang et al. 2021).

TABLE 6.4

Application of Immobilized Enzymes at Industrial Scale for Chemical Processing (Phitsuwan et al. 2013, Chapman et al. 2018)

Product	Immobilized Enzymes	Reaction Type	Type of Immobilization	Application
L-amino acid	Aminoacylase	Hydrolysis	Adsorption over ion exchange resins	Amino acids
Myristyl myristate	*Candida antarctica* lipase B (CALB)	Esterification	Adsorption over methacrylate or ion exchange copolymer	Sucrose caprylyl ester, lauryl ester, and myristyl ester.
Dimethenamid-P	*Candida antarctica* lipase B (CALB)	Transamidation	Adsorption over methacrylate or ion exchange copolymer	Outlook® Herbicide
Organosilicone polyamide	*Candida antarctica* lipase B (CALB)	Amidation	Adsorption over methacrylate or ion exchange copolymer	Polymers like polyamide-6
Methyl acrylate derivatives	*Candida antarctica* lipase B (CALB)	Transesterification	Adsorption over polyethylene	Polymer monomers like acrylate and methacrylate monomers

In reality, the production of allulose, is carried out in a continuous fixed bed reactor. During the process, fructose solution with a pH value of 8.0 is continuously pumped in with a recirculation rate of about 8-bed volumes per hour along with jacket columns heating to 57°C. The technology has now been industrialized, and Tate & Lyle introduced DolciaPrima® crystalline allulose in March 2017, utilizing the same manufacturing process.

6.7.5 IMMOBILIZED ENZYMES FOR THE CHEMICAL INDUSTRY

The chemical industry's attention has shifted in the last 20–30 years with an eye on implementing sustainable and green technologies to generate chemicals at bulk amounts as well as chiral compounds. Biocatalysts used in the chemical industry, like those in the food sector have to produce goods on a tons scale making recyclability combined with selectivity and fast reaction turnover a serious problem. On the contrary, in the pharmaceutical business biocatalysts have been utilized for the manufacture of finished products in the hundreds of kg. Table 6.4, provides an overall view of numerous large-scale commercial processes for bulk chemical synthesis by immobilized enzymes (Thompson et al. 2019).

6.7.6 LIPASE CALB FOR CHIRAL AMINES

In stereoselective organic synthesis, chiral amines play a significant role. They have been used as chiral auxiliaries, resolving agents, and building blocks. Biotechnological techniques offer the route to chiral amines production with a racemic resolution containing optically active acids. Enantioselective hydrolysis of amides has been described earlier utilizing dinitrogen tetroxide or chiral separation, both of which are not commercially feasible. Badische Anilin und SodaFabrik (BASF) published a unique methodology for the purification and production of chiral amines in 1997 that used an enzyme approach in contrast to previously described chemical methods (Basso and Serban 2019).

By resolving a racemate in three phases, it is possible to acquire both enantiomers with enantiopurity >99%. The first step involves employing immobilized CalB to chirally acylate rac-phenylethylamine with i-propyl methyloxyacetate in methyl tert-butyl ether (MTBE) to form a mixture containing S-phenylethylamine and R-phenylethylamine amide. Whereas in a continuous flow reactor, the biocatalytic transformation is carried out at an increased substrate concentration of 0.3 kg per liter of i-propyl methyloxyacetate and 1-phenylethylamine, resulting in a transformation rate at 50% (Ismail et al. 2008). The separation of unreacted amine is the second stage. Fractionation distillation can be used to extract S-phenylethylamine, which can be subsequently separated from the amide at 99% enantiopurity [1]. The next step is to hydrolyze the amide at 170°C with ethylene glycol in the presence of sodium hydroxide yielding 99% of R-phenylethylamine. BASF has scaled up the method for a range of rings containing amines like phenylethylamines, aliphatic, aryl, alkyl, bicyclic, and cycloaliphatic amines. ChiPros® is a brand of chiral amines that are sold commercially. In addition, BASF has looked into the possibility of using other enzymes instead of lipases to induce chirality into building blocks. Steroselectively the reduction of C=C bonds has been carried out by enoate reductases, whereas the alcohol dehydrogenases carry out ketones reduction (Wu et al. 2020, Kumar et al. 2021).

6.7.7 ACRYLATES AND ORGANOSILICONE ESTERS OR AMIDES PROCESSING BY LIPASE CALB

As indicated by Sumitomo's extensive research efforts in this field, acrylates play an important role in the manufacturing of polymeric-based products, including bio-polymers.(Sharmin et al. 2007). High temperatures are widely used in traditional chemical methods for generating acrylate esters, resulting in acrylic double bonds polymerization. To counter this, polymerization acid catalysts and inhibitors have also been designed, allowing processes to take place at temperatures between 80°C and 100°C. The enzymatic processes carried out at optimal temperature may be help-ful due to its selectivity, allowing additives such as inhibitors to be avoided, with the added benefit of energy savings and improved end product purity and quality. If the molecule of interest has numerous reactive functionalities (hydroxyls, amines, and thiols), enzyme selectivity is even more advantageous. As normal CalB reaction does not have the catalytic ability to perform acrylation reactions in terms of yield required for commercial applications. Hence, engineered CalB, that is mutant enzyme, has the ability to catalyze unnatural substrates, for instance, acrylates were found to be the key substrates translating a biocatalytic process into an effective commercial process. BASF's enzyme genetic manipulation turned the process into one that is more environmentally friendly and energy efficient (Wu et al. 2021). The mutant CalB enzyme has been immobilized by adsorption over polypropylene (Accurel® MP1000) in advance before its usage involving hydroxypropyl carbamate and methacrylate at a standard temperature of 40°C. This protocol yielded a conversion rate of 97% within six hours duration. However, in this example, the free soluble enzyme (i.e., non-immobilized) was inactive because of its substrate inaccessibility and instability. The biocatalytic method has the potential to create mixes or enantiomeric pure acrylates with greater yields. Immobilized CalB was found to be useful in the industrial manufacturing of polymers known as oleamides, which are utilized as slip and anti-blocking agents in plastic processing. The enzymatic ammonolysis of triolein in the presence of ammonia has been catalyzed by immobilized CalB over polymeric beads (Novozym 435) to produce oleamide. Dow Corning disclosed a process, in partnership with Professor Gross, for manufacturing polymers containing amides or organosilicone esters generated through enzyme bio-catalysis. Using immobilized enzyme at 10% w/v, the ester of diacid siloxane with hexanol has been produced in equimolar quantities. pon employing immobilized CalB several other groups have reported the synthesis of a polymer that is similar to the polymer nylon with substrates diamine disiloxane and dimethyl adipate to achieve a molecular weight of around 2100 (Wu et al. 2021).

6.7.8   ROLE OF ENZYMES IMMOBILIZATION IN PHARMACEUTICAL INDUSTRY

Biocatalysis and its benefits in industrial processes such as selectivity, sustainability, and safety are grabbing the attention of the pharmaceutical industry. Unnatural molecules have been proven to be compatible with enzyme substrates functioning in an organic solvent thereby resulting in high-yielding enantioselective, chemoselective, and regioselective chemicals. The creation of commercially high-value molecules through conventional chemical synthesis might be extremely difficult, which is why the pharmaceutical industry has made significant investments in high-speed manufacturing.

6.7.9   LIPASE CALB FOR ODANACATIB

Lipase B of *Candida antarctica* (CalB) is employed to catalyze the synthesis of active pharmaceutical components, cosmetic additives for personal care, and food components because of their extraordinary regioselectivity, chemoselectivity, enantioselectivity, and their ability to catalyze the conversion of fatty acids into fatty acids. CalB is widely used in industrial reactions because of its excellent resistance to organic solvents, a wide range of selectivity, and thermal stability in its immobilized form. The synthesis of Odanacatib utilizing immobilized CalB, disclosed by Merck in 2011 (O'Shea et al. 2009), is one notable API created through biocatalysis. In 2008, Odanacatib has been identified to be a strong cathepsin K inhibitor that was tested for treating osteoporosis in postmenopausal women and later discontinued (Stoch et al. 2009).

The most important stage in Merck's technique was reported as being the full ring opening via ethanolysis of azlactone. Using a high substrate concentration (i.e., azlactone yield of 200g/L) at 60°C in continuous plug flow reactor the immobilized CalB has the ability to catalyze 95% transformation of the required (S)-fluoroleucine ethyl ester in methyl tert-butyl ether (MTBE). With yields >90% and an ee of 88% the transformation has been ramped up to a 100 kg scale. In comparison to the commercial Novozym 435 (CalB immobilized on a divinylbenzene/methacrylate support matrix) the success of the process has been mainly due to the use of a carrier other than the one generally used for CalB immobilization also obtained a cost reduction of 99.9%. However, CalB's stability and activity were significantly increased when immobilized on an octadecyl functionalized methacrylate resin in comparison to the commercial preparation Novozym 435 reported as Purolite. This has been linked to the hydrophobic octadecyl groups' optimal interaction with lipases. According to Merck, the octadecyl functionalized carrier is also effective at immobilizing transaminase. (Neto et al. 2015).

6.7.10   Sofosbuvir Bio-catalysis by Lipase CalB

Hepatitis C infection is caused by the infectious agent Hepatitis C virus (HCV) that belongs to the family of viruses known as hepacivirus. Hepatitis C is reported to be the most common reason for chronic liver disease globally. Although the infection is mostly asymptomatic, it can lead to severe liver damage and eventually cirrhosis that is diagnosed after several years. Cirrhosis of the liver could lead to liver failure, malignancy, and oesophageal and gastric varics in some people. HCV infection typically spreads through direct contact with infected blood. A lot of interest lies in finding a better treatment against hepatitis C which could overcome HCV's high mutagenicity leading to the generation of diverse subtypes as well as genotypes s (Chandra et al. 2020). Gilead patented a novel drug in 2014 that inhibits the NS3 protease of the hepatitis C virus that has the advantage of blocking numerous virus genotypes. This drug has been marketed under the commercial name Sofosbuvir and reported by Sandoz with a novel crystalline formulation that is effective against HCV infection and the symptoms that accompany it. Immobilized in the Gilead patent, CalB is used to enantioselectively hydrolyze an acetate ester into chiral alcohol. Sofosbuvir synthesis takes numerous chemical steps to finally obtaining the desired product. An immobilized polymer of divinylbenzene/methacrylate (Novozym 435) has been used in MTBE upon saturation with phosphate buffer solution with pH 7 at 10°C; the racemate-to-target-enantiomer conversion was about 40%. In the year 2014, the Chemelectiva-HC-Pharma published a new biocatalytic technique describing the procedure for making sofosbuvir intermediate. The procedure uses immobilized lipase CalB with the reaction occurring at a different site on the molecule. This is accomplished at 60°C using an immobilized CalB catalyst to create the matching alcohol from the sofosbuvir intermediate in a protic organic solvent (Bull et al. 2016). Table 6.5. shows the different enzymes immobilized in the pharmaceutical industry to manufacture the different active chemical compounds.

6.7.11   Importance of Enzyme Immobilization in Biomedical Devices and Biosensors

6.7.11.1   Lipases in Biomedical Devices

Fat malabsorption found due to exocrine pancreatic insufficiency in cystic fibrosis patients can result in reduced caloric intake including fatty acid deficits like eicosapentaenoic acid (EPA) and docosahexaenoic acid (DHA) that are critical for

TABLE 6.5

Application of Enzyme Immobilization Techniques in the Pharmaceutical Industry (Basso and Serban 2019, Hassan et al. 2019)

API	Immobilized Enzyme	Type of Immobilization	Reaction Type	Target Application
Sitagliptin	Transaminases or aminotransferases	Adsorption	Transamination	Antidiabetic
Ampicillin and amoxicillin	Penicillin G acylase (PGA)-syn	Covalent on epoxy	Amidation	Beta-lactams
6-APA (6-aminopenicillanic acid)	Penicillin G acylase (PGA)-hyd	Covalent on epoxy	Hydrolysis	Beta-lactams
Sofosbuvir	*Candida antarctica* lipase B (CALB)	Adsorption over methacrylate/ divinylbenzene copolymer	Hydrolysis	Treatment for hepatitis C
Odanacatib	*Candida antarctica* lipase B (CALB)	Adsorption over octadecyl polymethacrylate resin	Esterification	Osteoporosis treatment

development and growth. Pancreatic enzyme replacement treatment has been used along with meals by cystic fibrosis patients to enhance fat and other nutritional absorption. Thousands of cystic fibrosis patients in the United States rely on enteral feeding to meet their development and weight objectives. Because hydrolyzed fats are poorly stabilized, current formulae comprise triglycerides rather than fatty acids (Loli et al. 2015).

The single-use cylindrical cartridge was developed by Alcresta Therapeutics in partnership with ChiralVision and features a chamber sealed by frits comprising digestive immobilized enzyme. The cartridge comprises polymeric beads with covalently immobilized lipase from *Pseudomonas fluorescens, Chromobacterium viscosum, Rhizopus oryzae*, that hydrolyze 90% of the lipids in the enteral formula, which passes through the cartridge. Patients' lives are extended as a result of the device's ability to boost fat absorption that improves chronic lung disease and cognitive ability, and also results in shorter times for parenteral feeding (Nunes et al. 2011).

6.7.11.2 Urease in Medical Devices

The urease enzyme catalyzes the hydrolysis of urea to produce ammonium and CO_2. Its applications are broader ranging from the removal and hydrolysis of urea in foods, beverages, and wastewater effluents to advanced applications like the removal of urea from the dialysate regeneration system for artificial kidneys or the blood extracorporeal detoxification (Agrawal and Verma, 2022). In patients with renal failure, the elimination of urea has been the main challenge. In order to eliminate urea, immobilized urease could be utilized in the artificial kidney's dialysis system., Portable dialysis equipment, which is much smaller in size depending on sorbent regenerative dialysis devices, was made possible with the development of immobilized urease. Several different procedures have been described for the elimination of urea in the past and currently, there is an increasing interest in innovative alternate approaches (Svane et al. 2020).

Several technologies are available when it comes to the urease-cation exchanger approach: the sorb system is a traditional divalent-selective ion exchanger, basically a cation exchanger, devised by Sorb, Renal Solutions Inc., while another is the zeolite (aluminosilicate) monovalent-selective cation exchanger by Fresenius Medical Care, and the zirconium silicate: where crystalline monovalent-s urease and liquid membrane acid-filled micro-capsules are used in other ways, such as at Exxon. The sorbent column is the only hands-on urease device that has been developed for medical application, despite all of the previous efforts. Portable dialysis devices, like Automated Wearable Artificial Kidney (AWAK) sorbent systems, Allient, and Regenerative Dialysis (REDY) use urease to convert urea into ammonium ions and bicarbonate ions. Unlike other metabolic wastes like uric acid and creatinine that are not readily absorbed by activated carbon (Zhao et al. 2020). Such immobilized urease dialysis devices were commercially available in 1980, but production ceased in 1990. Since 2000, similar systems have resurfaced as a viable technical alternative attracting considerable interest from industries with the AWAK system serving as the best example of commercially accessible technology.

6.8   CONCLUSIONS

In the industrial processes where they are now used, immobilized enzymes give significant benefits. When compared to chemical synthesis the advantages include lower environmental impact, and process simplification in a more sustainable method. In the industrial sector, the food industry uses the largest t amount of immobilized enzymes in established continuous biocatalysis like the synthesis of amino acids, cocoa butter analogs, and HFCS utilized as significant food ingredients globally. The chemical industry has significantly expanded its biocatalytic processes recently by improvised synthesis of complex chiral chemicals at low temperatures, like herbicides and chiral amines. These require naturally compatible substrates mainly in aqueous environments, for which immobilized enzymes are employed after modifying them by genetic engineering protocol for attaining desired characteristics.

However, the situation in the pharmaceutical industry is quite different. This industry has been putting the majority of its efforts into developing biocatalytic processes and enhancing the efficiency of currently available enzymes. Many enzymes used in the pharmaceutical industry must function on synthetic molecules that are quite different from standard enzyme substrates, and they must do so under unique circumstances, for instance, the presence of high substrate concentrations or organic solvents. This has encouraged the industry to build and develop novel enzymes or modify the existing ones to expand the spectrum of chemical substances for which they can be exploited.

The need for more sustainable and cost-effective processes together with significant advances in enzyme discovery, immobilized enzyme technology, and protein engineering undoubtedly demonstrates that biocatalysis in industrial processes has a promising future.

REFERENCES

Adrio, J.L. and Demain, A.L., 2014. Microbial enzymes: Tools for biotechnological processes. *Biomolecules*, 4 (1), 117–139.

Agostinelli, E., Belli, F., Tempera, G., Mura, A., Floris, G., Toniolo, L., Vavasori, A., Fabris, S., Momo, F., and Stevanato, R., 2007. Polyketone polymer: A new support for direct enzyme immobilization. *Journal of Biotechnology*, 127 (4), 670–678.

Agrawal, K., and Verma, P. (2020) The interest in nanotechnology: A step towards bioremediation. In: *Removal of Emerging Contaminants Through Microbial Processes*. Springer, Singapore 265–282.

Agrawal, K, and Verma, P. (2021) Applications of biomolecules of endophytic fungal origin and its future prospect. In: *Fungi Bio-Prospects in Sustainable Agriculture, Environment and Nano-technology. Vol 3:Fungal Metabolites and Nano-technology*. Academic Press, Elsevier, USA 207–230.

Agrawal, K., and Verma, P. (2022) An Overview of Wastewater Treatment Facilities in Asian and European Countries. Wastewater Treatment, CRC Press, Boca Raton 1–12.

Al-Adhami, A.J.H., Bryjak, J., Greb-Markiewicz, B., and Peczyńska-Czoch, W., 2002. Immobilization of wood-rotting fungi laccases on modified cellulose and acrylic carriers. *Process Biochemistry*, 37 (12), 1387–1394.

Alsafadi, D. and Paradisi, F., 2014. Covalent immobilization of alcohol dehydrogenase (ADH2) from *Haloferax volcanii*: How to maximize activity and optimize performance of halophilic enzymes. *Molecular Biotechnology*, 56 (3), 240–247.

Andler, S.M. and Goddard, J.M., 2018. Transforming food waste: How immobilized enzymes can valorize waste streams into revenue streams. *NPJ Science of Food*, 2 (1), 19.

Ashly, P.C., Joseph, M.J., and Mohanan, P.V., 2011. Activity of diastase α-amylase immobilized on polyanilines (PANIs). *Food Chemistry*, 127 (4), 1808–1813.

Atadashi, I.M., Aroua, M.K., and Aziz, A.A., 2010. High quality biodiesel and its diesel engine application: A review. *Renewable and Sustainable Energy Reviews*, 14 (7), 1999–2008.

Bai, X., Gu, H., Chen, W., Shi, H., Yang, B., Huang, X., and Zhang, Q., 2014. Immobilized laccase on activated poly(vinyl alcohol) microspheres for enzyme thermistor application. *Applied Biochemistry and Biotechnology*, 173 (5), 1097–1107.

Basso, A. and Serban, S., 2019. Industrial applications of immobilized enzymes—A review. *Molecular Catalysis*, 479, 110607.

Betigeri, S.S. and Neau, S.H., 2002. Immobilization of lipase using hydrophilic polymers in the form of hydrogel beads. *Biomaterials*, 23 (17), 3627–3636.

Bhardwaj, N., Agrawal, K., Kumar, B., and Verma, P. (2021a) Role of enzymes in deconstruction of waste biomass for sustainable generation of value-added products. In: Thatoi H., Mohapatra S., Das S.K. (eds) *Bioprospecting of Enzymes in Industry, Healthcare and Sustainable Environment*. 219–250; Springer, Singapore.

Bhardwaj, N., Kumar, B., Agrawal, K., and Verma, P. (2021b) Current perspective on production and applications of microbial cellulases: a review. *Bioresources and Bioprocessing*, 8: 1–34.

Bhardwaj, N., and Verma, P. (2021). Xylanases: A helping module for the enzyme biorefinery platform. In: *Bioenergy Research: Revisiting Latest Development Bioprospecting*, 7, 161–179, Springer, Singapore.

Blanco, R.M., Terreros, P., Muñoz, N., and Serra, E., 2007. Ethanol improves lipase immobilization on a hydrophobic support. *Journal of Molecular Catalysis B: Enzymatic*, 47 (1–2), 13–20.

Boller, Thomas, Meier, Christian, and Menzler, S., 2002. EUPERGIT oxirane acrylic beads: how to make enzymes fit for biocatalysis. *Organic Process Research and Development*, 6 (4), 509–519.

Brady, D. and Jordaan, J., 2009. Advances in enzyme immobilisation. *Biotechnology Letters*, 31 (11), 1639–1650.

Bull, J.A., Croft, R.A., Davis, O.A., Doran, R., and Morgan, K.F., 2016. Oxetanes: Recent advances in synthesis, reactivity, and medicinal chemistry. *Chemical Reviews*, 116 (19), 12150–12233.

Cahyaningrum, S.E., Herdyastusi, N., and Maharani, D.K., 2014. Immobilization of glucose isomerase in surface-modified chitosan gel beads. *Research Journal of Pharmaceutical. Biological and Chemical Science*, 5 (2), 104–111.

Chandra, P., Enespa, Singh R., and Arora, P.K., 2020. Microbial lipases and their industrial applications: A comprehensive review. *Microbial Cell Factories*, 19 (1), 169.

Chang, M.Y. and Juang, R.S., 2007. Use of chitosan–clay composite as immobilization support for improved activity and stability of β-glucosidase. *Biochemical Engineering Journal*, 35 (1), 93–98.

Chanitnun, K. and Pinphanichakarn, P., 2012. Glucose(xylose) isomerase production by *Streptomyces* SP. CH7 grown on agricultural residues. *Brazilian Journal of Microbiology*, 43 (3), 1084–1093.

Chapman, J., Ismail, A.E., and Dinu, C.Z., 2018. Industrial applications of enzymes: Recent advances, techniques, and outlooks. *Catalysts*, 8 (6), 20–29.

Chen, Z., Shen, Y.R., and Somorjai, G.A., 2002. Studies of polymer surfaces by sum frequency generation vibrational spectroscopy. *Annual Review of Physical Chemistry*, 53 (1), 437–465.

Chrisnasari, R., Wuisan, Z.G., Budhyantoro, A., and Widi, R.K., 2015. Glucose oxidase immobilization on TMAH-modified bentonite. *Indonesian Journal of Chemistry*, 15 (1), 22–28.

Cooper, G.M., 2000. The central role of enzymes as biological catalysts.

Daoud, F.B.O., Kaddour, S., and Sadoun, T., 2010. Adsorption of cellulase *Aspergillus niger* on a commercial activated carbon: Kinetics and equilibrium studies. *Colloids and Surfaces B: Biointerfaces*, 75 (1), 93–99.

Datta, S., Christena, L.R., and Rajaram, Y.R.S., 2013. Enzyme immobilization: An overview on techniques and support materials. *3 Biotech*, 3 (1), 1.

Demetrius, L., 2002. Thermodynamics and kinetics of protein folding: An evolutionary perspective. *Journal of Theoretical Biology*, 217 (3), 397–411.

Dezott, M., Innocentini-Mei, L.H., and Durán, N., 1995. Silica immobilized enzyme catalyzed removal of chlorolignins from eucalyptus kraft effluent. *Journal of Biotechnology*, 43 (3), 161–167.

DiCosimo, R., McAuliffe, J., Poulose, A.J., and Bohlmann, G., 2013. Industrial use of immobilized enzymes. *Chemical Society Reviews*, 42 (15), 6437–6474.

Ebrahimi, M., Placido, L., Engel, L., Ashaghi, K.S., and Czermak, P., 2010. A novel ceramic membrane reactor system for the continuous enzymatic synthesis of oligosaccharides. *Desalination*, 250 (3), 1105–1108.

Falahati, M., Ma'Mani, L., Saboury, A.A., Shafiee, A., Foroumadi, A., and Badiei, A.R., 2011. Aminopropyl-functionalized cubic Ia3d mesoporous silica nanoparticle as an efficient support for immobilization of superoxide dismutase. *Biochimica et Biophysica Acta - Proteins and Proteomics*, 1814 (9), 1195–1202.

Flores-Maltos, A., Rodríguez-Durán, L.V., Renovato, J., Contreras, J.C., Rodríguez, R., and Aguilar, C.N., 2011. Catalytical properties of free and immobilized *Aspergillus niger* tannase. *Enzyme Research*, 2011 (1), 1–6.

Fu, J., Reinhold, J., and Woodbury, N.W., 2011. Peptide-modified surfaces for enzyme immobilization. *PLoS ONE*, 6, 1–6.

Godjevargova, T., Nenkova, R., and Konsulov, V., 2006. Immobilization of glucose oxidase by acrylonitrile copolymer coated silica supports. *Journal of Molecular Catalysis B: Enzymatic*, 38 (2), 59–64.

Gómez, L., Ramírez, H.L., Neira-Carrillo, A., and Villalonga, R., 2006. Polyelectrolyte complex formation mediated immobilization of chitosan-invertase neoglycoconjugate on pectin-coated chitin. *Bioprocess and Biosystems Engineering*, 28 (6), 387–395.

Grosová, Z., Rosenberg, M., Rebroš, M., Šipocz, M., and Sedláčková, B., 2007. Entrapment of β-galactosidase in polyvinylalcohol hydrogel. *Biotechnology Letters*, 30 (4), 763–767.

Guisan, J.M., 2006. *Immobilization of Enzymes and Cells*. Springer. Humana Totowa, NJ.

Hamann, M.C.J. and Saville, B.A., 1996. Enhancement of tyrosinase stability by immobilization on nylon 66. *Food and Bioproducts Processing: Transactions of the Institution of Chemical Engineers, Part C*,74, 47–52.

Hanefeld, U., Gardossi, L., and Magner, E., 2009. Understanding enzyme immobilisation. *Chemical Society Reviews*, 38 (2), 453–468.

Hartmann, M., and Kostrov, X. 2013. Immobilization of enzymes on porous silicas--benefits and challenges. *Chemical Society Reviews*, 42 (15), 6277–6289.

Hassan, M.E., Yang, Q., Xiao, Z., Liu, L., Wang, N., Cui, X., and Yang, L., 2019. Impact of immobilization technology in industrial and pharmaceutical applications. *3 Biotech*, 9 (12), 440.

Hernandez, K. and Fernandez-Lafuente, R., 2011. Control of protein immobilization: Coupling immobilization and site-directed mutagenesis to improve biocatalyst or biosensor performance. *Enzyme and Microbial Technology*, 48 (2), 107–122.

Homaei, A.A., Sariri, R., Vianello, F., and Stevanato, R. 2013. Enzyme immobilization: an update. *Journal of Chemical Biology*, 6 (4), 185–205.

Hosseinkhani, S., Szittner, R., Nemat-Gorgani, M., and Meighen, E.A., 2003. Adsorptive immobilization of bacterial luciferases on alkyl-substituted Sepharose 4B. *Enzyme and Microbial Technology*, 32 (1), 186–193.

Huang, L. and Cheng, Z.M., 2008. Immobilization of lipase on chemically modified bimodal ceramic foams for olive oil hydrolysis. *Chemical Engineering Journal*, 144 (1), 103–109.

Iosin, M., Toderas, F., Baldeck, P.L., and Astilean, S., 2009. Study of protein–gold nanoparticle conjugates by fluorescence and surface-enhanced Raman scattering. *Journal of Molecular Structure*, 924–926, 196–200.

Ismail, H., Lau, R.M., van Langen, L.M., van Rantwijk, F., Švedas, V.K., and Sheldon, R.A., 2008. A green, fully enzymatic procedure for amine resolution, using a lipase and a penicillin G acylase. *Green Chemistry*, 10 (4), 415.

Jeong, J., Antonyraj, C.A., Shin, S., Kim, S., Kim, B., Lee, K.-Y., and Cho, J.K., 2013. Commercially attractive process for production of 5-hydroxymethyl-2-furfural from high fructose corn syrup. *Journal of Industrial and Engineering Chemistry*, 19 (4), 1106–1111.

Jiang, S., Xiao, W., Zhu, X., Yang, P., Zheng, Z., Lu, S., Jiang, S., Zhang, G., and Liu, J., 2020. Review on D-Allulose: In vivo metabolism, catalytic mechanism, engineering strain construction, bio-production technology. *Frontiers in Bioengineering and Biotechnology*, 8, 1–10.

Johnson, R.D., Wang, Z.-G., and Arnold, F.H., 1996. Surface site heterogeneity and lateral interactions in multipoint protein adsorption. *Journal of Physical Chemistry*, 100 (12), 5134–5139.

Kamori, M., Hori, T., Yamashita, Y., Hirose, Y., and Naoshima, Y., 2000. Immobilization of lipase on a new inorganic ceramics support, toyonite, and the reactivity and enantioselectivity of the immobilized lipase. *Journal of Molecular Catalysis B: Enzymatic*, 9 (4–6), 269–274.

Kilara, A., Shahani, K.M., and Shukla, T.P., 1979. The use of immobilized enzymes in the food industry: A review. *C R C Critical Reviews in Food Science and Nutrition*, 12 (2), 161–198.

Klein, M.P., Scheeren, C.W., Lorenzoni, A.S.G., Dupont, J., Frazzon, J., and Hertz, P.F., 2011. Ionic liquid-cellulose film for enzyme immobilization. *Process Biochemistry*, 46 (6), 1375–1379.

Klouda, L. and Mikos, A.G., 2008. Thermoresponsive hydrogels in biomedical applications. *European Journal of Pharmaceutics and Biopharmaceutics: Official Journal of Arbeitsgemeinschaft fur Pharmazeutische Verfahrenstechnik e.V*, 68 (1), 34–45.

Kołodziejczak-Radzimska, A., Zdarta, J., and Jesionowski, T., 2018. Physicochemical and catalytic properties of acylase I from *Aspergillus melleus* immobilized on amino- and carbonyl-grafted stöber silica. *Biotechnology Progress*, 34 (3), 767–777.

Kotal, M., Srivastava, S.K., and Maiti, T.K., 2011. Fabrication of gold nanoparticle assembled polyurethane microsphere template in trypsin immobilization. *Journal of Nanoscience and Nanotechnology*, 11 (11), 10149–10157.

Krajewska, B., 1991. Chitin and its derivative as supports for immobilization of enzymes. *Acta Biotechnologica*, 11 (3), 269–277.

Kramer, M., Cruz, J.C., Pfromm, P.H., Rezac, M.E., and Czermak, P., 2010. Enantioselective transesterification by *Candida antarctica* Lipase B immobilized on fumed silica. *Journal of Biotechnology*, 150 (1), 80–86.

Kreider, A., Sell, S., Kowalik, T., Hartwig, A., and Grunwald, I., 2014. Influence of immobilization protocol on the structure and function of surface bound proteins. *Colloids and Surfaces B: Biointerfaces*, 116, 378–382.

Kumar, B. and Verma, P. (2020). Application of hydrolytic enzymes in biorefinery and its future prospects. In: *Microbial Strategies for Techno-economic Biofuel Production.* Springer, Singapore 59–83.

Kumar, P., Kumar, A., Pandey, J.K., Chen, W., Patel, A., and Ashokkumar, V., 2021. Pyrolysis of sewage sludge for sustainable biofuels and value-added biochar production. *Journal of Environmental Management*, 298 (July), 113450.

Kumari, A. and Kayastha, A.M., 2011. Immobilization of soybean (Glycine max) α-amylase onto Chitosan and Amberlite MB-150 beads: Optimization and characterization. *Journal of Molecular Catalysis B: Enzymatic*, 69 (1–2), 8–14.

Loli, H., Narwal, S., Saun, N., and Gupta, R., 2015. Lipases in medicine: An overview. *Mini-Reviews in Medicinal Chemistry*, 15 (14), 1209–1216.

Mannen, T., Yamaguchi, S., Honda, J., Sugimoto, S., Kitayama, A., and Nagamune, T., 2001. Observation of charge state and conformational change in immobilized protein using surface plasmon resonance sensor. *Analytical Biochemistry*, 293 (2), 185–193.

Martins de Oliveira, S., Moreno-Perez, S., Romero-Fernández, M., Fernandez-Lorente, G., Rocha-Martin, J., and Guisan, J.M., 2018. Immobilization and stabilization of commercial β-1,4-endoxylanase Depol™ 333MDP by multipoint covalent attachment for xylan hydrolysis: Production of prebiotics (xylo-oligosaccharides). *Biocatalysis and Biotransformation*, 36 (2), 141–150.

Matto, M. and Husain, Q., 2009. Calcium alginate–starch hybrid support for both surface immobilization and entrapment of bitter gourd (*Momordica charantia*) peroxidase. *Journal of Molecular Catalysis B: Enzymatic*, 57 (1–4), 164–170.

Michota-Kaminska, A., Wrzosek, B., and Bukowska, J., 2006. Resonance raman evidence of immobilization of laccase on self-assembled monolayers of thiols on Ag and Au surfaces. *Applied Spectroscopy*, 60 (7), 752–757.

Minteer, S., 2014. Enzyme immobilization in biotechnology oxygen biocathodes view project hybrid electrocatalytic systems view project.

Mittal, S. and Singh, L.R., 2013. Denatured state structural property determines protein stabilization by macromolecular crowding: A thermodynamic and structural approach. *PLOS ONE*, 8 (11), e78936.

Mohamad, N.R., Marzuki, N.H.C., Buang, N.A., Huyop, F., and Wahab, R.A., 2015. An overview of technologies for immobilization of enzymes and surface analysis techniques for immobilized enzymes. *Biotechnology & Biotechnological Equipment*, 29 (2), 205–220.

Mohiuddin, M., Arbain, D., Islam, A.K.M.S., Ahmad, M.S., and Ahmad, M.N., 2016. Alpha-glucosidase enzyme biosensor for the electrochemical measurement of antidiabetic potential of medicinal plants. *Nanoscale Research Letters*, 11 (1), 1–12.

Moslemi, M., 2021. Reviewing the recent advances in application of pectin for technical and health promotion purposes: From laboratory to market. *Carbohydrate Polymers*, 254 (117324): 1–17.

Nagasawa, T., Sato, K., Shimada, Y., and Kasumi, T., 2016. Efficient conversion of d-glucose to d-fructose in the presence of organogermanium compounds. *Journal of Applied Glycoscience*, 63 (2), 39–45.

Namdeo, M. and Bajpai, S.K., 2009. Immobilization of α-amylase onto cellulose-coated magnetite (CCM) nanoparticles and preliminary starch degradation study. *Journal of Molecular Catalysis B: Enzymatic*, 59 (1–3), 134–139.

Neto, W., Schürmann, M., Panella, L., Vogel, A., and Woodley, J.M., 2015. Immobilisation of ω-transaminase for industrial application: Screening and characterisation of commercial ready to use enzyme carriers. *Journal of Molecular Catalysis B: Enzymatic*, 117, 54–61.

Nunes, P.A., Pires-Cabral, P., Guillén, M., Valero, F., Luna, D., and Ferreira-Dias, S., 2011. Production, purification, characterization, and applications of lipases. *Journal of the American Oil Chemists' Society*, 88 (4), 627–662.

Nygren, P., Lundqvist, M., Broo, K., and Jonsson, B.-H., 2008. Fundamental design principles that guide induction of helix upon formation of stable peptide–nanoparticle complexes. *Nano Letters*, 8 (7), 1844–1852.

O'Shea, P.D., Chen, C., Gauvreau, D., Gosselin, F., Hughes, G., Nadeau, C., and Volante, R.P., 2009. A practical enantioselective synthesis of odanacatib, a potent cathepsin k inhibitor, via triflate displacement of an α-trifluoromethylbenzyl triflate. *The Journal of Organic Chemistry*, 74 (4), 1605–1610.

Olempska-Beer, Z.S., Merker, R.I., Ditto, M.D., and DiNovi, M.J., 2006. Food-processing enzymes from recombinant microorganisms-a review. *Regulatory Toxicology and Pharmacology*, 45 (2), 144–158.

Ottolina, G., Carrea, G., Riva, S., Sartore, L., and Veronese, F.M., 1992. Effect of the enzyme form on the activity, stability and enantioselectivity of lipoprotein lipase in toluene. *Biotechnology Letters*, 14 (10), 947–952.

Parmar, A., Kumar, H., Marwaha, S.S., and Kennedy, J.F., 2000. Advances in enzymatic trans-formation of penicillins to 6-aminopenicillanic acid (6-APA). *Biotechnology Advances*, 18 (4), 289–301.

Phitsuwan, P., Laohakunjit, N., Kerdchoechuen, O., Kyu, K.L., and Ratanakhanokchai, K., 2013. Present and potential applications of cellulases in agriculture, biotechnology, and bioenergy. *Folia Microbiologica*, 58 (2), 163–176.

Pogorilyi, R.P., Siletskaya, E.Y., Goncharik, V.P., Kozhara, L.I., and Zub, Y.L., 2007. Immobilization of urease on the silica gel surface by sol-gel method. *Russian Journal of Applied Chemistry*, 80 (2), 330–334.

Prakash, O. and Jaiswal, N., 2011. Immobilization of a thermostable • -amylase on agarose and agar matrices and its application in starch stain removal. *World Applied Sciences Journal*, 13 (3), 572–577.

Rani, A.S., Das, M.L.M., and Satyanarayana, S., 2000. Preparation and characterization of amyloglucosidase adsorbed on activated charcoal. *Journal of Molecular Catalysis - B Enzymatic*, 10 (5), 471–476.

Rosa, C.C., Cruz, H.J., Vidal, M., and Oliva, A.G., 2002. Optical biosensor based on nitrite reductase immobilised in controlled pore glass. *Biosensors and Bioelectronics*, 17 (1–2), 45–52.

Saifuddin, N., Raziah, A.Z., and Junizah, A.R., 2013. Carbon nanotubes: A review on structure and their interaction with proteins. *Journal of Chemistry*, 2013, 1–18.

Schartner, J., Güldenhaupt, J., Mei, B., Rögner, M., Muhler, M., Gerwert, K., and Kötting, C., 2013. Universal method for protein immobilization on chemically functionalized germanium investigated by ATR-FTIR difference spectroscopy. *Journal of the American Chemical Society*, 135 (10), 4079–4087.

Serralha, F.N., Lopes, J.M., Lemos, F., Prazeres, D.M.F., Aires-Barros, M.R., Cabral, J.M.S., and Ramôa Ribeiro, F., 2001. Kinetics and modelling of an alcoholysis reaction catalyzed by cutinase immobilized on NaY zeolite. *Journal of Molecular Catalysis B: Enzymatic*, 11 (4–6), 713–718.

Sharmin, F., Rakshit, S.K., and Jayasuriya, H.P.W., 2007. Enzyme immobilization on glass surface for the development of phosphate detection biosensors. *Agricultural Engineering International*, 9, 1–12.

Sheldon, R.A. 2007a. Cross-linked enzyme aggregates (CLEAs): Stable and recyclable biocatalysts. *Biochemical Society Transactions*, 35 (6), 1583–1587.

Sheldon, R.A., 2007b. Enzyme immobilization: The quest for optimum performance. *Advanced Synthesis & Catalysis*, 349 (8–9), 1289–1307.

Shen, Q., Yang, R., Hua, X., Ye, F., Zhang, W., and Zhao, W., 2011. Gelatin-templated biomimetic calcification for β-galactosidase immobilization. *Process Biochemistry*, 46 (8), 1565–1571.

Shi, L.E., Tang, Z.X., Yi, Y., Chen, J.S., Xiong, W.Y., and Ying, G.Q., 2011. Immobilization of nuclease p1 on chitosan micro-spheres. *Chemical and Biochemical Engineering Quarterly*, 25 (1), 83–88.

Silva, M.F., Rigo, D., Mossi, V., Dallago, R.M., Henrick, P., Kuhn, G.D.O., Rosa, C.D., Oliveira, D., Oliveira, J.V., and Treichel, H., 2013. Evaluation of enzymatic activity of commercial inulinase from *Aspergillus niger* immobilized in polyurethane foam. *Food and Bioproducts Processing*, 91 (1), 54–59.

Silva, V.D.M., De Marco, L.M., Delvivo, F.M., Coelho, J.V., and Silvestre, M.P.C., 2005. Immobilization of pancreatin in activated carbon and in alumina for preparing whey hydrolysates. *Acta Scientiarum - Health Sciences*, 27 (2), 163–169.

Stoch, S.A., Zajic, S., Stone, J., Miller, D.L., Van Dyck, K., Gutierrez, M.J., De Decker, M., Liu, L., Liu, Q., Scott, B.B., Panebianco, D., Jin, B., Duong, L.T., Gottesdiener, K., and Wagner, J.A., 2009. Effect of the cathepsin K inhibitor odanacatib on bone resorption biomarkers in healthy postmenopausal women: Two double-blind, randomized, placebo-controlled phase I studies. *Clinical Pharmacology & Therapeutics*, 86 (2), 175–182.

Svane, S., Sigurdarson, J.J., Finkenwirth, F., Eitinger, T., and Karring, H., 2020. Inhibition of urease activity by different compounds provides insight into the modulation and association of bacterial nickel import and ureolysis. *Scientific Reports*, 10 (1), 8503.

Thompson, M.P., Peñafiel, I., Cosgrove, S.C., and Turner, N.J., 2019. Biocatalysis using immobilized enzymes in continuous flow for the synthesis of fine chemicals. *Organic Process Research & Development*, 23 (1), 9–18.

Tonini, D. and Astrup, T., 2012. Life-cycle assessment of a waste refinery process for enzymatic treatment of municipal solid waste. *Waste Management*, 32 (1), 165–176.

Tümtürk, H., Karaca, N., Demirel, G., and Şahin, F., 2007. Preparation and application of poly(N,N-dimethylacrylamide-co-acrylamide) and poly(N-isopropylacrylamide-co-acrylamide)/κ-Carrageenan hydrogels for immobilization of lipase. *International Journal of Biological Macromolecules*, 40 (3), 281–285.

Vaillant, F., Millan, A., Millan, P., Dornier, M., Decloux, M., and Reynes, M., 2000. Co-immobilized pectinlyase and endocellulase on chitin and Nylon supports. *Process Biochemistry*, 35 (9), 989–996.

Vartiainen, J., Rättö, M., and Paulussen, S., 2005. Antimicrobial activity of glucose oxidase-immobilized plasma-activated polypropylene films. *Packaging Technology and Science*, 18 (5), 243–251.

Wang, A., Wang, H., Zhu, S., Zhou, C., Du, Z., and Shen, S., 2008. An efficient immobilizing technique of penicillin acylase with combining mesocellular silica foams support and p-benzoquinone cross linker. *Bioprocess and Biosystems Engineering*, 31 (5), 509–517.

Wang, W., Li, Z., Liu, W., and Wu, J., 2012. Horseradish peroxidase immobilized on the silane-modified ceramics for the catalytic oxidation of simulated oily water. *Separation and Purification Technology*, 89, 206–211.

Wang, W., Zhou, W., Li, J., Hao, D., Su, Z., and Ma, G., 2015. Comparison of covalent and physical immobilization of lipase in gigaporous polymeric microspheres. *Bioprocess and Biosystems Engineering*, 38 (11), 2107–2115.

Wang, Y., Ravikumar, Y., Zhang, G., Yun, J., Zhang, Y., Parvez, A., Qi, X., and Sun, W., 2020. Biocatalytic synthesis of D-allulose using novel d-tagatose 3-epimerase from christensenella minuta. *Frontiers in Chemistry*, 8.

Wu, K., Yang, Z., Meng, X., Chen, R., Huang, J., and Shao, L., 2020. Engineering an alcohol dehydrogenase with enhanced activity and stereoselectivity toward diaryl ketones: Reduction of steric hindrance and change of the stereocontrol element. *Catalysis Science & Technology*, 10 (6), 1650–1660.

Wu, S., Snajdrova, R., Moore, J.C., Baldenius, K., and Bornscheuer, U.T., 2021. Biocatalysis: Enzymatic synthesis for industrial applications. *Angewandte Chemie International Edition*, 60 (1), 88–119.

Xie, T., Wang, A., Huang, L., Li, H., Chen, Z., Wang, Q., and Yin, X., 2009. Recent advance in the support and technology used in enzyme immobilization. *African Journal of Biotechnology*, 8 (19).

Xu, L.J., Zong, C., Zheng, X.S., Hu, P., Feng, J.M., and Ren, B., 2014. Label-free detection of native proteins by surface-enhanced Raman spectroscopy using iodide-modified nanoparticles. *Analytical Chemistry*, 86 (4), 2238–2245.

Youn, S.-C., Chen, L.-Y., Chiou, R.-J., Lai, T.-J., Liao, W.-C., Mai, F.-D., and Chang, H.-M., 2015. Comprehensive application of time-of-flight secondary Ion mass spectrometry (TOF-SIMS) for ionic imaging and bio-energetic analysis of club drug-induced cognitive deficiency. *Scientific Reports*, 5 (1), 18420.

Zdarta, J., Sałek, K., Kołodziejczak-Radzimska, A., Siwinska-Stefanska, K., Szwarc-Rzepka, K., Norman, M., Klapiszewski, Ł., Bartczak, P., Kaczorek, E., and Jesionowski, T., 2015. Immobilization of Amano Lipase A onto Stöber silica surface: Process characterization and kinetic studies. *Open Chemistry*, 13 (1), 138–148.

Zhang, J., Zhang, J., Zhang, F., Yang, H., Huang, X., Liu, H., and Guo, S., 2010. Graphene oxide as a matrix for enzyme immobilization. *Langmuir*, 26 (9), 6083–6085.

Zhang, W., Chen, J., and Mu, W., 2021. Recent Advances in ketose 3-epimerase and its application for D-allulose production. In: *Novel enzymes for functional carbohydrates production*. Singapore: Springer Singapore, 17–42.

Zhao, Q., Seredych, M., Precetti, E., Shuck, C.E., Harhay, M., Pang, R., Shan, C.X., and Gogotsi, Y., 2020. Adsorption of uremic toxins using $Ti_3C_2T_x$ xMXene for dialysate regeneration. *ACS Nano*, 14 (9), 11787–11798.

Zittan, L., Poulsen, P.B., and Hemmingsen, S.H., 1975. Sweetzyme - A new immobilized glucose isomerase. *Starch - Stärke*, 27 (7), 236–241.

7 Nanobiotechnology in Enzyme-based Biorefinement and Valorization of Waste

John M. Pisciotta and Azar E. Saikali

West Chester University, West Chester, USA

CONTENTS

7.1 OVERVIEW OF NANOBIOTECHNOLOGY

Nanotechnology utilizes structures and molecules on the scales of nanometers ranging between 1 and 100 nm (Bayda et al., 2019). Most of the commonly studied nanoparticles (NPs) that cause biological effects fall between 1 and 100 nm in at least one dimension (Ghosh et al., 2021; Qu et al., 2013). Abiotic NPs may be created through geological processes whereas biotic nanomaterials are naturally synthesized by organisms. Recently, genetic engineering has given rise to novel

DOI: 10.1201/9781003187721-7

biotic nanomaterials that can be produced using readily cultivated expression systems (Kolinko et al., 2014). Other nanomaterials are created and released as by-products of human activities, such as construction and mining, or are purposely synthesized to perform specialized functions. NPs may be produced using various methods including biological, electrochemical, or purely chemical methods and their production is increasing. Over 550 tons of NPs are produced annually in Europe alone and the long-term ecological effect is of concern (Tran, 2019). Chemical production methods often suffer from drawbacks including high input chemical costs and toxic by-products; aspects that may be reduced using biological production routes. Various microorganisms biologically produce an array of nanomaterials from inexpensive feedstocks, such as elements found in disparate anthropogenic waste streams. Microbially synthesized nanomaterials are currently being investigated for applications ranging from medical imaging to environmental bioremediation and renewable energy conversion (Ghosh et al., 2021; Agrawal & Verma, 2020).

7.1.1 HISTORICAL BASIS

The historical foundation of nanotechnology can be traced back to ancient Greek philosophers who speculated about the fundamental composition of matter. In the 5th century BCE, humans began to develop the idea of nanoscience and to consider the question of whether matter is continuous and thus infinitely divisible into smaller pieces, or composed of small, indivisible, and indestructible particles (Bayda et al., 2019). Indeed, the Greek prefix "nano" is translated to mean "dwarf" (Bayda et al., 2019). Throughout history, scientists have speculated about how small useful machines could be made. The age-old question of how many angels could dance on the head of a pin is one example of the concept of *reductio ad absurdum* (Naudet & Falissard, 2014). While throughout most of recorded history thinking about machines and other entities at the smallest of scales has been ridiculed, scientific progress during the latter half of the 20th century has revealed the endeavor to be full of merit and practical utility.

The term nanotechnology was originally coined by Norio Taniguchi in the mid-1970s; however, the father of modern nanotechnology is generally considered to be the theoretical physicist and Nobel laureate Richard Phillips Feynman (Mirkovic et al., 2010; Sun and Gupta, 2019). In 1959, Feynman presented a lecture "There's Plenty of Room at the Bottom: An Invitation to Enter a New Field of Physics" in which he envisioned a world in which tiny machines allowed an entire encyclopedia to be written upon the head of a pin (Pierotti et al., 2008). This concept of nanotechnology describes the characterization, fabrication, and manipulation of structures, devices, or materials that have one or more dimensions that are smaller than 100 nanometers (Santos et al., 2015). Nanotechnology has grown to be incorporated in different fields including, agriculture, medicine, textiles, and waste management. As early as 1959, Feynman postulated that minute surgical robots might one day be ingested for treatment of internal diseases. Consequently, Feynman can also be considered the conceptual father of nanobiotechnology, which represents the combination of nanotechnology with biology (Mele and Pisignano, 2009).

Nanobiotechnology is a subdiscipline of biotechnology, which is the use of living organisms for the production of useful products and services (Gupta et al., 2016). Biotechnology dates backs millennia to before the dawn of agriculture. It is concerned with the cultivation or use of organisms ranging in size from microbes to trees. Even before humans knew microbes existed, they were being applied in fields such as cheese and yogurt production and wine making, also called viniculture. As early as 6,000 BC stone age humans in Georgia employed viniculture (McGovern et al., 2017). Nanobiotechnology, *per se*, is a much newer subdiscipline of biotechnology. It is concerned with the production of nanomaterials via biological routes and the application of these nanomaterials in biotic and abiotic systems. Nanobiotechnology is only about two decades old, yet interest in the topic has increased exponentially.

7.1.2 NANOMATERIALS: TYPES AND DEFINITION

Many nanomaterials exist naturally in geological formations while others are synthesized by organisms, are created as a by-product of human activity, or are purposely produced to perform specialized functions. Nanomaterials synthesized by microorganisms, such as diatomaceous biosilica, can be highly complex and useful for applications like the filtration of pest management (Kröger and Brunner, 2014). NPs can be given types of protective coatings that provide for some beneficial function, such as to prevent agglomeration of particles in suspension or a synthetic polymer that acts as a stabilizing agent for metal NPs (Leynen et al., 2019). Silver NPs may be given a coating of polyvinylpyrrolidone (PVP), a synthetic polymer that chemically stabilizes the NP against oxidation (Liu et al., 2013). PVP is generally tissue-compatible and has been shown to be nonirritating to skin, eye, and mucosal membranes (Liu et al., 2013). NPs can be further classified as carbon-based, ceramic, semiconducting, or polymeric.

The identification of NPs has inspired researchers to find uses to which such NPs can be applied to help create a more sustainable world. NPs can be made out of virtually anything and have been applied to the mechanical industry, the food industry, used in flame retardants, and have been used medically for more efficient drug delivery methods (Prato et al., 2008). Because of the widespread use of anthropogenic nanoparticles, some of these synthetic NPs will inevitably leak into the environment, for example in waste streams (Agrawal et al., 2021). It is therefore important to understand NPs in waste and the environment and to recognize the effects that they may have on animals, microbes, and the broader environment. While some nanomaterials, like colloidal silver have been around for decades, many nanomaterial-based products are new and regulation is a challenging and ever-changing landscape (Rezvani et al., 2019).

Liposomes are soft, biocompatible nanostructures that are made of phospholipid vesicles and typically range from 50 to 100 nm for liposomes (Arsalan & Younus, 2018). Monolayered micelles are smaller and typically less than 50 nm (Cordero et al., 2018; Feng et al., 2014). Liposomal NPs possess versatile entrapment and self-assembly capabilities and can function as efficient drug carriers, diagnostics, vaccine components, nutrients, and as other biological agents (El Bissati et al., 2014). Liposomes can be classified into three basic types: multilamellar vesicles, consisting of several lipid bilayers; small unilamellar vesicles, consisting of a single bilayer that has a diameter that is less than 100 nm; and large unilamellar vesicles, consisting of a single lipid bilayer that has a diameter that is greater that 100 nm (Bhatia, 2016). Polymeric nanomaterials also have certain characteristics that make them ideal as drug carriers, such as controlled drug release, the ability to protect the drug and other molecules with biological activity against the environment, and improve their bioavailability and therapeutic index (Zielińska et al., 2020). Polymeric NPs possess two structures that are associated with drug delivery: nanocapsules and nanospheres. Nanocapsules are composed of an oily core in which the drug is usually dissolved and surrounded by a polymeric shell which controls the release profile of the drug from the core (Zielińska et al., 2020). Nanospheres are based on a continuous polymeric network in which the drug can be retained inside or adsorbed onto their surfaces (Guterres et al., 2007). The polymeric nanomaterials used are biodegradable and provide good pharmacokinetic control, are stable, nontoxic, noninflammatory, nonimmunogenic, do not activate neutrophils, and avoid reticuloendothelial clearance (Ahlawat et al., 2018). The polymer coating is used on some NPs to help prevent agglomeration, but the behavior and toxicity of NPs in biological systems may be affected. Industries such as food and textile often require NPs to be comprised of, or possessing a coating, of polymers or metals, such as silver (Ag) and copper (Cu), that in high amounts may be toxic (Chaturvedi & Verma, 2015). Waterways examined for AgNPs *Schmidtea mediterranea*, a type of flatworm, demonstrate adverse effects when exposed to a 10 ug/ml mixture of AgNPs (Bijnens et al., 2021: Leynen et al., 2019). AgNPs are classified as a moderately toxic substance from a study that included using AgBion-2, which is a commercial product used in Russia and represents a colloidal solution of AgNPs and E480, a food additive (Kustov et al., 2014).

Nanomaterials can be synthesized from a variety of products such as carbon, metals, and lipids. They may be comprised of single elements, such as precipitated metals, or a mixture of different molecules, both organic and inorganic. They can be crystalline or amorphous in their arrangement while the size distribution can vary substantially, or it may be strictly controlled via natural or imposed physicochemical conditions. Not all metal-containing nanoparticles (MCNPs) are acutely toxic. TiO_2 NPs unlike many other MeO NPs do not generate ROS in bacterial or animal cells, instead, they react with cellular structures that are involved in the process of respiration and block ROS formation (Musial et al., 2020). TiO_2 NPs cause genotoxic effects when injected in Winstar rats for 60 days, to simulate long-term, low dose ingestion of the FDA-approved food additive E171 in humans (Musial et al., 2020). Intragastric E171 exposure increased tumor progression markers and enhanced tumor formation in the colon in a murine model (Urrutia-Ortega et al., 2016). Although it should be noted that TiO_2 NPs did not induce tumor formation directly, they led to dysplastic changes in colonic epithelium and a decrease in goblet cells (Musial et al., 2020). Figure 7.1 depicts commonly produced types of inorganic and organic nanoparticles.

In nature, nanomaterials serve various physiological roles that help ensure survival by unicellular prokaryotes to multicellular eukaryotes. Bacteria use nanomaterials for internal energy reserves as well as to facilitate directional motility in the case of magnetotactic microbes. Archaea can accumulate intracellular 20–100 nm silver

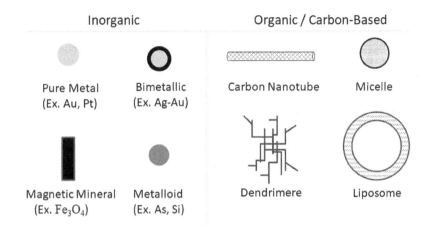

FIGURE 7.1 Representative inorganic (left) and organic carbon-based (right) NPs.

NPs as an adaptive means of metal tolerance and detoxification (Buda et al., 2019). Microbially synthesized polyphosphate granules provide intracellular energy reserves and facilitate the removal of nutrients from wastewater streams while concurrently creating a sustainable alternative to mined phosphates. Manufactured nanomaterials have a growing number of applications throughout society. MCNPs can be used for energy transduction and storage. Organic microbially-derived NPs based on unsaturated polyhydroxyalkanoates show promise as relatively inert and biodegradable drug delivery vehicles to promote or restore health (Pignatello et al., 2019). Other types of NPs directly or indirectly adversely influence pathogens; a phenomenon that can be usefully employed in the medical field.

Nanomaterials are widely used in paints and coatings. Ships, offshore rigs, and marine platforms are under constant attack from the elements of the marine environment, such as pounding surf, corrosive saltwater, biological species, and temperature fluctuations. Ceramic, metal, carbon-based, and other polymer coatings are used on marine equipment to prevent biofouling and corrosion (Fan et al., 2019). Commercially available NPs, such as ZnO and SiO_2 improve the compatibility with epoxy matrix and promote biofouling effects (Santos et al., 2015). Coated samples with and without NPs, exposed to a maritime environment for 45 days demonstrated antifouling properties while noncoated samples showed corrosion pits and additional effects (Santos et al., 2015). Such NP-containing coatings can protect metals from oxidation reactions and metal nanomaterials may be directly inhibitory to microbes and other organisms.

7.1.3 Effect of Nanomaterials on Organisms

Nanoparticles, depending on their chemical composition and morphology, possess properties that may cause lethal effects to microbes. Silver NPs have antimicrobial mechanisms that work individually or in conjunction with one another as they release Ag^+ ions causing oxidative stress, facilitating the destruction of bacterial cell membranes, and inhibiting DNA replication (Li et al., 2018). NPs that demonstrate antimicrobial activities are of interest for use as antimicrobials in medicine as well as on surfaces. Some such NPs are either entirely or partially composed of metal or a metal alloy Na nanoparticle (MeO-NPs). MeO-NPs have become attractive to combat microbes that are resistant to various classes of antibiotics. Their array of physicochemical properties enables MeO-NPs to act as antimicrobial agents through various mechanisms, such as the release of metal ions, the production of reactive oxygen species (ROS), induction of protein and enzyme dysfunction, genotoxicity, and the damaging, abrasive nature of MeO-NPs (Raghunath and Perumal, 2017). When exposed to air, water, or saline solutions they release metal ions that contribute primarily to the NPs toxicity (Li et al., 2018). MeO-NPs release metal ions through dissolution and hydrolysis (Wang et al., 2016). Metal ions, released from MeO-NPs, can cause surface defects to the cell wall and interact with thiol (-SH) groups in the peptidoglycan layer of bacteria, which disrupts the cell wall causing permeability of the cell envelope to increase and enabling additional MeO-NPs to enter the cell (Raghunath and Perumal, 2017). As NPs enter the bacterial cells, they increase oxidative stress by inducing the generation of ROS, including superoxide anion, hydrogen peroxide, and hydroxyl radicals. High levels of ROS are associated with molecular damage to cellular components and, consequently, cell death (Swanson et al., 2011).

Enzymatic dysfunction can be induced by metal oxide NPs and their metal ions. Metal ions catalyze the oxidation of amino acid side chains resulting in protein-bound carbonyls. This in turn can lead to a loss or reduction of catalytic activity in the case of many enzymes, ultimately triggering protein degradation (Raghunath & Perumal, 2017). AgNPs have the tendency to react with bases such as sulfur and phosphorus, which are the main constitutes of protein and DNA. This results in AgNPs causing damage to DNA and deactivating proteins which will ultimately lead to the demise of the cell (Hatchett & White, 1996). ROS production by MeO-NPs primarily inhibits respiratory enzymes, such as succinate hydrogenase and ATP synthase along the electron transport chain (Hajipour et al., 2012). Bacterial iron-sulfur (FeS) dehydratases are prone to inactivation by MeO-NPs, as seen in copper (Cu) as it depletes FeS dehydratases in *Escherichia coli* and eventually triggers bacteriostasis (Macomber and Imlay, 2009). Even though MCNPs can cause adverse effects, they demonstrate generally selective antimicrobial properties and promise for utilizing them to treat drug-resistant infectious diseases (Raghunath & Perumal, 2017).

Viruses too can be destroyed or otherwise inactivated by nanomaterials. Carbon-based nanomaterials (CBN) are emerging as promising options with potent antiviral activity demonstrated against a broad range of enveloped, positive-sense, single-stranded RNA viruses, including SARS-CoV-2, with low to no toxicity in humans (Serrano-Aroca et al., 2021). Fungi, including fungal pathogens, can be inhibited by nanoparticles. Immunocompromised individuals, such as transplant recipients and AIDS patients, are susceptible to infection by opportunistic fungal pathogens such as *Candida albicans*, which causes oral thrush, and *Aspergillum fumigatus*, which causes the respiratory infection known as aspergillosis (Karkowska-Kuleta et al., 2009). Various antifungals exist, but drugs like amphotericin B cause severe side effects and patients may elect to not complete their treatments (Johansen & Gøtzsche, 2014). Resistance to antifungals in emerging multi-drug resistant species such as *Candida auris* and *Candida glabrata* is a worsening problem, particularly in hospital settings (Colombo et al., 2017). For these reasons, novel agents are needed to combat fungal pathogens. Silver NPs ranging from 5 to 60 nm demonstrate direct active inhibition of *Candida* biofilms (Monteiro et al., 2012). MCNPs inhibit fungal cells in several ways. ZnO NPs increase levels of ROS, disrupt the functioning of the endoplasmic reticulum, and diminish the integrity of the chitin cell wall (Babele et al., 2018).

NPs can be used to carry potent antimicrobial drugs to sources of infection. Azole antifungal chemotherapeutics can be delivered to *Candida* biofilms via chitosan-coated iron oxide NPs and demonstrate greater efficacy than the drug alone (Arias et al., 2020). Carbon nanotubes, a type of CBN, are thin cylinders fabricated of rolled graphene sheets that can be monolayered or can have multiple layers that can reach lengths up to 1,000 nm (Eatemadi et al., 2014). Carbon nanotubes can enhance the solubility of some drugs and serve as a gene and peptide delivery system (Bhatia, 2016). Carbon nanotubes enhance drug delivery, efficacy and reduce toxicity as found in the case of amphotericin B-associated nanotubes (Bhatia, 2016). Mechanistically, amphotericin B-associated nanotubes enhanced drug delivery to the interior of fungal cells, increased selective toxicity while reducing toxicity to mammalian cells as compared to standard amphotericin B administration (Prato et al., 2008). The efficacy of amphotericin B-associated nanotubes was also heightened against fungal strains resistant to amphotericin B alone (Prato et al., 2008). Carbon nanofibers (CNF), which are similar to graphene, are another type of CBN that possesses electric conductivity that may be exploited to fabricate conductive composite biomaterials (Salesa et al., 2019).

Valorization of contaminated wastewater for crop irrigation purposes may be accomplished by the removal of diverse pathogens using nanomaterials. Sewage wastewater if left untreated can be a breeding ground for pathogenic bacteria, such as *Campylobacter jejuni*, *Vibrio cholerae* and *Salmonella* spp. (Chahal et al., 2016). Disease vectors, such as various fly and mosquito larvae, also thrive in wastewater. NPs are being used for applications in water and wastewater treatment plants, including adsorption, membranes, photocatalysis, disinfection, and microbial control, sensing, and monitoring (Qu et al., 2013). In adsorption, nanomaterials provided significant improvement with their extremely high specific surface area and associated sorption sites, short intraparticle diffusion distance, and tunable pore size and surface chemistry (Qu et al., 2013). The incorporation of functional nanomaterials into membranes offers an opportunity to improve the membrane permeability, fouling resistance, mechanical and thermal stability, as well as to render new functions for contaminant degradation and self-cleaning (Qu et al., 2013). The textile industry is one of the most water and chemical-intensive industries worldwide and the loss of dye in the effluents can reach up to 75% (Hussein, 2011). TiO_2 NPs displayed highly efficient photocatalytic detoxification of colored wastewater and when coupled with UV radiation the NPs can remove 100% of the color of water and 75% of the total organic carbon (Santos et al., 2015).

Protozoan parasites naturally found in water bodies can be particularly challenging for water treatment facilities. Oocysts of *Cryptosporidium parvum* and, to a lesser degree, *Giardia lamblia* (*Giardia duodenalis*) resist conventional wastewater treatments such as chlorination (Betancourt & Rose, 2004). *Toxoplasma gondii*, the most prevalent parasite of humans on earth, is extremely resistant to ozone (Wainwright et al., 2007). NPs are a promising alternative for the inactivation of such chlorine-resistant pathogens in waste streams. Chitosan NPs demonstrated activity against *C. parvum* oocysts at concentrations of 3 mg/ml (Ahmed et al., 2019). Giardia and other chlorine-resistant protozoan parasites are inexpensively inhibited by selenium and copper NPs (Malekifard et al., 2020). Silver nanoparticles, while more costly, can enter into oocytes of *C. parvum* and inactivate the parasite to enhance the water's value (Cameron et al., 2016).

Enzymatic disruption may be induced by nanoparticles. Respiratory enzymes including succinate dehydrogenase and ATP synthase are damaged by ROS generated from MeO-NPs, eventually resulting in cell death (Raghunath and Perumal, 2017). This oxidative stress is induced by various metal-containing NPs and in both eukaryotic cells and bacterial cells. Ag acts as a weak acid and has the tendency to react with bases such as sulfur and phosphorus of proteins and nucleotide bases, causing DNA damage and inactivating the proteins, ultimately leading to cell death (Raghunath and Perumal, 2017). Cytochrome P450 (CYP450) is an enzyme abundant in the brain and liver that metabolizes exogenous and endogenous compounds, such as antidepressants, opiates, steroids, arachidonic acid, dopamine, and serotonin (Wang et al., 2019). CYP450 expression can be regulated by oxidative stress via the activation of nuclear transport (Tolson and Wang, 2010). Copper NPs (CuNPs) increase ROS generation and decrease antioxidant enzyme activity (Zhang et al., 2018). When rats were exposed to CuNPs for 28 days, brain CYP450 protein expression levels decreased (Wang et al., 2019). Such findings reveal how NP exposure can enhance susceptibility to other toxins.

The morphology of NPs can influence how effectively the NP performs as an antimicrobial agent. AgNPs between the ranges of 1–10 nm can attach to the surface of cell membranes and drastically disturb proper functioning (Morones et al., 2005). NPs can cause direct damage to microbes by physically disrupting membranes and cell walls. AgNPs with diameters of 10nm, 75nm, and 110nm were studied for their bacterial toxicity and AgNP with a diameter of 10nm had the greatest bacterial toxicity on microbial populations and pronounced impact on gene expression (Williams et al., 2014). The optimal size for antimicrobial activity by Cu NPs ranges from 5 to 100 nm while other metal nanoparticles, such as 13 nm Fe nanoparticles, promote plant growth and cell metabolism (Tombuloglu et al., 2019). Different MCNPs can therefore either promote or inhibit cell growth and metabolism, depending on the metal(s) used, particle morphology, and the cell type targeted. Iron is often a limiting nutrient, so the application of iron NPs could enhance growth in some settings by overcoming a nutritional limitation.

Depending on the method used to create NPs, the morphologies displayed could be from a selection of varieties including nanocubes, nanoprisms, nanospheres, triangular nanoplates, and rod-like shapes. Comparative studies reveal differential antibacterial toxicity of divergent AgNPs morphologies. Triangular Ag nanoprisms exerted the greatest bacterial toxicity due to the sharp edges of the triangular shape while truncated triangular AgNps displayed stronger biocidal action than spherical and rod-shaped NPs. The morphology of AgNps influences antibacterial activity in terms of specific surface areas and facet reactivity as those with larger effective contact areas and higher reactive facets exhibit stronger antibacterial activity (Li et al., 2018). Microbial biofilms that develop on host surfaces, such as medical implants, typically demonstrate 100 to 1,000 times higher tolerance to antibiotics compared to bacteria in suspension and are therefore particularly difficult to eradicate using conventional treatment (Olsen, 2015).

7.1.3.1 NPs Effect on Biofilms

NPs affect microbial biofilms in various ways. A biofilm is an organized aggregate of microorganisms living within a self-produced matrix of extracellular polymeric substances (EPS) that is attached to a surface (Yin et al., 2019). EPS is composed of polysaccharides with associated proteins, lipids, and nucleic acids, which form a highly hydrated polar mixture that contributes to the overall scaffold and three-dimensional structure of a biofilm (Yin et al., 2019). Biofilms form on surfaces ranging from teeth to medical implants and several mechanisms are involved in the biofilm-specific tolerance and resistance to antibiotics. The heavily hydrated biofilm matrix serves as a physical barrier and their thickness and chemical composition can reduce the transfer of antibiotics to bacteria (Dunne et al., 1993). If antibiotics penetrate and enter the biofilm, the polar EPS molecules bind to charged antibiotics, allowing the microbes associated inside the biofilm to establish a tolerance to the antibiotics (Nadell et al., 2015). The PA1874-1877 efflux pump in *Pseudomonas aeruginosa* is upregulated in the biofilm state compared to the planktonic state and exports antibiotics to prevent toxic accumulation (Poole, 2001, Zhang & Mah, 2008). Accordingly, novel methods and materials are needed to disrupt antibiotic-resistant microbial biofilm communities that develop around keystone species. *Streptococcus mutans* is one facultatively-anaerobic, sucrose-fermenting bacterium important for dental plaque biofilm formation and cavity development that can be effectively inhibited using MCNPs (Hernández-Sierra et al., 2008).

MCNPs such as AgNPs induce significant decreases in the biomass of 24-hour *Pseudomonas putida* biofilms (Fabrega et al., 2009). *P. aeruginosa* biofilm exhibited Ag NPs resistance properties when exposed to a starting concentration value of 2 µg/mL, and at 18 µg/mL the growth of the biofilm was completely inhibited (Guo et al., 2019). Ag NPs generally prevented biofilm formation of *P. aeruginosa* and *S. epidermidis* by impeding the initial step of bacterial adhesion to the surface (Kalishwaralal et al., 2010). Ag NPs with an average diameter of 25.2 nm effectively precluded the formation of *P. aeruginosa* biofilms and killed bacteria in established biofilm structures (Martinez-Gutierrez et al., 2013). Stabilized NPs are used to slow the aggregation of NPs in solution. Ag NPs, stabilized by hydrolyzed casein peptides strongly inhibited biofilm formation by some gram negative bacteria, such as *E. coli*, *P. aeruginosa* and *Serratia spp.* (Radzig et al., 2013). Starch-stabilized nanoparticles, at concentrations of 1–2 mM, decreased *P. aeruginosa* biofilm formation by 65% and decreased *Staphylococcus aureus* biofilm formation by 88% (Mohanty et al., 2012). One of the most promising applications of NPs is for the destruction of antibiotic-resistant bacteria and MCNPs do hold potential as antimicrobial drugs (Almatar et al., 2018). However, their use appears best suited as a disinfectant or a component of surface coatings since many insoluble particles cannot match the pharmacokinetic and pharmacodynamic performance of small-molecule antibiotics.

7.1.3.2   NPs Effect on Host Microbiota

Some nanoparticles, such as ZnO, possess antimicrobial properties that are effective against a broad spectrum of microbes (Jones et al., 2008). This makes some NPs useful as preservatives in the food and medical industries. The whitening imparted by ZnO is further utilized as a processed food coloring agent, for example in frostings. However, the unregulated use of NPs in such industries may allow NPs to be leaked into the environment and pose a potential ecological threat to humans, animals, plants, insects, and microbes. Certain NPs pose a threat to living organisms and can disrupt the gut microbiome and the intestinal epithelial barrier of living organisms (Li et al., 2018). The microbiome is a complex community of microorganisms consisting of bacteria, protozoa, and fungi that reside on or inside an organism (Pascale et al., 2018). Over 70% of the human microbiota lives in the gastrointestinal (GI) tract, steadily increasing in species diversity from the gastric lumen to the large intestine (Pascale et al., 2018). The microbiome of the GI tract has co-evolved with the host to form an intricate mutualistic symbiotic relationship. The development of the human gut microbiome begins immediately after birth and its composition is strongly influenced by the type of birth, whether natural or cesarean, the individual's genetics, and environmental factors (Bäckhed et al., 2015). The gut microbiota is heavily influenced by host nutrition whether it be artificial, heavily processed, or natural products (Pascale et al., 2018). The microbiome helps the host in maintaining the integrity of the mucosal barrier, providing nutrients such as vitamins including short-chain fatty acids, B vitamins, and vitamin K, protecting against pathogens, and regulating host immunity (Thursby & Juge, 2017).

Bacteria of the gut microbiome can promote cell renewal and wound healing in the case of *Lactobacillus rhamnosus* (Thursby & Juge, 2017). Several other species have also been implicated in promoting epithelial integrity and bacteria modulate mucosal properties and turnover (Thursby & Juge, 2017). Mice raised in a germ-free environment had an extremely thin adherent colonic mucus layer, which can result in colonic bacteria invading the mucosa and causing inflammation. Bacterial products such as peptidoglycan and lipopolysaccharides increase the thickness of the adherent mucus layer which was restored to levels observed in conventionally reared mice (Petersson et al., 2011). Germ-free mice have reduced expansion of CD4+ T-cell populations, a major immune deficiency. However, this deficiency was reversed by the treatment of germ-free mice with polysaccharide A from the capsule of *Bacteroides fragilis* (Mazmanian et al., 2005). Such microbial effectors mediated processes can ameliorate certain inflammatory gut disorders and help differentiate between beneficial and pathogenic bacteria and bolster cellular immunity (Hevia et al., 2015). *B. fragilis* can also secrete membrane vesicles that reduce gut mucosal inflammation via regulatory T cell-independent mechanisms (Chu et al., 2016).

The gut microbiome facilitates the strengthening of the epithelial wall and mucosal membrane and contributes to proper nutrition. The major fermentation products in healthy adults are gases and organic acids, particularly the three short-chain fatty acids (SCFAs) acetate, propionate, and butyrate (Louis et al., 2014). These three SCFAs are typically found in a 3:1:1 ratio, and they play a role in the maintenance of gut and metabolic health (Blaak et al., 2020, Louis et al., 2014). These SCFAs are absorbed by the epithelial cells of the GI tract that are involved in maintaining the mucosal barrier and they use the SCFAs in the regulation of cellular processes such as gene expression, chemotaxis, differentiation, proliferation, and apoptosis (Corrêa-Oliveira et al., 2016). Acetate is produced by most gut anaerobes whereas propionate and butyrate are produced by different subsets of gut bacteria such as *Faecalibacterium prausnitzii* and *Eubacterium rectale* (Louis & Flint, 2016). Butyrate is formed from acetyl-CoA whereas propionate, depending on the nature of the sugar, has two production pathways: the succinate or the propanediol pathway (Louis & Flint, 2016). Butyrate is known for its anti-inflammatory properties and can attenuate bacterial translocation and enhance gut barrier function by affecting tight-junction assembly and mucin synthesis (Morrison & Preston, 2016). There are other gut microbial products helpful to the survival of the host. For example, lactic acid bacteria are probiotic microorganisms commonly found in yogurt that are involved in the production of vitamin B12, which cannot be synthesized by higher organisms. *Bifidobacteria* are producers of folate, a vitamin involved in vital host metabolic processes including DNA synthesis and repair (Thursby & Juge, 2017). Colonic bacteria can also metabolize bile acids that are not reabsorbed for biotransformation to secondary bile acids. Gut microbiota additionally synthesizes vitamin K, riboflavin, biotin, nicotinic acid, pantothenic acid, pyridoxine, and thiamine (Leblanc et al., 2013).

Modification in the microbiome, as can be caused by some NPs, may lead to several diseases, including metabolic diseases, such as obesity, diabetes, and cardiovascular disorders or invasion by pathogens like *Clostridium difficile* (Pascale et al., 2018). Recent research indicates the microbiome, shaped largely by diet, plays a role in the behavior and regeneration of an organism. NPs are increasingly playing a role in the food industry and becoming a component of processed foods that could affect the host animal microbiome. A disruption of the gut microbiome, also known as dysbiosis, is linked to medical conditions including, colitis, inflammatory bowel disease, diabetes, and metabolic syndrome (Williams et al., 2014). A better understanding of NPs, in particular metal-containing nanoparticles, on the microbiota of animals will provide knowledge to avoid the negative adverse effects that some NPs may possess. Copper oxide NPs (CuO-NPs) disrupt the microbiome of collembolan, primitive hexapods known as spring tails. Cu exposure has a profound influence on the gut microbiota of collembolans decreasing their microbial diversity and shifting their microbial community structure (Ding et al., 2020). Hens exposed to zinc oxide (ZnO) NPs and had their ileal microbiota sequenced and the bacterial community richness of the ileum was reduced as the dosage of NPs increased (Ghebretatios et al., 2021). *Lactobacillus* is the predominant bacterium in animal and human ileum, and members of this genus had a negative correlation with ZnO NP exposure. This change to the composition in the avian intestinal microbiota can be problematic (Ghebretatios et al., 2021). Different species may react differently to nanoparticles and the morphology of NPs may induce differential effects on the vertebrate gut microbiome. Sprague-Dawley rats were dosed with AgNps for 14 days and the researchers found that cuboidal AgNPs decreased populations of *Bacteroides uniformis, Clostridium spp., Christensenellaceae* and *Coprococcus eutactus* when spherical AgNPs decreased populations of *Oscillospira* spp., *Dehalobacterium* spp., *Peptococcaceae, Corynebacterium* spp. and *Aggregatibacter pneumotropica* (Javurek et al., 2017).

TABLE 7.1
Reported Effects of Various MCNPs On Diverse Species

MCNP	Host	NP Size (nm)	Effect	References
Ag	*P. putida* biofilm	65 ± 30	Decrease in biomass of biofilm	(Fabrega et al., 2009)
CuO	Collembolans	< 50	Altered microbial community structure of microbiome	(Ding et al., 2020)
FeO	Planarian	11	Did not inhibit regeneration or inhibit stem cell population	(Tran, 2019)
Ni	Zebrafish	30, 60, 90	Increased mortality and induced malformations	(Ispas et al., 2009)
Au	Mice	5	Disrupted cytoskeleton actin, degradation of clathrin	(Coradeghini et al., 2013)
TiO$_2$	Rats	5–12	Genotoxic effects	(Grissa et al., 2016)
ZnO	Hens	30	Reduced bacterial community richness of the ileum	(Feng et al., 2017)

MCNPs are typically inhibitory to the health of diverse cell types (Table 7.1). In some circumstances, MCNPs may be beneficial in terms of their effect on the animal microbiome. Administration of AgNPs in the drinking water of experimental mouse models with ulcerative disease and Crohn's disease effectively alleviated the colitis in mice (Li et al., 2018). ZnO NPs have similarly been indicated to have positive impacts on the gut microbiota of some host animals. The bacterial richness and diversity of the microbiota in the ileum of piglets were increased when exposed to ZnO NPs (Xia et al., 2017). In contrast to the case with chickens, when administered to piglets, ZnO NPs have a growth promoting effect (Pei et al., 2019). Copper containing nanoparticles, such as tribasic copper chloride (TBCC), are commonly used for growth promotion in animal farming and can bolster aspartate transaminase (AST) enzyme and the levels of antioxidant enzymes such as super oxide dismutase (SOD) (Zheng et al., 2018). In humans and other animals, NPs could play a role in the future treatment of gastrointestinal diseases. However, many of the mechanisms involved, particularly those related to perturbation of host microbiome, are largely unknown and research is ongoing to assess the effect of disparate NPs on various model organisms (Ding et al., 2020; Ghebretatios et al., 2021).

Planaria are flatworms with a flattened body architecture in the phylum Platyhelminthes (Ivankovic et al., 2019). Planaria have long been known to possess astonishing regenerative capabilities. If a planarian worm is chopped into three pieces, each of the piece's regenerates back into a complete and perfectly proportioned animal within ~2 weeks (Ivankovic et al., 2019). Hundreds of planarian species exist worldwide in marine, freshwater, or terrestrial habitats and the regenerative abilities vary greatly among different species (Ivankovic et al., 2019). The planarian species *Schmidtea mediterranea* can undergo whole-body regeneration while species like *Dendrocelum lacteum* possess more anatomically limited regenerative abilities (Ivankovic et al., 2019). Planarian flatworms have attracted interest in toxicology and pharmacology as an ideal model organism with which to study the toxicity of novel materials, such as nanoparticles, because of their many useful properties. Planaria possesses a primitive brain that has many features in common with vertebrate nervous systems, such as multipolar neurons and dendritic spines, they also possess nearly every neurotransmitter that has been found in mammals (Pagán et al., 2012). Planaria display behaviors associated with drug abuse; specifically withdrawal-like behaviors similar to humans. Consequently, planaria are ideal organisms to study the effect of NPs on regeneration and neurological diseases (Pagán, 2017).

In contrast to IONPs, toxicity of silver NPs to intact planaria and regenerating planaria had lethal effects on both intact and decapitated planaria (Leynen et al., 2019). Decapitated planaria exposed to polyvinylpyrrolidone (PVP) coated NPs had significantly smaller blastemas compared to nonexposed worms (Leynen et al., 2019). Regenerating planaria showed increased sensitivity to AgNPs compared to homeostatic planaria. Both PVP and nonPVP coated NPs decreased the motility of regenerating planaria and changed the type of motility behavior, with the PVP-coated NPs having a more pronounced effect. Prominent change in motility shifted normal gliding motility to a dragging motion (Leynen et al., 2019). Iron oxide NPs (IONPs) have limited effects on many animals. Exposure of IONPs to planaria indicated that IONPs did not affect the stem cell population dynamics, nor induce substantial changes in either homeostatic or regenerating planaria (Tran, 2019). Therefore, variables such as the metal(s) involved in conjunction with the morphology of MCNPs influence their effect on diverse host organisms.

7.2   WASTE FEEDSTOCKS FOR VALORIZATION

7.2.1   ORGANIC WASTES

Food waste, by its very nature, offers a rich source of nutrients and essential growth factors for various microbial processes, including the biosynthesis of useful nanoparticles. Food wastes are both abundant and, depending on the location and season, highly varied with regard to the diversity of constituent materials. The scale of food waste production varies considerably from large farms, food distributors and providers down to the restaurant and even household level (Ghosh et al., 2017). A large single crop industrial farm may have a homogeneous waste stream of many tons that can be effectively processed by a single highly specialized microbial strain. For example, corn cobs can be efficiently converted into medically useful, 40–70 nm silica NPs by the cellulolytic fungi *Fusarium culmorum* (Pieła et al., 2020). In contrast, the typical restaurant tends to have a tremendously varied food waste stream that includes animal, plant and artificial or highly processed wastes and therefore might require a diverse consortium of microbial species for successful bioprocessing.

Nanoparticles in certain food wastes may need to be considered when managing food wastes since some can be inhibitory to particular microbes. NPs made of titanium dioxide or zinc oxide are highly recalcitrant antimicrobial additives used in or with various processed food products (Urrutia-Ortega et al., 2016; Xia et al., 2017). Antimicrobials, such as ZnO, in food packaging play an important role in reducing the risk of pathogen contamination and extending the shelf life of food (Espitia et al., 2012). ZnO NPs exhibit significant antibacterial activity against *S. aureus* even when incorporated in polymeric matrices (Espitia et al., 2012; Lallo et al., 2019). In the food industry TiO_2 (E171) is applied as an additive to enhance the white color of certain products, such as sweets or milk-based products (Musial et al., 2020). Nanomaterials have proven to be useful materials in the processed food industry; however, because of their widespread use, it is inevitable that some of them will leak into the environment. When nanomaterials are leaked into the environment, for example in waste streams, they are generally released in uncontrolled amounts and may have negative downstream impacts on organisms. Accordingly, use of processes that can both synthesize *de novo* NPs and or capture and recover formed NPs present in wastes are of great interest (Biswas & Wu, 2005).

Sewage wastewater is an organic rich waste material common to human civilization that must be processed for numerous reasons including public health and odor control. Urban areas in developed nations often use advanced systems including sewers and wastewater treatment plants to process sewage wastewater to an acceptable level for environmental release into downstream water way. Sewage wastes contain a plethora of organic molecules in addition to inorganic phytonutrients including phosphorous and nitrogen that can cause downstream environmental issues. The water energy nexus present at sewage treatment facilities represents a largely untapped bioprocessing opportunity for the microbially mediated recovery of such resources. Biosynthesis of NPs by microorganisms naturally present in or tolerant to sewage waste is a particularly cost effective option. Naturally-occurring aquatic and oceanic coastal organic wastes can be collected and used for nanoparticle biosynthesis. In tropical coastal regions eukaryotic macroalgae such as *Sargassum muticum* can form vast blooms that wash up on beaches, to the detriment of tourism. These blooms of unsightly seaweed are exacerbated by eutrophication along the coasts and consequently have become more prominent in recent years (Wan et al., 2017). From a biotechnological standpoint this biomass can be viewed as a low cost, underutilized resource (Barbot et al., 2016). Indeed, *S. muticum* can be utilized for the production of valuable silver NPs (Azizi et al., 2013). In more northerly latitudes, metals As, Cu, Cr, Pb and Sn are taken up by the sea lettuce Ulva rigida (Wan et al., 2017). Various elements, including some toxic or valuable metals, can be removed and accumulated from waste contaminated soils by plants. The use of such plants for bioremediation through the environmental uptake of toxic elements from the environment is known as phytoremediation (DalCorso et al., 2019). Leaf biomass of agricultural wastes, such as soy leaves, can directly be used to directly produce metal NPs (Vivekanandhan, 2009).

7.2.2 INORGANIC WASTES

Industrial waste waters, such as those encountered at mining sites, are typically rich in various metals (Kumar et al., 2021). Some metal and metalloids including lead, uranium, arsenic, and cadmium are highly toxic to humans and other organisms (Stojsavljević et al., 2019). Lead heavy metal pollution causes neurological and developmental problems, particularly in children (Vorvolakos et al., 2016). Cadmium is used in various electronics and batteries and exposure to this heavy metal can cause kidney and brain damage as well as adverse effects to reproductive organs

(Rinaldi et al., 2017). Arsenic toxicity afflicts millions of people and may stem from geological contamination of drinking water as well as industrial pollution. Arsenic toxicity is a result of this metalloid's inhibition of hundreds of proteins including various enzymes needed for cellular energy metabolism (Ratnaike, 2003). Both chronic and acute forms of arsenic toxicity may occur in a dose and prior exposure dependent manner. Human populations in environments with elevated arsenic levels, such as are seen in the South American Andes mountains, appear to have evolved elevated tolerance to arsenic (Apata & Pfeifer, 2019). For other populations, proper water safeguarding is needed to protect against the toxic effects of metals and metalloids like arsenic. Electronic wastes, or e-wastes, are another growing point source of metals in the environment. Microorganisms including filamentous fungi conduct microbially mediated processes that can be employed to bioaccumulate such metals. (Bindschedler et al., 2017).

Bioremediation of contaminated natural and industrial soils and waste waters is possible using microbial nanobiotechnology using bacteria. For instance, sulfate reducing bacteria (SRBs) grown in conjunction with iron reducing bacteria facilitate removal of toxic arsenic from acidic mine waters as particles of nanoscale FeS coated limestone (Liu et al., 2017). Eukaryotes can also provide for removal. *Aspergillus* fungi facilitate absorption of arsenic from contaminated sources via a nonenzymatic mechanism that involves integration of silver NPs with carbonized fungal cells (Mukherjee et al., 2017). Culture supernatants of *Aspergillus terreus* were able catalyze the biosynthesize of silver NPs from an aqueous silver solution (Li et al., 2011b). Since reduced nicotinamide adenine dinucleotide (NADH) was consumed, the authors ascribed silver NP formation to be an enzymatically-driven process (Li et al., 2011a). Indeed, NADH and NADPH (*i.e.*, NAD(P)H) can serve as the source of reducing equivalents for reductive nanoparticle formation by microorganisms (Ovais et al., 2018). Different microbial redox enzymes, such as NADH-dependent nitrate reductase, can catalyze reduction of soluble metal ions into insoluble metal-containing NPs or pure metal NPs as indicted by Figure 7.2. (Rai et al., 2021).

Inexpensive, relatively reduced substrates are needed to supply electrons for reduction of soluble metal by heterotrophic species. *Aspergillus* is a common fungus that can be grown on various wastes, including inexpensive lignocellulosic biomass such as forestry wastes, suggesting how removal of organic woody biomass wastes can power reductive detoxification of arsenic from water in remote settings (Emtiazi et al., 2001). Magnetic chitosan beads adsorb and recover 99% of arsenic in water samples (Ayub et al., 2020). Leaching of accumulated adsorbed recalcitrant toxins from disposal sites is a longterm concern. Due to the low permissible level of arsenic set by the Environmental Protection Agency (EPA) of 10 ppb, leaching of the adsorbed toxin is a concern that may partly be addressed via stabilization through addition of arsenic containing bio-adsorbed materials into cement (Mondal & Garg, 2017). Toxic cadmium can be recovered as cadmium sulfide NPs via bioprecipitation as mediated by the purple non-sulphur (PNS) bacterium *Rhodopseudomonas palustris* TN110 (Sakpirom et al., 2019). Valuable nickel ions can be extracted microbiologically from electroplating factory waste waters as 40–90 nm $Ni_3(PO_4)_2 \cdot 8H_2O$ NPs using the common gram positive, spore forming bacterium *Bacillus subtilis* (Yu and Jiang, 2019).

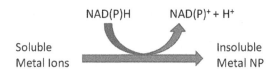

FIGURE 7.2 Enzyme-mediated, reductive formation of insoluble metal NPs from aqueous metal ions requires a source of reducing equivalents (*i.e.*, electrons) which are commonly provided to redox enzymes by NAD(P)H electron carriers. In heterotrophic microorganisms, enzymatically catabolized organic nutrients serve as the original electron donor.

7.3 MICROBIALLY SYNTHESIZED NANOMATERIALS

Nanomaterials are produced by various routes using both biological and abiotic methods. Abiotic chemical and physical methods are used to produce numerous types of NPs but, unlike cellular process, are not capable of replication or self-repair (Ghosh et al., 2021). One of the promising realms of nanotechnology in general and nanobiotechnology in particular pertains to the valorization, or value upcycling, of wastes. Certain microbial products, such as polyhydroxyalkanoates, can serve as feedstocks for bioplastic production (Agrawal and Verma, 2021). Others, such as polyphosphate granules, can elegantly recapture nutrients from waste streams while repurposing them to act as delayed release fertilizers for agricultural application. MCNPs have a range of uses and their microbiological production can concurrently provide for environmental bioremediation of heavy metal and toxic metalloid polluted sites.

7.3.1 METAL CONTAINING NANOPARTICLES

MCNPs, such as metal oxide NPs (MeO-NPs), like all other NPs range from 1 to 100nm and possess unique physical and quantum chemical characteristics linked to their nanometer size. Metallic NPs offer versatility with regard to ability of control size, shape, structure, composition, assembly, encapsulation, and tunable optical properties (Bhatia, 2016). Consequently, MeO-NPs are key constituents in catalysis, diagnosis, drug delivery, semiconductors, sensing, and solid oxide fuel cells (Raghunath & Perumal, 2017). In the medical field, AuNPs have shown great promise in the realm of cancer therapy (Cobley et al., 2011). AuNPs were utilized as ultrasensitive fluorescent probes to detect cancer biomarkers in human blood and the NP can be further controlled and modulated for the treatment and diagnosis of the disease which makes it more desirable as a medical tool (Bhatia, 2016). Metallic NPs are also used in the food and pharmaceutical industries when it comes to enzyme immobilization. Nanomaterials have been used in enzyme immobilization to improve the material properties such as stability, enzyme activity, and multicompartmentalization (Delcea et al., 2010). Even with their promise in the fields mentioned, MeO-NPs are of great interest due to their observed antimicrobial properties and have shown potential for being used as antimicrobial agents (Raghunath and Perumal, 2017).

Metal and metalloid precipitation from various types of wastewater is possible using diverse microbes. Selenium in mining wastewater is reduced to insoluble Se NP via glutathione-mediated reduction by *Streptomycetes* bacteria (Tan et al., 2016). Such microorganisms capable of precipitating metal and their products can also provide valuable goods and services by removing phytonutrients such as nitrogen and phosphorous. Precious metals can be accumulated and recovered microbiologically from wastes. Palladium is widely used in automotive catalytic converters and has increased significantly in value over the past five years. Palladium can be precipitated from solution as palladium NPs by bacteria via a hydrogenase-independent mechanism (Bunge et al., 2010). Platinum is broadly used in the jewelry industry and also has industrial applications in hydrogen fuel cells. Enzymes may also be involved in microbial precipitation of metals. Enzymes used for the precipitation of metals may be localized intracellularly or extracellularly. Additionally, in the case of certain gram negative bacteria, periplasmically localized enzymes may facilitate metal ion reduction and MCNP formation. For instance, nitrate reductase in the periplasm of *Achromobacter* denitrifying bacteria catalyzes biosynthesis of gold NPs with concurrent nitrate removal (Eltarahony et al., 2018). This demonstrates how microbially mediated MCNP synthesis can be coupled to bioremediation of metal-contaminated eutrophic wastewaters.

Cobalt is a metal increasingly needed for rapidly developing computer and battery-based societal applications, such as electric vehicles. The majority of the world's cobalt is mined in the Democratic Republic of Congo, largely with child labor (Banza Lubaba et al., 2018). Cobalt is genotoxic and is an environmental toxin that can be recovered from polluted waters using magnetotactic bacteria (Tajer et al., 2016). Lithium batteries have transformed many areas of human society over the past quarter century, from computing to large-scale renewable energy storage. As lithium batteries of ever-increasing size are produced, methods are needed to capture the light metal lithium from waste streams. Molybdenum is yet another metal with industrial applications that can be microbiologically recovered from waste streams (Kasra-Kermanshahi et al., 2020). The anaerobic bacterium *Clostridium pasteurianum* is capable of reducing nearly 90% of $Mo6^+$ ions from solution to produce molybdenum NPs of approximately 15 nm (Nordmeier et al., 2018). In turn, these NPs are able to degrade azo dyes, illustrating how various pollutants might be profitably bioremediated from waste streams co-contaminated with multiple pollutants (Nordmeier et al., 2018). Nanospheres of molybdenum were recently found capable of photocatalytically reducing CO_2 to CH_4 (Huang et al., 2020). It remains to be tested if microbially synthesized molybdenum NPs are capable of reducing CO_2 to industrially useful products like methane.

Metals can be coprecipitated with other metals or nonmetals using microbes. *Geobacter* bioprecipitates sulfur and copper Cu(II) from wastewater to produce CuS NPs (Kimber et al., 2020). The ability of this common soil microbe is not limited to copper as it also mineralizes cobalt; apparently as part of the detoxification defense against this toxic metal (Dulay et al., 2020) Certain extremophilic microorganisms are useful for production of nanomaterials in unusually harsh settings. Halophilic archaea of the genus *Haloferax* are able to precipitate silver from high salt liquids (Abdollahnia et al., 2020). Many other microorganisms are able to produce MCNPs (Table 7.2.).

TABLE 7.2

Chemical Composition, Size and Morphology of Representative Metal and Metalloid-containing NPs Synthesized by Various Microbial Producers

MCNP	Organism	Size (In nm)	Morphology	References
Ag	*Trichoderma viride*	5–40	Spherical	(Fayaz et al., 2010)
AgCl	*Macrophomina phaseolina*	5–30	Spherical	(Spagnoletti at al., 2019)
Au	*Rhodopseudomonas capsulata*	10–20	Spherical	(He et al., 2007)
TiO_2	*Lactobacillus* sp.	8–35	Spherical	(Jha et al., 2009a)
ZnO	*Aspergillus terreus*	54.8–82.6	Spherical	(Baskar et al., 2013)
ZnS	*Rhodobacter sphaeroides*	10.35–10.65	Spherical	(Bai et al., 2009)
Fe_3O_4	*Shewanella oneidensis*	40–50	Rectangular, Rhombic, Hexagonal	(Perez-Gonzalez et al., 2010)
Fe_2O_3	*Shewanella oneidensis* MR-1	30–43	Pseudohexagonal	(Bose et al., 2009)
ZrO_2	*Fusarium oxysporum*	3–11	quasi-spherical	(Bansal et al., 2004)
Pt	*Penicillium chrysogenum*	5–40	Spherical	(Subramaniyan et al., 2018)
Hg	*Enterobacter* sp.	2–5	Spherical	(Sinha et al., 2013)
Cu	*Shewanella loihica*	10–16	Spherical	(Lv et al., 2018)
$SrCO_3$	*Fusarium oxysporum*	10–50	Needle-like	(Rautaray et al., 2004)
CoO	*Aspergillus nidulans*	20.29	Spinel	(Vijayanandan and Balakrishnan, 2018)
Pd	*Chlorella vulgaris*	5–20	Spherical	(Arsiya et al., 2017)
Al_2O_3	*Sargassum ilicifolium*	20	Spherical	(Koopi and Buazar, 2018)
Sb_2O_3	*Saccharomyces cerevisiae*	2–10	Spherical	(Jha et al., 2009b)
$BaTiO_3$	*Lactobacillus* sp.	20–80	Tetragonal	(Jha and Prasad 2009)
Se	*Shewanella* sp.	141–221	Spherical	(Lee et al., 2007)
CdSe	*Fusarium oxysporum*	9–15	Spherical	(Kumar et al., 2007)
CdS	*Escherichia coli*	2–5	Wurtzite crystal	(Sweeney et al., 2004)
CdTe	*Escherichia coli*	2–3.2	Spherical	(Bao et al., 2010)

7.3.2 BIOPLASTICS

Valorization of waste is possible using microorganisms that remove nutrients and/or toxins while concurrently producing feedstock material for bioplastics that originate at the nano scale. Bioplastics have many functional attributes over conventional plastics. They are generally biodegradable and do not depend upon a nonrenewable fossil sourced feedstock such as petroleum (Roibás-Rozas et al., 2020, Lopez et al., 2015). Bioplastics may meet with greater customer acceptance than petroleum-based plastics, provided they offer comparable performance. Bioplastics can increase the profitability of waste streams by upcycling wastes into useful products. Diverse waste streams can be harnessed for the production of PHAs. These biodegradable products demonstrate low recalcitrance and relatively reduced negative environmental impact compared to conventional plastics (Lopez et al., 2015). However, certain challenges are associated with bioplastics including inconsistent polymer properties and higher production costs than petroleum-based plastics (Strong et al., 2016).

Polyhydroxyalkanoates are a versatile class of biologically produced polyesters made of hydroxy fatty acids that can be processed into various types of thermoplastics and elastomeric bioplastics (Rehm, 2003). They are synthesized by various organic waste and wastewater associated microorganisms, including purple non sulfur bacteria (PNS), typically when carbon and energy are abundant but one or more macronutrients needed for growth are limiting (Lopez et al., 2015). Various waste substrates can be used for production of polyhydroxyalkanoates. Glycerol is one major organic waste product of the rapidly growing biodiesel industry. Glycerol is released from triacyl glycerides as a result of the transesterification reactor employed to make biodiesel from oilseeds like soybeans. The surplus stocks of glycerol from biodiesel production have caused prices for glycerol to plummet and conventional disposal of this surplus is challenging and expensive (Crosse et al., 2019). *Cupriavidus necator*, formerly *Ralstonia eutropha*, is a metabolically versatile gram negative bacterium that can convert waste glycerol into polyhydroxyalkanoates (Canadas et al., 2014). Gram positive spore-forming bacteria such as *Bacillus megatherium* are able to produce polyhydroxyalkanoate from wastes (Mohapatra et al., 2020). Spore-forming producer strains are advantageous in that they have a long shelf-life and can tolerate environmental stresses. Gram positive bacteria do not produce lipopolysaccharide (LPS) which is beneficial if polyhydroxyalkanoate is being produced for biomedical applications. LPS is recognized as an endotoxin that must be removed from drugs and medical devices. Similar to the advent of biodiesel over the last decade, new food trends, such as the popularization of Greek yogurt throughout the American marketplace, have created surplus dairy industry waste in the form of whey; a protein-rich waste that is also produced during cheese production (Camacho et al., 2019). Whey disposal can be expensive and challenging so microbial upcycling is an attractive option. A wide variety of microbes from halophilic *Archaea* to recombinant bacteria are able to produce different types of polyhydroxyalkanoates including poly-(3-hydroxybutyrate-co-3-hydroxyvalerate (PHBV) which is ductile and useful for industrial applications (Nielsen et al., 2017). Technoeconomic analysis of whey as feedstock for bioplastics production indicates a possible biorefinery payback period of less than four years (Chalermthai et al., 2020). Whey may be available in suitable quantities in dairy intensive regions, such as Wisconsin, USA, but may be lacking in many areas. Accordingly, regionally suitable feedstocks and producer strains must be identified for profitable production of bioplastics from diverse wastes.

Minimal consumption of energy and potable freshwater is required to further minimize operating and production costs, particularly in coastal or desert regions. Brackish or marine feedstocks wastes can be cost-effectively employed for the cultivation of polyhydroxyalkanoate producing organisms. *Halomonas* proteobacteria are able to synthesize up to two-thirds of their biomass as poly (3-hydroxybutyrate) (PHB) NPs when grown in salt water on algal hydrosylate as their organic energy source (El-Malek et al., 2021). In this way, photosynthetic algae first fix CO_2 and inorganics in sea water into cell biomass using solar energy. The ability to synthesize PHA is found in a diverse array of microbes including certain halophilic Archaea that can coproduce antibiotics and antioxidants (Kasirajan & Maupin-Furlow, 2020;

Tan et al., 2021). PHB is considered an ideal feedstock for biodegradable bioplastics production and is currently being developed and commercialized by various companies. Biomedical engineering applications include "Biopol" a biodegradable suture material that slowly breaks down over time (Ray & Kalia, 2017). The production process can be coupled to other waste treatment methods, such as anaerobic digestion, for more complete waste catabolism and conversion into valuable products (Strong et al., 2016). Methane from diverse sources, such as anaerobic digesters can be a feedstock for the production of bioplastics and the controlled heat from methane combustion can be used to warm bioreactors to optimal temperatures.

Polylactic Acid (PLA) is another biodegradable bioplastic that is gaining interest for waste valorization using microorganisms (Marmol et al., 2020). Both inorganic and organic waste streams can be upcycled into PLA. Genetically modified *Synechocystis* sp. PCC 6803 cyanobacteria are able to use light energy to fix CO_2 into lactic acid, the key precursor for PLA production (Varman et al., 2013). Agricultural organic wastes such as potato peels can be effectively converted into lactic acid for PLA production via fermentation under anaerobic conditions (Liang et al., 2015). Ammonium-treated corn stover field waste can similarly be fermented into lactic acid by *P. aeruginosa*, *Enterococcus faecalis* and *Acinetobacter calcoaceticus* strains (Liu et al., 2019). Base pretreated corn stover can be fermented to PLA by *Bacillus coagulans* strain LA204 which has the advantages of endospore formation and the ability to ferment pentoses and cellobiose (Hu et al., 2015). PLA has many promising applications other than consumer bioplastics, which are of relatively low value since they must compete with petroleum derived plastic. For instance, antibiotic laden NPs of PLA are effective against methicillin resistant *S. aureus* (MRSA) as well as enteropathogenic *E. coli* O157/H7 (Herrera et al., 2017). Biomedically engineered tissues can be created using 3-D printed PLA as a biocompatible scaffolding material for cell growth (Bodnárová et al., 2019).

7.3.3 BIOFERTILIZERS

Valorization of wastes may be accomplished through the microbially mediated recovery of nutrients as intracellular particles, including nanoparticles. Phosphate is a common nutrient waste that is essential to the growth of plants. Most of the phosphate used in crop fertilizers at present is mined from phosphate-rich geological deposits in areas including Florida (Mao et al., 2017). Since phosphorous in its solid form deposits over geological time scales, terrestrial deposits of this element are considered a nonrenewable resource and phosphorous is projected to become a limiting nutrient for agriculture sometime between 2040 and 2070 (Neset et al., 2016, Blackwell et al., 2019). Consequently, novel methods are need to recover phosphate from waste and run-off streams and to avert its entry into sensitive natural waterways.

Polyphosphates are synthesized by a wide variety of different microbes. These intracellular granules range in size from a few to 100 nm in the cyanobacterium *Synechococcus* sp. PCC 7002 and up to a few micrometers in certain heterotrophic bacteria (Gao et al., 2018; Koller & Braunegg, 2015). Eukaryotic green microalgae including *Chlamydomonas reinhardtii* are similarly able to uptake and accumulate phosphate as polyphosphate granules from waste streams (Slocombe et al., 2020).

In *C. reinhardtii*, polyphosphate forms inside of membrane-bound acidic vacuoles that also contain calcium and magnesium (Komine et al., 2000). These vacuoles are accordingly known as acidocalcisomes (Sanz-Luque et al., 2020). Interestingly, in some green algae the granules can be secreted through the process of exocytosis; suggesting nanoproduct producer cells need not be lysed or otherwise sacrificed in order to recover the nutrient-rich product (Komine et al., 2000). In the green micro-algae *Chlorella vulgaris*, high levels of soluble phosphorous significantly inhibited growth and mitochondrial activity, possibly via disruption of membrane permeability (Fu et al., 2019). Secretion of granules from algal cells via exocytosis could represent a mechanism for reducing the concentration of soluble phosphate to subinhibitory levels. Polyphosphate granules are a storehouse of intracellular phosphorous as well as a reserve energy source (Achbergerová & Nahálka, 2011). Polyphosphate granules can additionally help cells adapt to environmental stresses such as shifts in pH or osmotic fluctuations (Achbergerová & Nahálka, 2011). Polyphosphate tends to accumulate when energy and phosphate is in abundance, but other resources needed for typical cell growth and replication are limiting. Enzymatically, polyphosphate granules are synthesized through the activity of phosphate kinases in diverse microbial species when growing in phosphorous replete medium, such as sewage (Kornberg et al., 1956).

Nitrogen is a macronutrient required in high quantities for many economically vital, nonleguminous plants, such as corn. Currently, most of the bioavailable nitrogen used in fertilizers derives from the Haber Bosch process; arguably one of the greatest inventions in human history. However, this thermochemical process depends predominantly on nonrenewable natural gas for the steam reformation of methane to produce the requisite hydrogen needed for N_2 reduction to ammonia. In 2016 alone, an estimated 175 million metric tons of ammonia were produced using this process (Qing et al., 2020) Up to 5% of global natural gas is consumed to fix nitrogen (Song et al., 2018). Consequently, nitrogen fixation via the Haber Bosch process is a significant source of carbon emissions and an important contributor to global climate change (Song et al., 2018). Nitrogen-fixing diazotrophic bacteria and their enzymes can be used to fix nitrogen in the air into bioavailable form. However, nanoparticle forming bacteria, such as *R. palustris* TN110, have been described that concurrently fix atmospheric nitrogen (Sakpirom et al., 2019). Nanoparticle formation is associated with an increase in levels of the enzyme nitrogenase, which catalyzes formation of ammonia from nitrogen (Sakpirom et al., 2019). This indicates that nanomaterial production can be effectively coupled to green fertilizer production.

7.4   PROTEIN-MEDIATED NANOMATERIAL REFINEMENT

Enzymatic biotransformation of toxic heavy metal and metalloid species to less toxic and/or less water-soluble forms is a useful redox-based means of improving waste and water safety and value. Arsenite oxidase is an enzyme produced by the arsenic-resistant bacterial genus *Exiguobacterium* which converts highly toxic arsenite, As(III), to less toxic arsenate, As(V) (Pandey & Bhatt, 2018). Chitosan NPs were used to stabilize and immobilize arsenite oxidase for up to one month enabling around 90% biotransformation of toxic arsenite into the less toxic arsenate (Pandey & Bhatt,

2018). Still greater biotransformation may be possible using nanochitosan-enzyme NPs since nanochitosan has an approximately 12 fold greater ability to adsorb arsenite compared to typical chitosan (Kwok et al., 2018). Chitosan is a nitrogen-containing material produced from the waste shells of crustaceans, such as shrimp and crabs. Reductive biotransformation of other toxic heavy metals, such as chromium VI, may similarly be carried out by *Geobacter* species as well as certain other exoelectrogenic bacterial genera such as *Shewanella* (Belchik et al., 2011). This represents a form of anaerobic respiration in which extracellular metals serve as terminal electron acceptors in anaerobic environments. As such, metal-contaminated anaerobic soils and sediments are best suited for this form of bioremediation. Precipitation of metal ions to less mobile, insoluble nanomaterials can help to reduce downstream leaching of toxic heavy metals into waterways. Hexavalant chromium is converted into chromium (III) oxide NPs by *Schwanniomyces occidentalis* (Mohite et al., 2015). Even radioactive elements can be precipitated by certain microorganisms. Plutonium VI is precipitated as by exoelectrogenic *Geobacter* strains and *Shewanella* as both PuV and PuVI are reduced to PuIV extracellular NPs (Boukhalfa et al., 2007, Icopini et al., 2009). Uranium is precipitated to form insoluble uranyl phosphate mineral by acid and alkaline phosphatase enzyme expressing *Serratia* bacteria (Chandwadkar et al., 2018). This form of radioresistant nanobiotechnology might be of use for the enzyme-mediated bioremediation and/or recovery of radioactive elements from sites contaminated with radioactive wastes, such as nuclear weapon production or test sites (Boukhalfa et al. 2007).

Enzyme and cell-based systems for producing nanomaterials from wastes have advantages over alternative methods, such as purely chemical or electrochemical approaches. Potentially toxic or expensive chemicals need not be use. Rather they can be removed microbiologically. While electrochemical methods generally require a net input of energy, cell-based methods harness living microbes to transduce freely available organic energy in waste or solar energy. To date, prokaryotic nanomaterials have been the most extensively studied; however, investigators are also examining waste-catabolizing eukaryotic groups. Certain fungal cells produce useful materials from organic wastes. *Stropharia rugosoannulata* is a cellulolytic basidiomycetes species which grows well on nitrogen-deficient wood chips and other forestry waste products. This easy-to-grow basidiomycetes produces a large, purple edible mushroom informally called the wine cap. Below ground, its mycelia synthesize spiculated acanthocyctes that are able to kill nematode worms in the surroundings, probably to provide needed nitrogen and other elements (Luo et al. 2006, Yang et al., 2020). Since nematodes can cause various maladies, acanthocytes could prove to be a useful organic product or could be useful as an organic treatment for garden soils.

Some pathogens can produce metal-containing nanomaterials as a metabolic biproduct. For instance, blood-feeding human parasite *Schistosoma* trematode worms generate an iron containing nanocrystal called hemozoin (Oliveira et al. 2000). Malaria parasites in the genus *Plasmodium* parasitize the interior of red blood cells where they consume amino acid-rich hemoglobin protein and likewise produce hemozoin crystals. As hemoglobin is catabolized, the free heme is released which is swiftly crystallized into hemozoin nanocrystals and potent antimalaria drugs like chloroquine can disrupt this process (Orjih and Fitch, 1993). Free heme is highly hydrophobic and the hemozoin nanocrystal is thought to be formed within hydrophobic lipid nanospheres inside of the parasite's acidic digestive vacuole (Pisciotta et al., 2007). Diagnostically, this microbially produced nanomaterial is of value since uninfected humans generally do not contain hemozoin. The presence of hemozoin can be identified by a variety of methods for biosensing or to diagnose disease with high sensitivity and specificity (Hyeon et al., 2018). From an optical standpoint, the heme crystal rotates the plane of polarized light. This property, known as birefringence, can be utilized by microscopists to aid disease diagnosis.

Enzymatic involvement is being harnessed for an increasing number of nanomaterial applications. The interaction between enzymes and particles, whether organic or inorganic, can result in favorable changes in physicochemical properties of either the enzymes or nanomaterials (Chen et al., 2017). Nanomaterials are directly conjugated to enzymes to adjust enzymatic activity, and to affect enzymatic structures and functions. Some enzymes are capable of modifying nanoparticle properties to develop their conjugates (Chen et al., 2017). Natural enzymes present inherent drawbacks such as heat-induced denaturation or pH-mediated inactivation, requirement for cofactors and time-consuming extraction. Enzyme-linked nanomaterials are being fabricated to overcome these drawbacks (Chen et al., 2017). The food and pharmaceutical industries are particularly interested in nanoparticulate immobilized enzymes for these are generally more robust and resistant to environmental changes compared to free enzymes (Bhavaniramya et al., 2019). NP immobilization of enzymes may improve the binding affinity of enzymes to substrate significantly (Numanoğlu and Sungur, 2004) which can improve the yield of food processing and can minimize the process cost (Gupta et al., 2013). Immobilized enzymes, such as alpha-amylase, act as biocatalysts in the food industry mainly for starch processing while others aid cheese-making, fermentation, lipid hydrolysis, and food preservations. Amylase, a widely used saccharolytic enzyme in the food industry, is isolated from *Bacillus licheniformis* and is used for starch saccharification and liquification. The immobilization technique used should not alter the reaction mechanism of enzymes; however, NP immobilized amylase can suffer a lack of long-term operational stability (Bhavaniramya et al., 2019). To improve the enzyme activity, nanomaterials, with their unique physicochemical properties, are used as carriers for enzyme immobilization.

Enzymatically, microbial products such as polyhydroxyalkanoates, can be anabolically synthesized from inorganic wastes, like CO_2, by certain chemolithoautotrophic and mixotrophic bacteria (Shimizu et al., 2013). Accordingly, gaseous waste streams from breweries or even fossil-fueled power plants could serve as feedstock for polyhydroxyalkanoate production. Should a tax on carbon emissions be imparted, this avenue of bioplastics production could prove profitable for dual reasons. The PNS bacterium *R. eutropha* uses the enzyme ribulose 1,5-bisphosphate carboxylase oxygenase (Rubisco), which is a core component of the Calvin Benson cycle, to fix CO_2 into poly(3-hydroxybutyrate) (Shimizu et al., 2013). Cysteine desulfhydrase activity is present in the cytoplasm of nanomaterial producing *R. palustris* PNS bacteria that also efficiently export MCNPs out of the cell (Bai et al., 2009). If such pathways can be effectively harnessed for the export of bioplastic precursors, the producer cells need not be lysed and product formation rates can be enhanced (Tan et al., 2016).

7.4.1 Cell-free Systems and Magnetic Nanoparticles

Bioinspired crystalline products, such as synthetic hemozoin, called beta hematin, can be produced in high quantities using cell-free systems, possibly from the heme extracted from slaughterhouse wastes. This crystal is finding interesting new uses in diverse industries. One application for this heme crystal is to facilitate oxidation reactions in rocket fuel (Slocik et al., 2015). Archaeologists are interested in identifying the presence of hemozoin in skeletal remains to obtain paleo-epidemiological information about disease prevalence in ancient human populations (Setzer, 2014). Based on the magnetoelectric properties of iron-rich heme crystals, these crystals are of considerable interest for biosensing and microelectronic applications. Parasite infected cells have been successfully concentrated based on the paramagnetic properties of this iron rich nano-scale crystal, which may facilitate infection diagnosis (Bhakdi et al., 2010, Inyushin et al., 2016).

Waste-associated bacteria can synthesize useful magnetic nanoparticles. In 1975, Richard P. Blakemore described aquatic bacteria that were physiologically specialized to align to a magnetic field (Blakemore, 1975). These magnetotactic bacteria, as they came to be known, are commonly present in water bodies near the oxic-anoxic boundary. There, they produce different types of magnetic NPs that are proving extremely useful for various industrial and medical applications. Magnetotactic bacteria produce intracellular 35–120 nm ferrimagnetic nanocrystals of greigite (Fe_3S_4) in anaerobic species, or magnetite (Fe_3O_4) in micro-aerophilic representatives, that are sheathed inside of a protein-rich, membrane-bound intracellular compartment, termed the magnetosome (Bazylinski et al., 1994; Barber-Zucker & Zarivach, 2016). The presence of membranous magnetosomes in these motile, typically gram-negative prokaryotes challenges the notion that all bacteria lack membrane-bound organelles. Physiologically, these prokaryotic iron-rich magnetic crystals are normally arrayed in a linear relation with one another inside the cell forming a collective structure functionally akin to a compass needle (Müller et al., 2020). By aligning to the magnetic field of the earth, magnetosomes help magnetotactic bacteria thereby sense and respond to their position in the surrounding environment (Barber-Zucker & Zarivach, 2016; Lin et al., 2020). This form of magnetically guided motility is referred to as magnetotaxis (Kuzajewska et al., 2020).

Wastewater treatment and valorization is possible using magnetotactic bacteria which can be usefully employed for the recovery of iron as well as precious metals from waste streams (Yan et al., 2017). For instance, magnetotactic bacteria in the genus *Stenotrophomonas* are able to precipitate gold III and then facilitate its isolation from suspension via magnetic separation of the gold laden cells (Song et al., 2008). The genetic, biochemical and physiological mechanisms by which magnetosomes are produced is an area of active research. The magnetosome membrane-associated protein MmsF works in concert with at least 18 additional genes of the mamAB gene cluster to catalyze magnetosome formation (Murat et al., 2012). The protein-mediated crystallization occurs via a process of biomineralization that takes place inside the vacuolar membrane of the magnetosome (Singh and Ahmad, 2018). For this, significant iron is required and both ferric and ferrous iron transporter have been identified (Taoka et al., 2008). Magnetotactic bacteria accumulate iron from their surrounding within magnetosome vacuoles as phosphate-rich ferric hydroxide precursor used to synthesize magnetic particles (Baumgartner et al., 2013). Initiation of biomineralization in turn causes the dynamic expansion of the magnetosome vacuolar membrane with accumulation of additional magnetic particles (Cornejo et al., 2016). These magnetic NPs are randomly ordered but align to magnetic fields through interaction with the fibrous actin homologue protein MamK (Pradel et al., 2006). Gravity also seems to be required for, when cultivated in a 0 G environment on the Space Shuttle, ordered magnetosomes could not form (Urban, 2000). Following magnetic nanoparticle biosynthesis and protein-mediated alignment, bacteria initiate magnetotactic motility in either a north or south seeking manner, depending on species (Figure 7.3).

The ability to synthesize magnetic NPs can be genetically transferred to different cell types. Genetic modification of organic wastewater-associated purple non sulfur (PNS) bacteria with magnetotactic genes provides for magnetosome production in metabolically versatile, easy-to-cultivate, waste-catabolizing PNS strains (Kolinko et al., 2014). Now that a basic understanding of the genetic basic for magnetosome synthesis exists, future research aims to use genetically modified organisms as cellular factories to produce particles of customized sizes and dimensions. Modification of cultivation conditions is one way to accomplish this even in wild-type producer strains. When the model magnetotactic bacterium *Magnetospirillum gryphiswaldense* was starved of iron during early growth and was provided iron in the stationary phase of its growth cycle, magnetosomes of smaller, more uniform size were produced compared to when iron was provided during the logarithmic phase (Marcano et al., 2017). It is evident that genetic modification, modulation of culture conditions, and producer strain used all play important roles in the nature of the final magnetosome product.

Medical applications for magnetic NPs have increased considerably in recent years. An innovative proposed use for bacterial magnetosomes is as an injectable and low dose contrast enhancement agent for magnetic resonance imaging (MRI) (Mériaux et al., 2015). Commonly used MRI contrast enhancement dyes are based on gadolinium and are cleared from the blood via the kidney and can cause renal

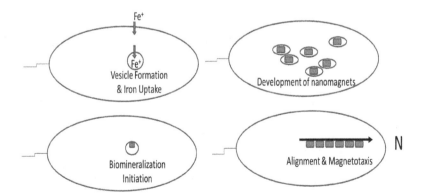

FIGURE 7.3 Mechanistic basis of magnetosome formation. Cell surface transporters import both ferrous and ferric iron from the surrounding wastewater (a). Iron, accumulated within intracellular vacuoles termed magnetosomes is biomineralized (b), to form particles of greigite (Fe_3S_4) or magnetite (Fe_3O_4) (c). The magnetic NPs are aligned via interaction with fibrous actin-like protein mamK, enabling sensing of magnetic fields and directional magnetotaxis (d).

damage in certain patients (Maripuri and Johansen, 2020). In contrast, bacterial magnetosomes show good biocompatibility with predominantly hepatic and splenic clearance in the mouse model (Nan et al., 2021). Magnetic NPs produced by *Magnetospirillum gryphiswaldense* can be taken up by human carcinoma cells and show promise as anticancer chemotherapeutics in *in vitro* tests and in animal studies (Mannucci et al., 2014). Magnetosomes, which tend to accumulate in the anoxic interior of tumors, can be loaded with chemotherapeutics to serve as anticancer drug delivery vehicles (Fdez-Gubieda et al., 2020). Magnetosomes, once delivered to the tumor, can even be electromagnetically vibrated and heated to destroy the tumor *in situ*; a form of treatment termed magnetic hyperthemia (Fdez-Gubieda et al., 2020). Bacterially derived magnetosomes are taken up by diverse eukaryotic cell lines and this can be exploited for magnetic cell separations; potentially of use for certain blood cancers (Mickoleit et al., 2021). Coupling of microbially synthesized magnetic NPs to antibodies could provide a highly specific means of directing NP effectors to transformed target cell types.

7.4.2 Nanowire Production by Electrically Active Bacteria

Valorization of organic wastes into useful nanomaterials can be achieved via biological conversion using bio electrochemically active microbes cultivated within bio electrochemical systems. Metals are widely used throughout human society largely because of their strength and high electrical conductivity. Certain microorganisms, including both photosynthetic oxygenic phototrophs as well as anaerobic heterotrophic bacteria synthesize electrically conductive, metal-containing nanoscale structures called microbial nanowires under certain environmental circumstances (Gorby et al., 2006). Diverse cyanobacteria interface with electrically conductive surfaces and can generate an electrical potential under illumination (Pisciotta et al., 2010). Microbial nanowires have been most extensively studied in the strict anaerobic gram negative bacterium *Geobacter sulfurreducens* (Pokkuluri et al., 2004). *Geobacter sulfurreducens* is a common gram negative bacteria found in soil that is able to generate electricity in sewage fed microbial fuel cells (Reguera et al., 2006). As it catabolizes organic wastes such as acetate under anaerobic condition in the MFC, *G. sulfurreducens* synthesizes these thin, pili like structures that incorporate iron in heme. Anaerobic exoelectrogenic bacteria, such as *Geobacter* species, primarily use nanowires to discharge organically derived electrons to their surrounding as a form of anaerobic respiration (Chin et al., 2004). Humans are interested in using nanowires and nanowire-producing microbes that can be grown inexpensively on organic wastes for an increasing array of applications. These range from toxic metal capture to bioenergy harvesting from waste to facilitation of electrically driven, industrial fermentation reactions (Koul et al., 2021, Reguera, 2018).

Mechanistically, nanowire-mediated electron transfer to the environment appears to be facilitated by stacked iron-containing multiheme cytochromes (Wang et al., 2019). Paradoxically, the porphyrin ring of heme, rather than its centrally coordinated metal, provides for efficient electron transfer via physical coherence (Agam et al. 2020). While nanowire mediated electron transfer in *Geobacter* biofilms is limited to around 50 um, relatively long-distance electron transfer also occurs in nature over centimeter scales, as revealed by the discovery of so-called cable bacteria (Pfeffer et al., 2012). These microorganisms bio electrochemically link the catabolism of reduced sulfur compounds in anaerobic sediments with oxidation reactions in the overlaying surface so as to accelerate microbial metabolism or organic wastes in sediments (Kjeldsen et al., 2019). Therefore, nanowire producing microbes can both be used to catabolize and treat organic wastes and also to use that waste as the production feedstock for a rapidly advancing microbial sector of nanobiotechnology (Zou et al., 2021). Organic waste treatment carried out by nanowire-producing microorganisms can provide valuable services under diverse environmental settings. Electrically active microbes can be put to use treating organic waste while producing crystals of struvite ($NH_4MgPO_4.6H_2O$), a phosphorous-rich mineral that can be used as a delayed release crop fertilizer (Hövelmann & Putnis, 2016). Microbial electrolysis cells (MECs) were efficiently producing struvite while generating hydrogen gas at the cathode electrode (Cusick et al., 2014).

7.4.3 PHOTOSYNTHETIC NP PRODUCTION

Photosynthetic microorganisms, such as some microalgae, grow effectively on waste streams and may be used to produce useful nanomaterials (Subramaniyam et al., 2016). Municipal wastewater is commonly rich in phytonutrients such as nitrogen and phosphorous that algae require to grow (Preisner et al., 2020). Downstream release of such nutrients can cause eutrophication-related damage to natural waterways in freshwater and marine setting (Smith, 2003). Dead zones at the mouths of rivers that feed into natural systems can cause a severe economic burden on natural habitats and fisheries. Vast dead zones at the mouth of the Mississippi river in the Gulf of Mexico and the Susquehanna river at the head of Chesapeake Bay are in part due to the large-scale release of upstream nutrients (Stow et al., 2004). Photosynthetic microorganisms grown on wastewater can provide value by performing a useful service through wastewater phytonutrient removal of nitrogen and phosphorous (Tang et al., 2018). Bioremediation of the water released plus upcycling of the captured waste pollutants via biochemical accumulation and removal may be possible. Photosynthetic microalgae and cyanobacteria are optimally suited for sunlight-driven bioconversion of wastes into other useful materials, such as polyphosphate granules. Polyphosphates have nutritional and nutraceutical applications for humans and other animals. For instance, the marine cyanobacterium *Synechococcus* sp. PCC 7002 was found to produce anti-inflammatory polyphosphate NPs (Feng et al., 2018).

Certain photosynthetic microorganisms are capable of precipitating a variety of metals from waste waters (Goswami et al., 2021). The filamentous cyanobacterium *Plectonema boryanum* reductively precipitates gold and silver NPs extracellularly and accumulates them intracellularly when exposed to solutions containing dissolved metal solutions (Lengke et al., 2007). Phototrophic biological recovery of soluble precious metals could provide a valuable material return when applied to contaminated waste waters. Protein-rich extracts of the green microalgae *C. vulgaris* can precipitate silver NPs (Xie et al., 2007). Algae or cyanobacteria can form symbiotic associations with certain fungi to form lichen. Lichens and their extracts are of growing interest for the production of various nanomaterials including nanocomposites and bimetallic alloys (Hamida et al., 2021). Lichen-derived NPs are being investigated for a variety of applications and broad-spectrum antimicrobial activity has been reported (Rattan et al., 2021). Anticancer activity towards certain *in vitro* cancer cell lines, including breast and colon cancers, has been reported using metal NPs produced using lichens (Alqahtani et al., 2020).

Cyanobacteria and microalgae use the Calvin Benson cycle to fix CO_2 and some species can synthesize bioplastic precursors from CO_2. For instance, *Chlorella* microalgae and *Spirulina* cyanobacteria have the capacity to synthesize bioplastics from this greenhouse gas (Onen et al., 2020). A key challenge associated with axenic (*i.e.*, pure culture) phototrophic wastewater treatment is the risk of contamination. Mixed cultures of algae and PNS bacteria can be better used to convert domestic wastewater and CO_2 to polyhydroxyalkanoates (Almeida et al., 2021). Biochemically, polyester synthases, or PHA synthases, convert substrate (R)-hydroxyacyl-CoA thioesters into polyhydroxyalkanoates with the concomitant release of CoA (Rehm, 2003; Zou et al., 2017). Genetic modification of PHA synthases may be possible using recombinant DNA technologies, such as CRISPR-Cas9, to produce and an array of novel bioplastics not otherwise found in nature (Zou et al., 2017). Disinfection of treated wastewater prior to environmental discharge into streams and rivers is required in many countries but common chemical treatments, such as ozone or chlorine addition, may be subject to price fluctuation. This cost consideration might be reduced using waste grown microorganisms and their NP products. Under illuminated conditions, titanium dioxide NPs, like those produced by *Lactobacillus* spp., absorb energy from sunlight and are useful for wastewater disinfection (Yusuf et al., 2021).

7.5 CONCLUSION

As Earth's population continues its inexorable surge towards eight billion hungry inhabitants, there exists a growing need to develop profitable means of converting humankind's diverse wastes into valuable products. Nanoscale materials ranging in size from 1 to 100 nm or more have been recognized as possessing many useful properties and applications over the past quarter century. Some useful nanomaterials are synthesized abiotically whereas many others are the products of protein and enzyme-mediated biochemical reactions. Chemically and morphologically diverse nanomaterials can be microbiologically synthesized by certain naturally occurring bacteria, eukaryotes and even archaea from elements in their surroundings using energy in reduced organic wastes or in sunlight.

Microbially synthesized nanomaterials offer an attractive option for the production of renewable resources such as bioplastics and biofertilizers at wastewater facilities. These sustainably sourced biological products can help humanity reduce the amount of nonrenewable fossil fuels consumed and combusted to fix atmospheric nitrogen or mine phosphate deposits. Polyhydroxyalkanoates are microbial energy storage compounds accumulated in diverse microbes when carbon and energy is in abundance, yet one or more other nutrients are limited. Metals present in industrial wastewaters located near mines are often rich in valuable metals that can be harvested following microbiologically mediated biomineralization or bioprecipitation. Biomineralization of iron in the form of magnetic NPs by magnetotactic bacteria can yield a variety of particles currently being explored for a range of promising applications. Medically, these magnetic NPs can be used for enhanced MRI imagining or may be useful for targeted anticancer therapy. Electrically conductive nanomaterials can similarly be synthesized by microorganisms naturally present in wastes such as sewage. *Geobacter* species produce microbial nanowires enriched in heme-containing cytochromes that demonstrate metal-like levels of electrical conductivity. Nanowires have a range of promising uses for renewable energy generation from organic wastes using microbial fuel cells. Nanowires could also have biomedical applications, for instance when integrated into biocompatible sensor devices, such as blood glucose meters. Interest in the production of nanomaterials from wastes using microorganisms and their enzymes is growing and this nascent field appears poised for future development.

REFERENCES

Abdollahnia, Maryam, Ali Makhdoumi, Mansour Mashreghi, and Hossein Eshghi. "Exploring the Potentials of Halophilic Prokaryotes from a Solar Saltern for Synthesizing Nanoparticles: The Case of Silver and Selenium". *PLOS ONE* 15, no. 3 (2020): e0229886. doi:10.1371/journal.pone.0229886.

Achbergerová, Lucia, and Jozef Nahálka. "Polyphosphate - An Ancient Energy Source and Active Metabolic Regulator". *Microbial Cell Factories* 10, no. 1 (2011): 63. doi:10.1186/1475-2859-10-63.

Agam, Yuval, Ramesh Nandi, Alexander Kaushansky, Uri Peskin, and Nadav Amdursky. "The Porphyrin Ring Rather than the Metal Ion Dictates Long-Range Electron Transport Across Proteins Suggesting Coherence-Assisted Mechanism". *Proceedings of The National Academy of Sciences* 117, no. 51 (2020): 32260–32266. doi:10.1073/pnas.2008741117.

Agrawal, Komal, Vijal Kumar Gupta, and Pradeep Verma. "Microbial Cell Factories a New Dimension in Bio-nanotechnology: Exploring the Robustness of Nature". *Critical Reviews in Microbiology* (2021): 1–31. doi:10.1080/1040841X.2021.1977779.

Agrawal, Komal, and Pradeep Verma. (2020) "The Interest in Nanotechnology: A Step Towards Bioremediation". In: *Removal of Emerging Contaminants Through Microbial Processes*. 265–282. Springer, Singapore. doi:10.1007/978-981-15-5901-3_13.

Agrawal, Komal, and Pradeep Verma. (2021). "Applications of Biomolecules of Endophytic Fungal Origin and its Future Prospect". In: *Fungi Bio-Prospects in Sustainable Agriculture, Environment and Nano-technology. Vol 3: Fungal Metabolites and Nanotechnology*. 207–230. Academic Press, Elsevier. doi:10.1016/B978-0-12-821734-4.00015-0.

Ahlawat, Jyoti, Gabriela Henriquez, and Mahesh Narayan. "Enhancing the Delivery of Chemotherapeutics: Role of Biodegradable Polymeric Nanoparticles". *Molecules* 23, no. 9 (2018): 2157. doi:10.3390/molecules23092157.

Ahmed, Shahira A., Heba S. El-Mahallawy, and Panagiotis Karanis. "Inhibitory Activity of Chitosan Nanoparticles Against *Cryptosporidium parvum* Oocysts". *Parasitology Research* 118, no. 7 (2019): 2053–2063. doi:10.1007/s00436-019-06364-0.

AlMatar, Manaf, Essam A. Makky, Isil Var, and Fatih Koksal. "The Role of Nanoparticles in The Inhibition of Multidrug-Resistant Bacteria and Biofilms". *Current Drug Delivery* 15, no. 4 (2018): 470–484. doi:10.2174/1567201815666171207163504.

Almeida, J.R., E. Serrano, M. Fernandez, J.C. Fradinho, A. Oehmen, and M.A.M. Reis. "Polyhydroxyalkanoates Production from Fermented Domestic Wastewater Using Phototrophic Mixed Cultures". *Water Research* 197 (2021): 117101. doi:10.1016/j.watres.2021.117101.

Alqahtani, Mona A., Monerah R. Al Othman, and Afrah E. Mohammed. "Bio Fabrication of Silver Nanoparticles with Antibacterial and Cytotoxic Abilities Using Lichens". *Scientific Reports* 10, no. 1 (2020). doi:10.1038/s41598-020-73683-z.

Apata, Mario, and Susanne P. Pfeifer. "Recent Population Genomic Insights into the Genetic Basis of Arsenic Tolerance in Humans: The Difficulties of Identifying Positively Selected Loci in Strongly Bottlenecked Populations". *Heredity* 124, no. 2 (2019): 253–262. doi:10.1038/s41437-019-0285-0.

Arias, Laís Salomão, Juliano Pelim Pessan, Francisco Nunes de Souza Neto, Bruno Henrique Ramos Lima, Emerson Rodrigues de Camargo, Gordon Ramage, Alberto Carlos Botazzo Delbem, and Douglas Roberto Monteiro. "Novel Nanocarrier of Miconazole Based on Chitosan-Coated Iron Oxide Nanoparticles as a Nanotherapy to Fight Candida Biofilms". *Colloids And Surfaces B: Biointerfaces* 192 (2020): 111080. doi:10.1016/j.colsurfb.2020.111080.

Arsalan, Abdullah, and Hina Younus. "Enzymes and Nanoparticles: Modulation of Enzymatic Activity Via Nanoparticles". *International Journal of Biological Macromolecules* 118 (2018): 1833–1847. doi:10.1016/j.ijbiomac.2018.07.030.

Arsiya, F., Sayadi, M.H., and Sobhani, S. "Green Synthesis of Palladium Nanoparticles Using *Chlorella vulgaris*". *Materials Letters* 186 (2017): 113–115. doi:10.1016/j.matlet.2016.09.101

Ayub, Asif, Zulfiqar Ali Raza, Muhammad Irfan Majeed, Muhammad Rizwan Tariq, and Ayesha Irfan. "Development of Sustainable Magnetic Chitosan Biosorbent Beads for Kinetic Remediation of Arsenic Contaminated Water". *International Journal Of Biological Macromolecules* 163 (2020): 603–617. doi:10.1016/j.ijbiomac.2020.06.287.

Azizi, Susan, Farideh Namvar, Mahnaz Mahdavi, Mansor Ahmad, and Rosfarizan Mohamad. "Biosynthesis of Silver Nanoparticles Using Brown Marine Macroalga, *Sargassum Muticum* Aqueous Extract". *Materials* 6, no. 12 (2013): 5942–5950. doi:10.3390/ma6125942.

Babele, Piyoosh Kumar, Pilendra Kumar Thakre, Ramesh Kumawat, and Raghuvir Singh Tomar. "Zinc Oxide Nanoparticles Induce Toxicity by Affecting Cell Wall Integrity Pathway, Mitochondrial Function and Lipid Homeostasis in *Saccharomyces cerevisiae*". *Chemosphere* 213 (2018): 65–75. doi:10.1016/j.chemosphere.2018.09.028.

Bäckhed, Fredrik, Josefine Roswall, Yangqing Peng, Qiang Feng, Huijue Jia, Petia Kovatcheva-Datchary, and Yin Li et al. "Dynamics and Stabilization of the Human Gut Microbiome During the First Year of Life". *Cell Host & Microbe* 17, no. 5 (2015): 690–703. doi:10.1016/j.chom.2015.04.004.

Bai, H.J., Z.M. Zhang, Y. Guo, and G.E. Yang. "Biosynthesis of Cadmium Sulfide Nanoparticles by Photosynthetic Bacteria *Rhodopseudomonas palustris*". *Colloids and Surfaces B: Biointerfaces* 70, no. 1 (2009): 142–146. doi:10.1016/j.colsurfb.2008.12.025.

Bansal, V., Rautaray, D., Ahmad, A., and Sastry, M. "Biosynthesis of Zirconia Nanoparticles Using the Fungus *Fusarium oxysporum*". *Journal of Materials Chemistry* 14, no. 22 (2004): 3303–3305. doi: 10.1039/B407904C

Banza Lubaba, Nkulu, Lidia Casas Célestin, Vincent Haufroid, Thierry De Putter, Nelly D. Saenen, Tony Kayembe-Kitenge, and Paul Musa Obadia et al. "Sustainability of Artisanal Mining of Cobalt in DR Congo". *Nature Sustainability* 1, no. 9 (2018): 495–504. doi:10.1038/s41893-018-0139-4.

Bao, H., Lu, Z., Cui, X., Qiao, Y., Guo, J., Anderson, J.M., and Li, C.M., "Extracellular Microbial Synthesis of Biocompatible CdTe Quantum Dots". *Acta Biomaterialia* 6, no. 9 (2010): 3534–3541. doi: 10.1016/j.actbio.2010.03.030

Barber-Zucker, Shiran, and Raz Zarivach. "A Look into the Biochemistry of Magnetosome Biosynthesis in Magnetotactic Bacteria". *ACS Chemical Biology* 12, no. 1 (2016): 13–22. doi:10.1021/acschembio.6b01000.

Barbot, Yann, Hashem Al-Ghaili, and Roland Benz. "A Review on the Valorization of Macroalgal Wastes for Biomethane Production". *Marine Drugs* 14, no. 6 (2016): 120. doi:10.3390/md14060120.

Baskar, G., Chandhuru, J., Fahad, K. S., and Praveen, A. S. "Mycological Synthesis, Characterization and Antifungal Activity of Zinc Oxide Nanoparticles". *Asian Journal of Pharmacy and Technology* 3, no. 4 (2013): 142–146. doi: https://ajptonline.com/AbstractView.aspx?PID=2013-3-4-2

Baumgartner, J., G. Morin, N. Menguy, T. Perez Gonzalez, M. Widdrat, J. Cosmidis, and D. Faivre. "Magnetotactic Bacteria form Magnetite From a Phosphate-Rich Ferric Hydroxide Via Nanometric Ferric (Oxyhydr)Oxide Intermediates". *Proceedings of the National Academy of Sciences* 110, no. 37 (2013): 14883–14888. doi:10.1073/pnas.1307119110.

Bayda, Samer, Muhammad Adeel, Tiziano Tuccinardi, Marco Cordani, and Flavio Rizzolio. "The History of Nanoscience and Nanotechnology: From Chemical–Physical Applications to Nanomedicine". *Molecules* 25, no. 1 (2019): 112. doi:10.3390/molecules25010112.

Bazylinski, Dennis A., Anthony J. Garratt-Reed, and Richard B. Frankel. "Electron Microscopic Studies Of Magnetosomes In Magnetotactic Bacteria". Microscopy Research And Technique 27, no. 5 (1994): 389–401. doi:10.1002/jemt.1070270505.

Belchik, Sara M., David W. Kennedy, Alice C. Dohnalkova, Yuanmin Wang, Papatya C. Sevinc, Hong Wu, Yuehe Lin, H. Peter Lu, James K. Fredrickson, and Liang Shi. "Extracellular Reduction of Hexavalent Chromium by Cytochromes Mtrc and Omca of *Shewanella oneidensis* MR-1". *Applied And Environmental Microbiology* 77, no. 12 (2011): 4035–4041. doi:10.1128/aem.02463-10.

Betancourt, Walter Q., and Joan B. Rose. "Drinking Water Treatment Processes for Removal of *Cryptosporidium* and *Giardia*". *Veterinary Parasitology* 126, no. 1–2 (2004): 219–234. doi:10.1016/j.vetpar.2004.09.002.

Bhakdi, Sebastian C, Annette Ottinger, Sangdao Somsri, Panudda Sratongno, Peeranad Pannadaporn, Pattamawan Chimma, Prida Malasit, Kovit Pattanapanyasat, and Hartmut PH Neumann. "Optimized High Gradient Magnetic Separation for Isolation of *Plasmodium*-Infected Red Blood Cells". *Malaria Journal* 9, no. 1 (2010). doi:10.1186/1475-2875-9-38.

Bhatia, Saurabh. "Nanoparticles Types, Classification, Characterization, Fabrication Methods and Drug Delivery Applications". *Natural Polymer Drug Delivery Systems* 2016, 33–93. doi:10.1007/978-3-319-41129-3_2.

Bhavaniramya, Sundaresan, Ramar Vanajothi, Selvaraju Vishnupriya, Kumpati Premkumar, Mohammad S. Al-Aboody, Rajendran Vijayakumar, and Dharmar Baskaran. "Enzyme Immobilization on Nanomaterials for Biosensor and Biocatalyst in Food and Biomedical Industry". *Current Pharmaceutical Design* 25, no. 24 (2019): 2661–2676. doi:10.2174/1381612825666190712181403.

Bijnens, Karolien, Sofie Thijs, Nathalie Leynen, Vincent Stevens, Breanne McAmmond, Jonathan Van Hamme, Jaco Vangronsveld, Tom Artois, and Karen Smeets. "Differential Effect of Silver Nanoparticles on the Microbiome of Adult and Developing Planaria". *Aquatic Toxicology* 230 (2021): 105672. doi:10.1016/j.aquatox.2020.105672.

Bindschedler, Saskia, Thi Quynh Trang Vu Bouquet, Daniel Job, Edith Joseph, and Pilar Junier. "Fungal Biorecovery of Gold From E-Waste". *Advances in Applied Microbiology* 2017, 53–81. doi:10.1016/bs.aambs.2017.02.002.

Biswas, Pratim, and Chang-Yu Wu "Nanoparticles and the Environment". *Journal of the Air & Waste Management Association* 55, no. 6 (2005): 708–746. doi:10.1080/10473289.2005.10464656.

Blaak, E.E., E.E. Canfora, S. Theis, G. Frost, A.K. Groen, G. Mithieux, and A. Nauta et al. "Short Chain Fatty Acids in Human Gut and Metabolic Health". *Beneficial Microbes* 11, no. 5 (2020): 411–455. doi:10.3920/bm2020.0057.

Blackwell, Martin, Tegan Darch, and Richard Haslam. "Phosphorus Use Efficiency and Fertilizers: Future Opportunities for Improvements". *Frontiers of Agricultural Science and Engineering* 6, no. 4 (2019): 332. doi:10.15302/j-fase-2019274.

Blakemore, R. "Magnetotactic Bacteria". *Science* 190, no. 4212 (1975): 377–379. doi:10.1126/science.170679.

Bodnárová, Simona, Sylvia Gromošová, Radovan Hudák, Ján Rosocha, Jozef Živčák, Jana Plšíková, and Marek Vojtko et al. "3D Printed Polylactid Acid Based Porous Scaffold for Bone Tissue Engineering: An in Vitro Study". *Acta of Bioengineering and Biomechanics* 21, no. 4 (2019). doi:10.37190/abb-01407-2019-02.

Bose, S., Hochella Jr, M. F., Gorby, Y. A., Kennedy, D. W., McCready, D. E., Madden, A. S., Lower, B. H. "Bioreduction of Hematite Nanoparticles by the Dissimilatory Iron Reducing Bacterium *Shewanella oneidensis* MR-1". *Geochimica et Cosmochimica Acta* 73, no. 4 (2009): 962–976. doi: 10.1016/j.gca.2008.11.031

Boukhalfa, Hakim, Gary A. Icopini, Sean D. Reilly, and Mary P. Neu. "Plutonium(IV) Reduction by the Metal-Reducing Bacteria *Geobacter metallireducens* GS15 and Shewanella Oneidensis MR1". *Applied And Environmental Microbiology* 73, no. 18 (2007): 5897–5903. doi:10.1128/aem.00747-07.

Buda, Doriana-Mădălina, Paul-Adrian Bulzu, Lucian Barbu-Tudoran, Alina Porfire, Laura Pătraş, Alina Sesărman, Septimiu Tripon, Marin Şenilă, Mihaela Ileana Ionescu, and Horia Leonard Banciu. "Physiological Response to Silver Toxicity in the Extremely Halophilic Archaeon *Halomicrobium mukohataei*". *FEMS Microbiology Letters* 366, no. 18 (2019). doi:10.1093/femsle/fnz231.

Bunge, Michael, Lina S. Søbjerg, Amelia-Elena Rotaru, Delphine Gauthier, Anders T. Lindhardt, Gerd Hause, Kai Finster, Peter Kingshott, Troels Skrydstrup, and Rikke L. Meyer. "Formation of Palladium(0) Nanoparticles at Microbial Surfaces". *Biotechnology and Bioengineering* 107, no. 2 (2010): 206–215. doi:10.1002/bit.22801.

Camacho, Flinois, Robin Dando Julie, and Olga I. Padilla-Zakour. "Yogurt Acid Whey Utilization for Production of Baked Goods: Pancakes and Pizza Crust". *Foods* 8, no. 12 (2019): 615. doi:10.3390/foods8120615.

Cameron, Pamela, Birgit K. Gaiser, Bidha Bhandari, Paul M. Bartley, Frank Katzer, and Helen Bridle. "Silver Nanoparticles Decrease the Viability of *Cryptosporidium parvum* Oocysts". *Applied and Environmental Microbiology* 82, no. 2 (2016): 431–437. doi:10.1128/aem.02806-15.

Canadas, Raphaël F., João M.B.T. Cavalheiro, João D.T. Guerreiro, M. Catarina, M.D. de Almeida, Eric Pollet, Cláudia Lobato da Silva, M.M.R. da Fonseca, and Frederico Castelo Ferreira. "Polyhydroxyalkanoates: Waste Glycerol Upgrade Into Electrospun Fibrous Scaffolds for Stem Cells Culture". *International Journal of Biological Macromolecules* 71 (2014): 131–140. doi:10.1016/j.ijbiomac.2014.05.008.

Chahal, C., B. van den Akker, F. Young, C. Franco, J. Blackbeard, and P. Monis. "Pathogen and Particle Associations in Wastewater". *Advances in Applied Microbiology*, 2016, 63–119. doi:10.1016/bs.aambs.2016.08.001.

Chalermthai, Bushra, Muhammad Tahir Ashraf, Juan-Rodrigo Bastidas-Oyanedel, Bradley D. Olsen, Jens Ejbye Schmidt, and Hanifa Taher. "Techno-Economic Assessment of Whey Protein-Based Plastic Production from a Co-Polymerization Process". *Polymers* 12, no. 4 (2020): 847. doi:10.3390/polym12040847.

Chandwadkar, Pallavi, Hari Sharan Misra, and Celin Acharya. "Uranium Biomineralization Induced By a Metal Tolerantserratiastrain Under Acid, Alkaline and Irradiated Conditions". *Metallomics* 10, no. 8 (2018): 1078–1088. doi:10.1039/c8mt00061a.

Chaturvedi, Venkatesh, and Pradeep Verma. "Fabrication of Silver Nanoparticles from Leaf Extract of *Butea monosperma* (Flame of Forest) and their Inhibitory Effect on Bloomforming Cyanobacteria". *Bioresources and Bioprocessing* 2 (2015): 18. doi: 10.1186/s40643-015-0048-6.

Chen, Ming, Guangming Zeng, Piao Xu, Cui Lai, and Lin Tang. "How Do Enzymes 'Meet' Nanoparticles and Nanomaterials?" *Trends in Biochemical Sciences* 42, no. 11 (2017): 914–930. doi:10.1016/j.tibs.2017.08.008.

Chin, Kuk-Jeong, Abraham Esteve-Núñez, Ching Leang, and Derek R. Lovley. "Direct Correlation Between Rates of Anaerobic Respiration and Levels of Mrna for Key Respiratory Genes in *Geobacter sulfurreducens*". *Applied And Environmental Microbiology* 70, no. 9 (2004): 5183–5189. doi:10.1128/aem.70.9.5183-5189.2004.

Chu, H., A. Khosravi, I. P. Kusumawardhani, A. H. K. Kwon, A. C. Vasconcelos, L. D. Cunha, and A. E. Mayer et al. "Gene-Microbiota Interactions Contribute to the Pathogenesis of Inflammatory Bowel Disease". *Science* 352, no. 6289 (2016): 1116–1120. doi:10.1126/science.aad9948.

Cobley, Claire M., Jingyi Chen, Eun Chul Cho, Lihong V. Wang, and Younan Xia. "Gold Nanostructures: A Class of Multifunctional Materials for Biomedical Applications". *Chemical Society Reviews* 40, no. 1 (2011): 44–56. doi:10.1039/b821763g.

Colombo, Arnaldo L., João N. de Almeida Júnior, and Jesus Guinea. "Emerging Multidrug-Resistant *Candida* Species". *Current Opinion in Infectious Diseases* 30, no. 6 (2017): 528–538. doi:10.1097/qco.0000000000000411.

Coradeghini, Rosella, Sabrina Gioria, César Pascual García, Paola Nativo, Fabio Franchini, Douglas Gilliland, Jessica Ponti, and François Rossi. "Size-Dependent Toxicity and Cell Interaction Mechanisms of Gold Nanoparticles on Mouse Fibroblasts". *Toxicology Letters* 217 (3): 205–216. (2013). doi:10.1016/j.toxlet.2012.11.022.

Cordero, Roselynn, Ali Jawaid, Ming-Siao Hsiao, Zoë Lequeux, Richard A. Vaia, and Christopher K. Ober. "Mini Monomer Encapsulated Emulsion Polymerization of PMMA Using Aqueous ARGET ATRP". *ACS Macro Letters* 7, no. 4 (2018): 459–463. doi:10.1021/acsmacrolett.8b00038.

Cornejo, Elias, Poorna Subramanian, Zhuo Li, Grant J. Jensen, and Arash Komeili. "Dynamic Remodeling of the Magnetosome Membrane is Triggered by the Initiation of Biomineralization". *Mbio* 7, no. 1 (2016). doi:10.1128/mbio.01898-15.

Corrêa-Oliveira, Renan, José Luís Fachi, Aline Vieira, Fabio Takeo Sato, and Marco Aurélio R Vinolo. "Regulation of Immune Cell Function by Short-Chain Fatty Acids". *Clinical & Translational Immunology* 5, no. 4 (2016): e73. doi:10.1038/cti.2016.17.

Crosse, Amanda Jane, Dean Brady, Nerve Zhou, and Karl Rumbold. "Biodiesel'S Trash is a Biorefineries' Treasure: The Use of "Dirty" Glycerol as an Industrial Fermentation Substrate". *World Journal of Microbiology and Biotechnology* 36, no. 1 (2019). doi:10.1007/s11274-019-2776-9.

Cusick, Roland D., Mark L. Ullery, Brian A. Dempsey, and Bruce E. Logan. "Electrochemical Struvite Precipitation from Digestate with a Fluidized Bed Cathode Microbial Electrolysis Cell". *Water Research* 54 (2014): 297–306. doi:10.1016/j.watres.2014.01.051.

DalCorso, Giovanni, Elisa Fasani, Anna Manara, Giovanna Visioli, and Antonella Furini. "Heavy Metal Pollutions: State of the Art and Innovation in Phytoremediation". *International Journal of Molecular Sciences* 20, no. 14 (2019): 3412. doi:10.3390/ijms20143412.

Delcea, Mihaela, Alexey Yashchenok, Kristina Videnova, Oliver Kreft, Helmuth Möhwald, and Andre G. Skirtach. "Multicompartmental Micro- and Nanocapsules: Hierarchy and Applications in Biosciences". *Macromolecular Bioscience* 10, no. 5 (2010): 465–474. doi:10.1002/mabi.200900359.

Ding, Jing, Jin Liu, Xian Bo Chang, Dong Zhu, and Simon Bo Lassen. "Exposure of Cuo Nanoparticles and Their Metal Counterpart Leads to Change in the Gut Microbiota and Resistome of Collembolans". *Chemosphere* 258 (2020): 127347. doi:10.1016/j.chemosphere.2020.127347.

Dulay, Hunter, Marcela Tabares, Kazem Kashefi, and Gemma Reguera. "Cobalt Resistance Via Detoxification and Mineralization in the Iron-Reducing Bacterium *Geobacter sulfurreducens*". *Frontiers in Microbiology* 11 (2020). doi:10.3389/fmicb.2020.600463.

Dunne, W. M., E. O. Mason, and S. L. Kaplan. "Diffusion of Rifampin and Vancomycin through a *Staphylococcus epidermidis* Biofilm". *Antimicrobial Agents And Chemotherapy* 37, no. 12 (1993): 2522–2526. doi:10.1128/aac.37.12.2522.

Eatemadi, Ali, Hadis Daraee, Hamzeh Karimkhanloo, Mohammad Kouhi, Nosratollah Zarghami, Abolfazl Akbarzadeh, Mozhgan Abasi, Younes Hanifehpour, and Sang Joo. "Carbon Nanotubes: Properties, Synthesis, Purification, and Medical Applications". *Nanoscale Research Letters* 9, no. 1 (2014): 393. doi:10.1186/1556-276x-9-393.

El Bissati, Kamal, Ying Zhou, Debleena Dasgupta, Drew Cobb, Jitender P. Dubey, Peter Burkhard, David E. Lanar, and Rima McLeod. "Effectiveness of a Novel Immunogenic Nanoparticle Platform for Toxoplasma Peptide Vaccine in HLA Transgenic Mice". *Vaccine* 32, no. 26 (2014): 3243–3248. doi:10.1016/j.vaccine.2014.03.092.

El-Malek, Fady Abd, Marian Rofeal, Aida Farag, Sanaa Omar, and Heba Khairy. "Polyhydroxyalkanoate Nanoparticles Produced by Marine Bacteria Cultivated on Cost Effective Mediterranean Algal Hydrolysate Media". *Journal of Biotechnology* 328 (2021): 95–105. doi:10.1016/j.jbiotec.2021.01.008.

Eltarahony, Marwa, Sahar Zaki, Zeinab Kheiralla, and Desouky Abd-El-haleem. "NAP Enzyme Recruitment in Simultaneous Bioremediation and Nanoparticles Synthesis". *Biotechnology Reports* 18 (2018): e00257. doi:10.1016/j.btre.2018.e00257.

Emtiazi, G., N. Naghavi, and A. Bordbar. "Biodegradation of lignocellulosic waste by *Aspergillus terreus*". *Biodegradation* 12, no. 4 (2001): 257–261. doi:10.1023/a:1013 155621336.

Espitia, Paula Judith Perez, Nilda de Fátima Ferreira Soares, Jane Sélia dos Reis Coimbra, Nélio José de Andrade, Renato Souza Cruz, and Eber Antonio Alves Medeiros. "Zinc Oxide Nanoparticles: Synthesis, Antimicrobial Activity and Food Packaging Applications". *Food And Bioprocess Technology* 5, no. 5 (2012): 1447–1464. doi:10.1007/s11947-012-0797-6.

Fabrega, Julia, Joanna C. Renshaw, and Jamie R. Lead. "Interactions of Silver Nanoparticles with *Pseudomonas Putida* Biofilms". *Environmental Science & Technology* 43, no. 23 (2009): 9004–9009. doi:10.1021/es901706j.

Fan, Xinfei, Yanming Liu, Xiaochen Wang, Xie Quan, and Shuo Chen. "Improvement of Antifouling and Antimicrobial Abilities on Silver–Carbon Nanotube Based Membranes Under Electrochemical Assistance". *Environmental Science & Technology* 53, no. 9 (2019): 5292–5300. doi:10.1021/acs.est.9b00313.

Fayaz, A. M., Balaji, K., Girilal, M., Yadav, R., Kalaichelvan, P. T., and Venketesan, R. "Biogenic Synthesis of Silver Nanoparticles and Their Synergistic Effect with Antibiotics: A Study Against Gram-Positive and Gram-Negative Bacteria". *Nanomedicine: Nanotechnology, Biology and Medicine* 6, no. 1 (2010): 103–109. doi: 10.1016/j.nano.2009.04.006

Fdez-Gubieda, M. L., J. Alonso, A. García-Prieto, A. García-Arribas, L. Fernández Barquín, and A. Muela. "Magnetotactic Bacteria for Cancer Therapy". *Journal of Applied Physics* 128, no. 7 (2020): 070902. doi:10.1063/5.0018036.

Feng, Guangxin, Shiyuan Dong, Min Huang, Mingyong Zeng, Zunying Liu, Yuanhui Zhao, and Haohao Wu. "Biogenic Polyphosphate Nanoparticles from a Marine Cyanobacterium *Synechococcus* Sp. PCC 7002: Production, Characterization, and Anti-Inflammatory Properties in Vitro". *Marine Drugs* 16, no. 9 (2018): 322. doi:10.3390/md16090322.

Feng, Shi-Ting, Jingguo Li, Yanji Luo, Tinghui Yin, Huasong Cai, Yong Wang, Zhi Dong, Xintao Shuai, and Zi-Ping Li. "Ph-Sensitive Nanomicelles for Controlled and Efficient Drug Delivery to Human Colorectal Carcinoma Lovo Cells". *Plos ONE* 9, no. 6 (2014): e100732. doi:10.1371/journal.pone.0100732.

Feng, Yanni, Lingjiang Min, Weidong Zhang, Jing Liu, Zhumei Hou, Meiqiang Chu, Lan Li, Wei Shen, Yong Zhao, and Hongfu Zhang. 2017. "Zinc Oxide Nanoparticles Influence Microflora in Ileal Digesta and Correlate Well with Blood Metabolites". *Frontiers In Microbiology* 8. doi:10.3389/fmicb.2017.00992.

Fu, Liang, Qingcheng Li, Ge Yan, Dandan Zhou, and John C. Crittenden. "Hormesis Effects of Phosphorus on the Viability of *Chlorella regularis* Cells Under Nitrogen Limitation". *Biotechnology for Biofuels* 12, no. 1 (2019). doi:10.1186/s13068-019-1458-z.

Gao, Fengzheng, Haohao Wu, Mingyong Zeng, Min Huang, and Guangxin Feng. "Overproduction, Purification, and Characterization of Nanosized Polyphosphate Bodies from *Synechococcus* Sp. PCC 7002". *Microbial Cell Factories* 17, no. 1 (2018). doi:10.1186/s12934-018-0870-6.

Ghebretatios, Merry, Sabrina Schaly, and Satya Prakash. "Nanoparticles in the Food Industry and their Impact on Human Gut Microbiome and Diseases". *International Journal of Molecular Sciences* 22, no. 4 (2021): 1942. doi:10.3390/ijms22041942.

Ghosh, Purabi, Derek Fawcett, Shashi Sharma, and Gerrard Poinern. "Production of High-Value Nanoparticles Via Biogenic Processes Using Aquacultural and Horticultural Food Waste". *Materials* 10, no. 8 (2017): 852. doi:10.3390/ma10080852.

Ghosh, Shubhrima, Razi Ahmad, Md. Zeyaullah, and Sunil Kumar Khare. "Microbial Nano-Factories: Synthesis and Biomedical Applications". *Frontiers in Chemistry* 9 (2021). doi:10.3389/fchem.2021.626834.

Gorby, Y. A., S. Yanina, J. S. McLean, K. M. Rosso, D. Moyles, A. Dohnalkova, and T. J. Beveridge et al. "Electrically Conductive Bacterial Nanowires Produced by *Shewanella oneidensis* Strain MR-1 and Other Microorganisms". *Proceedings of the National Academy of Sciences* 103, no. 30 (2006): 11358–11363. doi:10.1073/pnas.0604517103.

Goswami, R.K., K. Agrawal, M.P. Shah, and Pradeep Verma. "Bioremediation of Heavy Metals from Wastewater: A Current Perspective on Microalgae-based Future". *Letters in Applied Microbiology* (2021): 1–21. doi:10.1111/lam.13564.

Grissa, Intissar, Jaber Elghoul, Lobna Ezzi, Sana Chakroun, Emna Kerkeni, Mohsen Hassine, Lassaad El Mir, Meriem Mehdi, Hassen Ben Cheikh, and Zohra Haouas. "Anemia and Genotoxicity Induced by Sub-Chronic Intragastric Treatment of Rats with Titanium Dioxide Nanoparticles". *Mutation Research/Genetic Toxicology and Environmental Mutagenesis* 794: 25–31. (2016). doi:10.1016/j.mrgentox.2015.09.005.

Guo, Jingyang, Simin Qin, Yan Wei, Shima Liu, Hongzhen Peng, Qingnuan Li, Liqiang Luo, and Min Lv. "Silver Nanoparticles Exert Concentration-Dependent Influences on Biofilm Development and Architecture". *Cell Proliferation* 52, no. 4 (2019). doi:10.1111/cpr.12616.

Gupta, Kapish, Asim Kumar Jana, Sandeep Kumar, and Mithu Maiti. "Immobilization of α-Amylase and Amyloglucosidase Onto Ion-Exchange Resin Beads and Hydrolysis of Natural Starch at High Concentration". *Bioprocess and Biosystems Engineering* 36, no. 11 (2013): 1715–1724. doi:10.1007/s00449-013-0946-y.

Gupta, Varsha, Manjistha Sengupta, Jaya Prakash, and Baishnab Charan Tripathy. "An Introduction to Biotechnology". *Basic and Applied Aspects of Biotechnology*, 2016, 1–21. doi:10.1007/978-981-10-0875-7_1.

Guterres, Sílvia S., Marta P. Alves, and Adriana R. Pohlmann. "Polymeric Nanoparticles, Nanospheres and Nanocapsules, for Cutaneous Applications". *Drug Target Insights* 2 (2007): 117739280700200. doi:10.1177/117739280700200002.

Hajipour, Mohammad J., Katharina M. Fromm, Ali Akbar Ashkarran, Dorleta Jimenez de Aberasturi, Idoia Ruiz de Larramendi, Teofilo Rojo, Vahid Serpooshan, Wolfgang J. Parak, and Morteza Mahmoudi. "Antibacterial Properties of Nanoparticles". *Trends in Biotechnology* 30, no. 10 (2012): 499–511. doi:10.1016/j.tibtech.2012.06.004.

Hamida, Reham Samir, Mohamed Abdelaal Ali, Nabila Elsayed Abdelmeguid, Mayasar Ibrahim Al-Zaban, Lina Baz, and Mashael Mohammed Bin-Meferij. "Lichens—a Potential Source for Nanoparticles Fabrication: A Review on Nanoparticles Biosynthesis and their Prospective Applications". *Journal of Fungi* 7, no. 4 (2021): 291. doi:10.3390/jof7040291.

Hatchett, David W., and Henry S. White. "Electrochemistry of Sulfur Adlayers on the Low-Index Faces of Silver". *The Journal of Physical Chemistry* 100, no. 23 (1996): 9854–9859. doi:10.1021/jp953757z.

He, S., Guo, Z., Zhang, Y., Zhang, S., Wang, J., and Gu, N. "Biosynthesis of Gold Nanoparticles Using the Bacteria *Rhodopseudomonas capsulate*". *Materials Letters* 61, no. 18 (2007): 3984–3987. doi: 10.1016/j.matlet.2007.01.018

Hernández-Sierra, Juan Francisco, Facundo Ruiz, Diana Corina Cruz Pena, Fidel Martínez-Gutiérrez, Alberto Emilio Martínez, Amaury de Jesús Pozos Guillén, Humberto Tapia-Pérez, and Gabriel Martínez Castañón. "The Antimicrobial Sensitivity of *Streptococcus* Mutans to Nanoparticles of Silver, Zinc Oxide, and Gold". *Nanomedicine: Nanotechnology, Biology and Medicine* 4, no. 3 (2008): 237–240. doi:10.1016/j.nano.2008.04.005.

Herrera, Mónica Tatiana, Jhon Jhamilton Artunduaga, Claudia Cristina Ortiz, and Rodrigo Gonzalo Torres. "Synthesis of Antibiotic Loaded Polylactic Acid Nanoparticles and their Antibacterial Activity Against *Escherichia coli* O157:H7 and Methicillin-Resistant Staphylococcus Aureus". *Biomédica* 37, no. 1 (2017): 11. doi:10.7705/biomedica.v37i1.2995.

Hevia, Arancha, Susana Delgado, Borja Sánchez, and Abelardo Margolles. "Molecular Players Involved in the Interaction Between Beneficial Bacteria and the Immune System". *Frontiers in Microbiology* 6 (2015). doi:10.3389/fmicb.2015.01285.

Hövelmann, Jörn, and Christine V. Putnis. "In Situ Nanoscale Imaging of Struvite Formation During the Dissolution of Natural Brucite: Implications for Phosphorus Recovery from Wastewaters". *Environmental Science & Technology* 50, no. 23 (2016): 13032–13041. doi:10.1021/acs.est.6b04623.

Hu, Jinlong, Zhenting Zhang, Yanxu Lin, Shumiao Zhao, Yuxia Mei, Yunxiang Liang, and Nan Peng. "High-Titer Lactic Acid Production From Naoh-Pretreated Corn Stover by *Bacillus coagulans* LA204 Using Fed-Batch Simultaneous Saccharification and Fermentation Under Non-Sterile Condition". *Bioresource Technology* 182 (2015): 251–257. doi:10.1016/j.biortech.2015.02.008.

Huang, Shaolong, Huan Yi, Luhong Zhang, Zhengyuan Jin, Yaojia Long, Yiyue Zhang, and Qiufan Liao et al. "Non-Precious Molybdenum Nanospheres as a Novel Cocatalyst for Full-Spectrum-Driven Photocatalytic CO_2 Reforming To CH4". *Journal of Hazardous Materials* 393 (2020): 122324. doi:10.1016/j.jhazmat.2020.122324.

Hussein, F. H. (2011). Photochemical Treatments of Textile Industries Wastewater. In (Ed.), Advances in Treating Textile Effluent. IntechOpen. https://doi.org/10.5772/18902

Hyeon, Jeong Eun, Da Woon Jeong, Young Jin Ko, Seung Wook Kim, Chulhwan Park, and Sung Ok Han. "Biomimetic Magnetoelectric Nanocrystals Synthesized by Polymerization of Heme as Advanced Nanomaterials for Biosensing Application". *Biosensors and Bioelectronics* 114 (2018): 1–9. doi:10.1016/j.bios.2018.05.007.

Icopini, Gary A., Joe G. Lack, Larry E. Hersman, Mary P. Neu, and Hakim Boukhalfa. "Plutonium(V/VI) Reduction by the Metal-Reducing Bacteria *Geobacter metallireducens* GS-15 and *Shewanella oneidensis* MR-1". *Applied and Environmental Microbiology* 75, no. 11 (2009): 3641–3647. doi:10.1128/aem.00022-09.

Inyushin, M., Yu. Kucheryavih, L. Kucheryavih, L. Rojas, I. Khmelinskii, and V. Makarov. "Superparamagnetic Properties of Hemozoin". *Scientific Reports* 6, no. 1 (2016). doi:10.1038/srep26212.

Ispas, Cristina, Daniel Andreescu, Avni Patel, Dan V. Goia, Silvana Andreescu, and Kenneth N. Wallace. "Toxicity and Developmental Defects of Different Sizes and Shape Nickel Nanoparticles in Zebrafish". *Environmental Science & Technology* 43 (16): 6349–6356 (2009) doi:10.1021/es9010543.

Ivankovic, Mario, Radmila Haneckova, Albert Thommen, Markus A. Grohme, Miquel Vila-Farré, Steffen Werner, and Jochen C. Rink. "Model Systems for Regeneration: Planarians". *Development* 146, no. 17 (2019). doi:10.1242/dev.167684.

Javurek, Angela B., Dhananjay Suresh, William G. Spollen, Marcia L. Hart, Sarah A. Hansen, Mark R. Ellersieck, and Nathan J. Bivens et al. "Gut Dysbiosis and Neurobehavioral Alterations in Rats Exposed to Silver Nanoparticles". *Scientific Reports* 7, no. 1 (2017). doi:10.1038/s41598-017-02880-0.

Jha, A.K. and Prasad, K. "Ferroelectric $BaTiO_3$ Nanoparticles: Biosynthesis and Characterization". *Colloids and Surfaces B: Biointerfaces* 75, no. 1 (2009): 330–334. doi: 10.1016/j.colsurfb.2009.09.005

Jha, A. K., Prasad, K., and Kulkarni, A. R. "Synthesis of TiO_2 Nanoparticles Using Microorganisms". *Colloids and Surfaces B: Biointerfaces* 71, no. 2 (2009a): 226–229. doi: 10.1016/j.colsurfb.2009.02.007

Jha, A.K., Prasad, K., and Prasad, K., "A Green Low-Cost Biosynthesis of Sb_2O_3 Nanoparticles". *Biochemical Engineering Journal* 43, no. 3 (2009b): 303-306. doi: 10.1016/j.bej.2008.10.016.

Johansen, Helle Krogh, and Peter C Gøtzsche. "Amphotericin B Versus Fluconazole for Controlling Fungal Infections in Neutropenic Cancer Patients". *Cochrane Database of Systematic Reviews*, 2014. doi:10.1002/14651858.cd000239.pub2.

Jones, Nicole, Binata Ray, Koodali T. Ranjit, and Adhar C. Manna. "Antibacterial Activity of Zno Nanoparticle Suspensions on a Broad Spectrum of Microorganisms". *FEMS Microbiology Letters* 279, no. 1 (2008): 71–76. doi:10.1111/j.1574-6968.2007.01012.x.

Kalishwaralal, Kalimuthu, Selvaraj BarathManiKanth, Sureshbabu Ram Kumar Pandian, Venkataraman Deepak, and Sangiliyandi Gurunathan. "Silver Nanoparticles Impede the Biofilm Formation by *Pseudomonas aeruginosa* and *Staphylococcus epidermidis*". *Colloids and Surfaces B: Biointerfaces* 79, no. 2 (2010): 340–344. doi:10.1016/j.colsurfb.2010.04.014.

Karkowska-Kuleta, Justyna, Maria Rapala-Kozik, and Andrzej Kozik. "Fungi Pathogenic to Humans: Molecular Bases of Virulence of *Candida albicans*, *Cryptococcus neoformans* and *Aspergillus fumigatus*.". *Acta Biochimica Polonica* 56, no. 2 (2009). doi:10.18388/abp.2009_2452.

Kasirajan, Lakshmi, and Julie A. Maupin-Furlow. "Halophilic Archaea and their Potential to Generate Renewable Fuels and Chemicals". *Biotechnology and Bioengineering* 118, no. 3 (2020): 1066–1090. doi:10.1002/bit.27639.

Kasra-Kermanshahi, Rouha, Parisa Tajer-Mohammad-Ghazvini, and Marziyeh Bahrami-Bavani. "A Biotechnological Strategy for Molybdenum Extraction Using *Acidithiobacillus ferrooxidans*". *Applied Biochemistry and Biotechnology* 193, no. 3 (2020): 884–895. doi:10.1007/s12010-020-03468-7.

Kimber, Richard L., Heath Bagshaw, Kurt Smith, Dawn M. Buchanan, Victoria S. Coker, Jennifer S. Cavet, and Jonathan R. Lloyd. "Biomineralization of Cu_2S Nanoparticles by *Geobacter sulfurreducens*". *Applied And Environmental Microbiology* 86, no. 18 (2020). doi:10.1128/aem.00967-20.

Kjeldsen, Kasper U., Lars Schreiber, Casper A. Thorup, Thomas Boesen, Jesper T. Bjerg, Tingting Yang, and Morten S. Dueholm et al. "On the Evolution and Physiology of Cable Bacteria". *Proceedings of the National Academy of Sciences* 116, no. 38 (2019): 19116–19125. doi:10.1073/pnas.1903514116.

Kolinko, Isabel, Anna Lohße, Sarah Borg, Oliver Raschdorf, Christian Jogler, Qiang Tu, and Mihály Pósfai et al. "Biosynthesis of Magnetic Nanostructures in a Foreign Organism by Transfer of Bacterial Magnetosome Gene Clusters". *Nature Nanotechnology* 9, no. 3 (2014): 193–197. doi:10.1038/nnano.2014.13.

Koller, Martin, and Gerhart Brauuegg. "Potential and Prospects of Continuous Polyhydroxyalkanoate (PHA) Production". *Bioengineering* 2, no. 2 (2015): 94–121. doi:10.3390/bioengineering2020094.

Komine, Yutaka, Laura L. Eggink, Hyoungshin Park, and J. Kenneth Hoober. "Vacuolar Granules in *Chlamydomonas reinhardtii* Polyphosphate and a 70-Kda Polypeptide as Major Components". *Planta* 210, no. 6 (2000): 897–905. doi:10.1007/s004250050695.

Koopi, H. and Buazar, F. "A Novel One-Pot Biosynthesis of Pure Alpha Aluminum Oxide Nanoparticles Using the Macroalgae *Sargassum ilicifolium*: A Green Marine Approach". *Ceramics International* 44, no. 8 (2018): 8940–8945. doi: 10.1016/j. ceramint.2018.02.091.

Kornberg, Arthur, S.R. Kornberg, and Ernest S. Simms. "Metaphosphate Synthesis by an Enzyme from *Escherichia coli*". *Biochimica Et Biophysica Acta* 20 (1956): 215–227. doi:10.1016/0006-3002(56)90280-3.

Koul, Bhupendra, Anil Kumar Poonia, Dhananjay Yadav, and Jun-O Jin. "Microbe-Mediated Biosynthesis of Nanoparticles: Applications and Future Prospects". *Biomolecules* 11, no. 6 (2021): 886. doi:10.3390/biom11060886.

Kröger, Nils, and Eike Brunner. "Complex-Shaped Microbial Biominerals for Nanotechnology". *Wiley Interdisciplinary Reviews: Nanomedicine And Nanobiotechnology* 6, no. 6 (2014): 615–627. doi:10.1002/wnan.1284.

Kumar, S.A., Ansary, A.A., Ahmad, A., and Khan, M.I. (2007). "Extracellular Biosynthesis of CdSe Quantum Dots by the Fungus, *Fusarium oxysporum*". *Journal of Biomedical Nanotechnology* 3, no. 2: 2007: 190–194. doi: 10.1166/jbn.2007.027

Kumar, Bikash, Komal Agrawal, and Pradeep Verma. "Current Perspective and Advances of Microbe Assisted Electrochemical System as a Sustainable Approach for Mitigating Toxic Dyes and Heavy Metals from Wastewater". *ASCE's Journal of Hazardous, Toxic, and Radioactive Waste* 25, no. 2 (2021): 04020082. doi:10.1061/(ASCE)HZ.2153-5515.0000590.

Kustov, Leonid, Kharlampii Tiras, Souhail Al-Abed, Natalia Golovina, and Mikhail Ananyan. "Estimation of the Toxicity of Silver Nanoparticles by Using Planarian Flatworms". *Alternatives to Laboratory Animals* 42, no. 1 (2014): 51–58. doi:10.1177/026119291404200108.

Kuzajewska, Danuta, Agata Wszołek, Wojciech Zwierełło, Lucyna Kirczuk, and Agnieszka Maruszewska. "Magnetotactic Bacteria and Magnetosomes as Smart Drug Delivery Systems: A New Weapon on the Battlefield With Cancer?" *Biology* 9, no. 5 (2020): 102. doi:10.3390/biology9050102.

Kwok, Katrina C. M., Len Foong Koong, Tareq Al Ansari, and Gordon McKay. "Adsorption/Desorption of Arsenite and Arsenate on Chitosan and Nanochitosan". *Environmental Science and Pollution Research* 25, no. 15 (2018): 14734–14742. doi:10.1007/s11356-018-1501-9.

Lallo, da Silva, Bruna, Bruno Leonardo Caetano, Bruna Galdorfini Chiari-Andréo, Rosemeire Cristina Linhari Rodrigues Pietro, and Leila Aparecida Chiavacci. "Increased Antibacterial Activity of Zno Nanoparticles: Influence of Size and Surface Modification". *Colloids and Surfaces B: Biointerfaces* 177 (2019): 440–447. doi:10.1016/j.colsurfb.2019.02.013.

Leblanc, Jean Guy, Christian Milani, Graciela Savoy de Giori, Fernando Sesma, Douwe van Sinderen, and Marco Ventura. "Bacteria as Vitamin Suppliers to their Host: A Gut Microbiota Perspective". *Current Opinion in Biotechnology* 24, no. 2 (2013): 160–168. doi:10.1016/j.copbio.2012.08.005.

Lee, J.H., Han, J., Choi, H., and Hur, H.G. "Effects of Temperature and Dissolved Oxygen on Se(IV) Removal and Se(0) Precipitation by *Shewanella* Sp. HN-41". *Chemosphere* 68, no. 10 (2007): 1898–1905. doi: 10.1016/j.chemosphere.2007.02.062

Lengke, Maggy F., Michael E. Fleet, and Gordon Southam. "Biosynthesis of Silver Nanoparticles by Filamentous Cyanobacteria from a Silver(I) Nitrate Complex". *Langmuir* 23, no. 5 (2007): 2694–2699. doi:10.1021/la0613124.

Leynen, Nathalie, Frank G.A.J Van Belleghem, Annelies Wouters, Hannelore Bove, Jan-Pieter Ploem, Elsy Thijssen, and Sabine A.S. Langie et al. "In Vivo Toxicity Assessment of Silver Nanoparticles in Homeostatic Versus Regenerating Planarians". *Nanotoxicology* 13, no. 4 (2019): 476–491. doi:10.1080/17435390.2018.1553252.

Li, Guangquan, Dan He, Yongqing Qian, Buyuan Guan, Song Gao, Yan Cui, Koji Yokoyama, and Li Wang. "Fungus-Mediated Green Synthesis of Silver Nanoparticles Using *Aspergillus terreus*". *International Journal of Molecular Sciences* 13, no. 1 (2011a): 466–476. doi:10.3390/ijms13010466.

Li, Jiangyan, Meng Tang, and Yuying Xue. "Review of the Effects of Silver Nanoparticle Exposure on Gut Bacteria". *Journal of Applied Toxicology* 39, no. 1 (2018): 27–37. doi:10.1002/jat.3729.

Li, Xiangqian, Huizhong Xu, Zhe-Sheng Chen, and Guofang Chen. "Biosynthesis of Nanoparticles by Microorganisms and their Applications". *Journal of Nanomaterials* 2011b (2011): 1–16. doi:10.1155/2011/270974.

Liang, Shaobo, Armando G. McDonald, and Erik R. Coats. "Lactic Acid Production from Potato Peel Waste by Anaerobic Sequencing Batch Fermentation Using Undefined Mixed Culture". *Waste Management* 45 (2015): 51–56. doi:10.1016/j.wasman.2015.02.004.

Lin, Wei, Wensi Zhang, Greig A. Paterson, Qiyun Zhu, Xiang Zhao, Rob Knight, Dennis A. Bazylinski, Andrew P. Roberts, and Yongxin Pan. "Expanding Magnetic Organelle Biogenesis in the Domain Bacteria". *Microbiome* 8, no. 1 (2020). doi:10.1186/s40168-020-00931-9.

Liu, Jing, Lei Zhou, Faqin Dong, and Karen A. Hudson-Edwards. "Enhancing As(V) Adsorption and Passivation using Biologically Formed Nano-Sized Fes Coatings on Limestone: Implications for Acid Mine Drainage Treatment and Neutralization". *Chemosphere* 168 (2017): 529–538. doi:10.1016/j.chemosphere.2016.11.037.

Liu, Ling, Yuyuan Cai, Hong Li, Shumiao Zhao, Mingxiong He, Guo-quan Hu, Yunxiang Liang, Nan Peng, and Jinglong Hu. "Bio-Detoxification Bacteria Isolated from Dye-Polluted Soils Promote Lactic acid Production from Ammonia Pretreated Corn Stover". *Applied Biochemistry and Biotechnology* 189, no. 1 (2019): 129–143. doi:10.1007/s12010-019-02993-4.

López, Nancy I., M. Julia Pettinari, Pablo I. Nikel, and Beatriz S. Méndez. "Polyhydroxy-alkanoates". *Advances in Applied Microbiology* 2015, 73–106. doi:10.1016/bs.aambs.2015.06.001.

Louis, Petra, and Harry J. Flint. "Formation of Propionate and Butyrate by the Human Colonic Microbiota". *Environmental Microbiology* 19, no. 1 (2016): 29–41. doi:10.1111/1462-2920.13589.

Louis, Petra, Georgina L. Hold, and Harry J. Flint. "The Gut Microbiota, Bacterial Metabolites and Colorectal Cancer". *Nature Reviews Microbiology* 12, no. 10 (2014): 661–672. doi:10.1038/nrmicro3344.

Luo, Hong, Xuan Li, Guohong Li, Yanbo Pan, and Keqin Zhang. "Acanthocytes of *Stropharia rugosoannulata* Function as a Nematode-Attacking Device". *Applied And Environmental Microbiology* 72, no. 4 (2006): 2982–2987. doi:10.1128/aem.72.4.2982-2987.2006.

Lv, Q., Zhang, B., Xing, X., Zhao, Y., Cai, R., Wang, W., and Gu, Q. "Biosynthesis of Copper Nanoparticles Using *Shewanella loihica* PV-4 with Antibacterial Activity: Novel Approach and Mechanisms Investigation". *Journal of Hazardous Materials* 347 (2018): 141–149. doi: 10.1016/j.jhazmat.2017.12.070

Macomber, L., and J. A. Imlay. "The Iron-Sulfur Clusters of Dehydratases are Primary Intracellular Targets of Copper Toxicity". *Proceedings of the National Academy of Sciences* 106, no. 20 (2009): 8344–8349. doi:10.1073/pnas.0812808106.

Malekifard, Farnaz, Mousa Tavassoli, and Kiana Vaziri. "In Vitro Assessment Antiparasitic Effect of Selenium and Copper Nanoparticles on *Giardia deodenalis* Cyst". *Iranian Journal of Parasitology* 2020. doi:10.18502/ijpa.v15i3.4206.

Mannucci, Silvia, Leonardo Ghin, Giamaica Conti, Stefano Tambalo, Alessandro Lascialfari, Tomas Orlando, and Donatella Benati et al. "Magnetic Nanoparticles from *Magnetospirillum gryphiswaldense* Increase the Efficacy of Thermotherapy in a Model of Colon Carcinoma". *Plos ONE* 9, no. 10 (2014): e108959. doi:10.1371/journal. pone.0108959.

Mao, Xiaoyun, Qin Lu, Wei Mo, Xiaoping Xin, Xian Chen, and Zhenli He. "Phosphorus Availability and Release Pattern from Activated Dolomite Phosphate Rock in Central Florida". *Journal of Agricultural And Food Chemistry* 65, no. 23 (2017): 4589–4596. doi:10.1021/acs.jafc.7b01037.

Marcano, L., A. García-Prieto, D. Muñoz, L. Fernández Barquín, I. Orue, J. Alonso, A. Muela, and M.L. Fdez-Gubieda. "Influence of the Bacterial Growth Phase on the Magnetic Properties of Magnetosomes Synthesized by *Magnetospirillum gryphiswaldense*". *Biochimica Et Biophysica Acta (BBA) - General Subjects* 1861, no. 6 (2017): 1507–1514. doi:10.1016/j.bbagen.2017.01.012.

Maripuri, Saugar, and Kirsten L. Johansen. "Risk Of Gadolinium-Based Contrast Agents in Chronic Kidney Disease—is Zero Good Enough?". *JAMA Internal Medicine* 180, no. 2 (2020): 230. doi:10.1001/jamainternmed.2019.5278.

Mármol, Gonzalo, Christian Gauss, and Raul Fangueiro. "Potential of Cellulose Microfibers for PHA and PLA Biopolymers Reinforcement". *Molecules* 25, no. 20 (2020): 4653. doi:10.3390/molecules25204653.

Martinez-Gutierrez, Fidel, Laura Boegli, Alessandra Agostinho, Elpidio Morales Sánchez, Horacio Bach, Facundo Ruiz, and Garth James. "Anti-Biofilm Activity of Silver Nanoparticles Against Different Microorganisms". *Biofouling* 29, no. 6 (2013): 651–660. doi:10.1080/08927014.2013.794225.

Mazmanian, Sarkis K., Cui Hua Liu, Arthur O. Tzianabos, and Dennis L. Kasper. "An Immunomodulatory Molecule of Symbiotic Bacteria Directs Maturation of the Host Immune System". *Cell* 122, no. 1 (2005): 107–118. doi:10.1016/j.cell.2005.05.007.

McGovern, Patrick, Mindia Jalabadze, Stephen Batiuk, Michael P. Callahan, Karen E. Smith, Gretchen R. Hall, and Eliso Kvavadze et al. "Early Neolithic Wine of Georgia in the South Caucasus". *Proceedings of the National Academy of Sciences* 114, no. 48 (2017): E10309–E10318. doi:10.1073/pnas.1714728114.

Mele, Elisa, and Dario Pisignano. "Nanobiotechnology: Soft Lithography". *Biosilica in Evolution, Morphogenesis, and Nanobiotechnology*, 2009, 341–358. doi:10.1007/978-3-540-88552-8_15.

Mériaux, Sébastien, Marianne Boucher, Benjamin Marty, Yoann Lalatonne, Sandra Prévéral, Laurence Motte, and Christopher T. Lefèvre et al. "Magnetosomes, Biogenic Magnetic Nanomaterials for Brain Molecular Imaging with 17.2 T MRI Scanner". *Advanced Healthcare Materials* 4, no. 7 (2015): 1076–1083. doi:10.1002/adhm.201400756.

Mickoleit, Frank, Cornelia Jörke, Stefan Geimer, Denis S. Maier, Jörg P. Müller, Johanna Demut, Christine Gräfe, Dirk Schüler, and Joachim H. Clement. "Biocompatibility, Uptake and Subcellular Localization of Bacterial Magnetosomes in Mammalian Cells". *Nanoscale Advances* 3, no. 13 (2021): 3799–3815. doi:10.1039/d0na01086c.

Mirkovic, Tihana, Nicole S. Zacharia, Gregory D. Scholes, and Geoffrey A. Ozin. "Fuel for Thought: Chemically Powered Nanomotors Out-Swim Nature'S Flagellated Bacteria". *ACS Nano* 4, no. 4 (2010): 1782–1789. doi:10.1021/nn100669h.

Mohanty, Soumitra, Saswati Mishra, Prajna Jena, Biju Jacob, Biplab Sarkar, and Avinash Sonawane. "An Investigation on the Antibacterial, Cytotoxic, and Antibiofilm Efficacy of Starch-Stabilized Silver Nanoparticles". *Nanomedicine: Nanotechnology, Biology And Medicine* 8, no. 6 (2012): 916–924. doi:10.1016/j.nano.2011.11.007.

Mohapatra, S., S. Pattnaik, S. Maity, S. Mohapatra, S. Sharma, J. Akhtar, S. Pati, D.P. Samantaray, and Ajit Varma. "Comparative Analysis of Phas Production by *Bacillus megaterium* OUAT 016 Under Submerged and Solid-State Fermentation". *Saudi Journal of Biological Sciences* 27, no. 5 (2020): 1242–1250. doi:10.1016/j.sjbs.2020.02.001.

Mohite, Pallavi T., Ameeta Ravi Kumar, and Smita S. Zinjarde. "Biotransformation of Hexavalent Chromium into Extracellular Chromium(III) Oxide Nanoparticles Using *Schwanniomyces occidentalis*". *Biotechnology Letters* 38, no. 3 (2015): 441–446. doi:10.1007/s10529-015-2009-8.

Mondal, Monoj Kumar, and Ravi Garg. "A Comprehensive Review on Removal of Arsenic Using Activated Carbon Prepared from Easily Available Waste Materials". *Environmental Science and Pollution Research* 24, no. 15 (2017): 13295–13306. doi:10.1007/s11356-017-8842-7.

Monteiro, D.R., S. Silva, M. Negri, L.F. Gorup, E.R. de Camargo, R. Oliveira, D.B. Barbosa, and M. Henriques. "Silver Nanoparticles: Influence of Stabilizing Agent and Diameter on Antifungal Activity Against *Candida albicans* and *Candida glabrata* Biofilms". *Letters in Applied Microbiology* 54, no. 5 (2012): 383–391. doi:10.1111/j.1472-765x.2012.03219.x.

Morrison, Douglas J., and Tom Preston. "Formation of Short Chain Fatty Acids by the Gut Microbiota and their Impact on Human Metabolism". *Gut Microbes* 7, no. 3 (2016): 189–200. doi:10.1080/19490976.2015.1134082.

Morones, J.R., Elechiguerra, J.L., Camacho, A., Holt, K., Kouri, J.B., Ramírez, J.T., Yacaman, M.J., "The Bactericidal Effect of Silver Nanoparticles". *IOP Sciences Nanotechnology* 16, no. 10 (2005): 2346. doi: 10.1088/0957-4484/16/10/059

Mukherjee, Triparna, Shatarupa Chakraborty, Aksar Ali Biswas, and Tapan Kumar Das. "Bioremediation Potential Of Arsenic By Non-Enzymatically Biofabricated Silver Nanoparticles Adhered to the Mesoporous Carbonized Fungal Cell Surface of *Aspergillus foetidus* MTCC8876". *Journal of Environmental Management* 201 (2017): 435–446. doi:10.1016/j.jenvman.2017.06.030.

Müller, Frank D., Dirk Schüler, and Daniel Pfeiffer. "A Compass to Boost Navigation: Cell Biology of Bacterial Magnetotaxis". *Journal of Bacteriology* 202, no. 21 (2020). doi:10.1128/jb.00398-20.

Murat, Dorothée, Veesta Falahati, Luca Bertinetti, Roseann Csencsits, André Körnig, Kenneth Downing, Damien Faivre, and Arash Komeili. "The Magnetosome Membrane Protein, Mmsf, is a Major Regulator of Magnetite Biomineralization in *Magnetospirillum magneticum* AMB-1". *Molecular Microbiology* 85, no. 4 (2012): 684–699. doi:10.1111/j.1365-2958.2012.08132.x.

Musial, Joanna, Rafal Krakowiak, Dariusz T. Mlynarczyk, Tomasz Goslinski, and Beata J. Stanisz. "Titanium Dioxide Nanoparticles in Food and Personal Care Products—What Do We Know About Their Safety?". *Nanomaterials* 10, no. 6 (2020): 1110. doi:10.3390/nano10061110.

Nadell, Carey D, Knut Drescher, Ned S Wingreen, and Bonnie L Bassler. "Extracellular Matrix Structure Governs Invasion Resistance in Bacterial Biofilms". *The ISME Journal* 9, no. 8 (2015): 1700–1709. doi:10.1038/ismej.2014.246.

Nan, Xiaohui, Wenjia Lai, Dan Li, Jiesheng Tian, Zhiyuan Hu, and Qiaojun Fang. "Biocompatibility of Bacterial Magnetosomes as MRI Contrast Agent: A Long-Term in Vivo Follow-Up Study". *Nanomaterials* 11, no. 5 (2021): 1235. doi:10.3390/nano11051235.

Naudet, Florian, and Bruno Falissard. "Does Reductio Ad Absurdumhave a Place in Evidence-Based Medicine?". *BMC Medicine* 12, no. 1 (2014). doi:10.1186/1741-7015-12-106.

Neset, Tina-Simone, Dana Cordell, Steve Mohr, Froggi VanRiper, and Stuart White. "Visualizing Alternative Phosphorus Scenarios for Future Food Security". *Frontiers in Nutrition* 3 (2016). doi:10.3389/fnut.2016.00047.

Nielsen, Chad, Asif Rahman, Asad Ur Rehman, Marie K. Walsh, and Charles D. Miller. "Food Waste Conversion to Microbial Polyhydroxyalkanoates". *Microbial Biotechnology* 10, no. 6 (2017): 1338–1352. doi:10.1111/1751-7915.12776.

Nordmeier, Akira, Augustus Merwin, Donald F. Roeper, and Dev Chidambaram. "Microbial Synthesis of Metallic Molybdenum Nanoparticles". *Chemosphere* 203 (2018): 521–525. doi:10.1016/j.chemosphere.2018.02.079.

Numanoğlu, Yasemin, and Sibel Sungur. "B-Galactosidase from Kluyveromyces Lactis Cell Disruption and Enzyme Immobilization Using a Cellulose–Gelatin Carrier System". *Process Biochemistry* 39, no. 6 (2004): 705–711. doi:10.1016/s0032-9592(03) 00183-3.

Oliveira, Marcus F, Joana C.P d'Avila, Christiane R Torres, Pedro L Oliveira, Antônio J Tempone, Franklin D Rumjanek, and Cláudia M.S Braga et al. "Haemozoin in *Schistosoma mansoni*". *Molecular and Biochemical Parasitology* 111, no. 1 (2000): 217–221. doi:10.1016/s0166-6851(00)00299-1.

Olsen, I. "Biofilm-Specific Antibiotic Tolerance and Resistance". *European Journal of Clinical Microbiology & Infectious Diseases* 34, no. 5 (2015): 877–886. doi:10.1007/ s10096-015-2323-z.

Onen, Cinar, Zhi Kai Chong Senem, Mehmet Ali Kucuker, Nils Wieczorek, Ugur Cengiz, and Kerstin Kuchta. "Bioplastic Production from Microalgae: A Review". *International Journal of Environmental Research and Public Health* 17, no. 11 (2020): 3842. doi:10.3390/ijerph17113842.

Orjih, Augustine U., and Coy D. Fitch. "Hemozoin Production by *Plasmodium falciparum*: Variation with Strain and Exposure to Chloroquine". *Biochimica Et Biophysica Acta (BBA)-General Subjects* 1157, no. 2 (1993): 270–274. doi:10.1016/0304-4165(93)90109-l.

Ovais, Muhammad, Ali Khalil, Muhammad Ayaz, Irshad Ahmad, Susheel Nethi, and Sudip Mukherjee. "Biosynthesis of Metal Nanoparticles Via Microbial Enzymes: A Mechanistic Approach". *International Journal of Molecular Sciences* 19, no. 12 (2018): 4100. doi:10.3390/ijms19124100.

Pagán, Oné R. "Planaria: An Animal Model that Integrates Development, Regeneration and Pharmacology". *The International Journal of Developmental Biology* 61, no. 8–9 (2017): 519–529. doi:10.1387/ijdb.160328op.

Pagán, Oné R., Debra Baker, Sean Deats, Erica Montgomery, Matthew Tenaglia, Clinita Randolph, and Dharini Kotturu et al. "Planarians in Pharmacology: Parthenolide is a Specific Behavioral Antagonist of Cocaine in the Planarian Girardia Tigrina". *The International Journal of Developmental Biology* 56, no. 1-2-3 (2012): 193–196. doi:10.1387/ijdb.113486op.

Pandey, Neha, and Renu Bhatt. "Improved Biotransformation of Arsenic by Arsenite Oxidase – Chitosan Nanoparticle Conjugates". *International Journal of Biological Macromolecules* 106 (2018): 258–265. doi:10.1016/j.ijbiomac.2017.08.021.

Pascale, Alessia, Nicoletta Marchesi, Cristina Marelli, Adriana Coppola, Livio Luzi, Stefano Govoni, Andrea Giustina, and Carmine Gazzaruso. "Microbiota and Metabolic Diseases". *Endocrine* 61, no. 3 (2018): 357–371. doi:10.1007/s12020-018-1605-5.

Pei, X., Xiao, Z., Liu, L., Wang, G., Tao, W., Wang, M., Zou, J., and Leng, D. "Effects of Dietary Zinc Oxide Nanoparticles Supplementation on Growth Performance, Zinc Status, Intestinal Morphology, Microflora Population, and Immune Response in Weaned Pigs". *Journal of the Science of Food and Agriculture* 99, no. 3 (2019): 1366–1374. doi: 10.1002/jsfa.9312

Perez-Gonzalez, T., Jimenez-Lopez, C., Neal, A. L., Rull-Perez, F., Rodriguez-Navarro, A., Fernandez-Vivas, A., and Iañez-Pareja, E. (2010). "Magnetite Biomineralization Induced by *Shewanella oneidensis*". *Geochimica et Cosmochimica Acta* 74, no. 3 (2010): 967–979. doi: 10.1016/j.gca.2009.10.035

Petersson, J., O. Schreiber, G. C. Hansson, S. J. Gendler, A. Velcich, J. O. Lundberg, S. Roos, L. Holm, and M. Phillipson. "Importance and Regulation of the Colonic Mucus Barrier in a Mouse Model of Colitis". *American Journal of Physiology-Gastrointestinal and Liver Physiology* 300, no. 2 (2011): G327–G333. doi:10.1152/ ajpgi.00422.2010.

Pfeffer, Christian, Steffen Larsen, Jie Song, Mingdong Dong, Flemming Besenbacher, Rikke Louise Meyer, and Kasper Urup Kjeldsen et al. "Filamentous Bacteria Transport Electrons Over Centimetre Distances". *Nature* 491, no. 7423 (2012): 218–221. doi:10.1038/nature11586.

Piela, Aleksandra, Ewa Zymańczyk-Duda, Małgorzata Brzezińska-Rodak, Maciej Duda, Jakub Grzesiak, Agnieszka Saeid, Małgorzata Mironiuk, and Magdalena Klimek-Ochab. "Biogenic Synthesis of Silica Nanoparticles from Corn Cobs Husks. Dependence of the Productivity on the Method of Raw Material Processing". *Bioorganic Chemistry* 99 (2020): 103773. doi:10.1016/j.bioorg.2020.103773.

Pierotti, Marco A, Claudio Lombardo, and Camillo Rosano. "Nanotechnology: Going Small for a Giant Leap in Cancer Diagnostics and Therapeutics". *Tumori Journal* 94, no. 2 (2008): 191–196. doi:10.1177/030089160809400210.

Pignatello, Rosario, Giuseppe Impallomeni, Sarha Cupri, Giuseppe Puzzo, Claudia Curcio, Maria Rizzo, Salvatore Guglielmino, and Alberto Ballistreri. "Unsaturated Poly(Hydroxyalkanoates) for the Production of Nanoparticles and the Effect of Cross-Linking on Nanoparticle Features". *Materials* 12, no. 6 (2019): 868. doi:10.3390/ma12060868.

Pisciotta, John M., Isabelle Coppens, Abhai K. Tripathi, Peter F. Scholl, Joel Shuman, Sunil Bajad, Vladimir Shulaev, and David J. Sullivan. "The Role of Neutral Lipid Nanospheres in *Plasmodium falciparum* Haem Crystallization". *Biochemical Journal* 402, no. 1 (2007): 197–204. doi:10.1042/bj20060986.

Pisciotta, John M., Yongjin Zou, and Ilia V. Baskakov. "Light-Dependent Electrogenic Activity of Cyanobacteria". *Plos ONE* 5, no. 5 (2010): e10821. doi:10.1371/journal.pone.0010821.

Pokkuluri, P. Raj, Yuri Y. Londer, Norma E. C. Duke, W. Chris Long, and Marianne Schiffer. "Family of Cytochrome C7-Type Proteins from *Geobacter Sulfurreducens*: Structure of one Cytochrome C7 At 1.45 Å Resolution,". *Biochemistry* 43, no. 4 (2004): 849–859. doi:10.1021/bi0301439.

Poole, K. "Multidrug Resistance in Gram-Negative Bacteria". *Current Opinion in Microbiology* 4, no. 5 (2001): 500–508. doi:10.1016/s1369-5274(00)00242-3.

Pradel, N., C.-L. Santini, A. Bernadac, Y. Fukumori, and L.-F. Wu. "Biogenesis of Actin-Like Bacterial Cytoskeletal Filaments Destined for Positioning Prokaryotic Magnetic Organelles". *Proceedings of the National Academy of Sciences* 103, no. 46 (2006): 17485–17489. doi:10.1073/pnas.0603760103.

Prato, Maurizio, Kostas Kostarelos, and Alberto Bianco. "Functionalized Carbon Nanotubes in Drug Design and Discovery". *Accounts of Chemical Research* 41, no. 1 (2008): 60–68. doi:10.1021/ar700089b.

Preisner, M., E. Neverova-Dziopak, and Z. Kowalewski. "Analysis of Eutrophication Potential of Municipal Wastewater". *Water Science and Technology* 81, no. 9 (2020): 1994–2003. doi:10.2166/wst.2020.254.

Qing, Geletu, Reza Ghazfar, Shane T. Jackowski, Faezeh Habibzadeh, Mona Maleka Ashtiani, Chuan-Pin Chen, Milton R. Smith, and Thomas W. Hamann. "Recent Advances and Challenges of Electrocatalytic N_2 Reduction to Ammonia". *Chemical Reviews* 120, no. 12 (2020): 5437–5516. doi:10.1021/acs.chemrev.9b00659.

Qu, Xiaolei, Pedro J.J. Alvarez, and Qilin Li. "Applications of Nanotechnology in Water and Wastewater Treatment". *Water Research* 47, no. 12 (2013): 3931–3946. doi:10.1016/j.watres.2012.09.058.

Radzig, M.A., V.A. Nadtochenko, O.A. Koksharova, J. Kiwi, V.A. Lipasova, and I.A. Khmel. "Antibacterial Effects of Silver Nanoparticles on Gram-Negative Bacteria: Influence on the Growth and Biofilms Formation, Mechanisms of Action". *Colloids and Surfaces B: Biointerfaces* 102 (2013): 300–306. doi:10.1016/j.colsurfb.2012.07.039.

Raghunath, A., and Perumal, E. "Metal Oxide Nanoparticles as Antimicrobial Agents: A Promise for the Future". *International Journal of Antimicrobial Agents* 49, no.2 (2017): 137–152. doi: 10.1016/j.ijantimicag.2016.11.011

Rai, Mahendra, Shital Bonde, Patrycja Golinska, Joanna Trzcińska-Wencel, Aniket Gade, Kamel A. Abd-Elsalam, Sudhir Shende, Swapnil Gaikwad, and Avinash P. Ingle. "Fusarium as a Novel Fungus for the Synthesis of Nanoparticles: Mechanism and Applications". *Journal of Fungi* 7, no. 2 (2021): 139. doi:10.3390/jof7020139.

Ratnaike, R. N. "Acute and Chronic Arsenic Toxicity". *Postgraduate Medical Journal* 79, no. 933 (2003): 391–396. doi:10.1136/pmj.79.933.391.

Rattan, Rohit, Sudeep Shukla, Bharti Sharma, and Mamta Bhat. "A Mini-Review on Lichen-Based Nanoparticles and their Applications as Antimicrobial Agents". *Frontiers In Microbiology* 12 (2021). doi:10.3389/fmicb.2021.633090.

Rautaray, D., Sanyal, A., Adyanthaya, S.D., Ahmad, A., and Sastry, M. "Biological Synthesis of Strontium Carbonate Crystals Using the Fungus *Fusarium oxysporum*". *Langmuir* 20, no. 16 (2004): 6827–6833. doi: 10.1021/la049244d

Ray, Subhasree, and Vipin Chandra Kalia. "Biomedical Applications of Polyhydroxy-alkanoates". *Indian Journal of Microbiology* 57, no. 3 (2017): 261–269. doi:10.1007/s12088-017-0651-7.

Reguera, Gemma, Kelly P. Nevin, Julie S. Nicoll, Sean F. Covalla, Trevor L. Woodard, and Derek R. Lovley. "Biofilm and Nanowire Production Leads to Increased Current in *Geobacter sulfurreducens* Fuel Cells". *Applied And Environmental Microbiology* 72, no. 11 (2006): 7345–7348. doi:10.1128/aem.01444-06.

Reguera, Gemma. "Harnessing the Power of Microbial Nanowires". *Microbial Biotechnology* 11, no. 6 (2018): 979–994. doi:10.1111/1751-7915.13280.

Rehm, Bernd H. A. "Polyester Synthases: Natural Catalysts for Plastics". *Biochemical Journal* 376, no. 1 (2003): 15–33. doi:10.1042/bj20031254.

Rezvani, Ehsan, Aran Rafferty, Cormac McGuinness, and James Kennedy. "Adverse Effects of Nanosilver on Human Health and the Environment". *Acta Biomaterialia* 94 (2019): 145–159. doi:10.1016/j.actbio.2019.05.042.

Rinaldi, Mariagrazia, Antonio Micali, Herbert Marini, Elena Bianca Adamo, Domenico Puzzolo, Antonina Pisani, Vincenzo Trichilo, Domenica Altavilla, Francesco Squadrito, and Letteria Minutoli. "Cadmium, Organ Toxicity and Therapeutic Approaches: A Review on Brain, Kidney and Testis Damage". *Current Medicinal Chemistry* 24, no. 35 (2017). doi:10.2174/0929867324666170801101448.

Roibás-Rozas, Alba, Anuska Mosquera-Corral, and Almudena Hospido. "Environmental Assessment of Complex Wastewater Valorisation by Polyhydroxyalkanoates Production". *Science of the Total Environment* 744 (2020): 140893. doi:10.1016/j.scitotenv.2020.140893.

Sakpirom, Jakkapan, Duangporn Kantachote, Sumana Siripattanakul-Ratpukdi, John McEvoy, and Eakalak Khan. "Simultaneous Bioprecipitation of Cadmium to Cadmium Sulfide Nanoparticles and Nitrogen Fixation by *Rhodopseudomonas palustris* TN110". *Chemosphere* 223 (2019): 455–464. doi:10.1016/j.chemosphere.2019.02.051.

Salesa, Beatriz, Miguel Martí, Belén Frígols, and Ángel Serrano-Aroca. "Carbon Nanofibers IN Pure form and in Calcium Alginate Composites Films: New Cost-Effective Antibacterial Biomaterials Against the Life-Threatening Multidrug-Resistant *Staphylococcus epidermidis*". *Polymers* 11, no. 3 (2019): 453. doi:10.3390/polym11030453.

Santos, Cátia S.C., Barbara Gabriel, Marilys Blanchy, Olivia Menes, Denise García, Miren Blanco, Noemí Arconada, and Victor Neto. "Industrial Applications of Nanoparticles – A Prospective Overview". *Materials Today: Proceedings* 2, no. 1 (2015): 456–465. doi:10.1016/j.matpr.2015.04.056.

Sanz-Luque, Emanuel, Devaki Bhaya, and Arthur R. Grossman. "Polyphosphate: A Multifunctional Metabolite in Cyanobacteria and Algae". *Frontiers in Plant Science* 11 (2020). doi:10.3389/fpls.2020.00938.

Serrano-Aroca, Ángel, Kazuo Takayama, Alberto Tuñón-Molina, Murat Seyran, Sk. Sarif Hassan, Pabitra Pal Choudhury, and Vladimir N. Uversky et al. "Carbon-Based Nanomaterials: Promising Antiviral Agents to Combat COVID-19 in the Microbial-Resistant Era". *ACS Nano* 15, no. 5 (2021): 8069–8086. doi:10.1021/acsnano.1c00629.

Setzer, Teddi J. "Malaria Detection in the Field of Paleopathology: A Meta-Analysis of the State of The Art". *Acta Tropica* 140 (2014): 97–104. doi:10.1016/j.actatropica.2014.08.010.

Shimizu, Rie, Kenta Chou, Izumi Orita, Yutaka Suzuki, Satoshi Nakamura, and Toshiaki Fukui. "Detection of Phase-Dependent Transcriptomic Changes and Rubisco-Mediated CO2 Fixation Into Poly (3-Hydroxybutyrate) Under Heterotrophic Condition in *Ralstonia eutropha* H16 Based on RNA-Seq and Gene Deletion Analyses". *BMC Microbiology* 13, no. 1 (2013): 169. doi:10.1186/1471-2180-13-169.

Sinha, A, Kumar, S., Khare, S.K. "Biochemical Basis of Mercury Remediation and Bioaccumulation by *Enterobacter* Sp. EMB21". *Applied Biochemistry and Biotechnology* 169, no. 1 (2013): 256–267. doi: 10.1007/s12010-012-9970-7.

Singh, Renu, and Tanzeel Ahmad. "Expression and Localization of Gene Encoding Biomineralization in Magnetotactic Bacteria". *International Journal of Life-Sciences Scientific Research* 4, no. 1 (2018). doi:10.21276/ijlssr.2018.4.1.10.

Slocik, Joseph M., Lawrence F. Drummy, Matthew B. Dickerson, Christopher A. Crouse, Jonathan E. Spowart, and Rajesh R. Naik. "Bioinspired High-Performance Energetic Materials Using Heme-Containing Crystals". *Small* 11, no. 29 (2015): 3539–3544. doi:10.1002/smll.201403659.

Slocombe, Stephen P., Tatiana Zúñiga-Burgos, Lili Chu, Nicola J. Wood, Miller Alonso Camargo-Valero, and Alison Baker. "Fixing The Broken Phosphorus Cycle: Wastewater Remediation by Microalgal Polyphosphates". *Frontiers in Plant Science* 11 (2020). doi:10.3389/fpls.2020.00982.

Smith, Val H. "Eutrophication of Freshwater and Coastal Marine Ecosystems a Global Problem". *Environmental Science and Pollution Research* 10, no. 2 (2003): 126–139. doi:10.1065/espr2002.12.142.

Spagnoletti, F. N., Spedalieri, C., Kronberg, F., and Giacometti, R. "Extracellular Biosynthesis of Bactericidal Ag/AgCl Nanoparticles for Crop Protection Using the Fungus *Macrophomina phaseolina*". *Journal of Environmental Management* 231 (2019): 457–466. doi: 10.1016/j.jenvman.2018.10.081

Song, Hui-Ping, Xin-Gang Li, Jin-Sheng Sun, Shi-Min Xu, and Xu Han. "Application of a Magnetotactic Bacterium, *Stenotrophomonas* sp. to the Removal of Au(III) from Contaminated Wastewater With a Magnetic Separator". *Chemosphere* 72, no. 4 (2008): 616–621. doi:10.1016/j.chemosphere.2008.02.064.

Song, Yang, Daniel Johnson, Rui Peng, Dale K. Hensley, Peter V. Bonnesen, Liangbo Liang, and Jingsong Huang et al. "A Physical Catalyst for the Electrolysis of Nitrogen to Ammonia". *Science Advances* 4, no. 4 (2018): e1700336. doi:10.1126/sciadv.1700336.

Stojsavljević, Aleksandar, Slavica Borković-Mitić, Ljiljana Vujotić, Danica Grujičić, Marija Gavrović-Jankulović, and Dragan Manojlović. "The Human Biomonitoring Study in Serbia: Background Levels for Arsenic, Cadmium, Lead, Thorium and Uranium in the Whole Blood of Adult Serbian Population". *Ecotoxicology and Environmental Safety* 169 (2019): 402–409. doi:10.1016/j.ecoenv.2018.11.043.

Stow, Craig A., Song S. Qian, and J. Kevin Craig. "Declining Threshold for Hypoxia in the Gulf of Mexico". *Environmental Science & Technology* 39, no. 3 (2004): 716–723. doi:10.1021/es049412o.

Strong, Peter, Bronwyn Laycock, Syarifah Mahamud, Paul Jensen, Paul Lant, Gene Tyson, and Steven Pratt. "The Opportunity for High-Performance Biomaterials from Methane". *Microorganisms* 4, no. 1 (2016): 11. doi:10.3390/microorganisms4010011.

Subramaniyan, S. A., Sheet, S., Vinothkannan, M., Yoo, D. J., Lee, Y. S., Belal, S. A., and Shim, K. S. "One-Pot Facile Synthesis of Pt Nanoparticles Using Cultural Filtrate of Microgravity Simulated Grown *P. chrysogenum* and Their Activity on Bacteria and Cancer Cells". *Journal of Nanoscience and Nanotechnology* 18, no. 5 (2018): 3110–3125. doi: 10.1166/jnn.2018.14661

Subramaniyam, Vidhyasri, Suresh Ramraj Subashchandrabose, Vimalkumar Ganeshkumar, Palanisami Thavamani, Zuliang Chen, Ravi Naidu, and Mallavarapu Megharaj. "Cultivation of *Chlorella* on Brewery Wastewater and Nano-Particle Biosynthesis by its Biomass". *Bioresource Technology* 211 (2016): 698–703. doi:10.1016/j.biortech.2016.03.154.

Sun, Michael, and Anirban Sen Gupta. "Vascular Nanomedicine: Current Status, Opportunities, and Challenges". *Seminars in Thrombosis and Hemostasis* 46, no. 05 (2019): 524–544. doi:10.1055/s-0039-1692395.

Swanson, P. A., A. Kumar, S. Samarin, M. Vijay-Kumar, K. Kundu, N. Murthy, J. Hansen, A. Nusrat, and A. S. Neish. "Enteric Commensal Bacteria Potentiate Epithelial Restitution Via Reactive Oxygen Species-Mediated Inactivation of Focal Adhesion Kinase Phosphatases". *Proceedings of the National Academy of Sciences* 108, no. 21 (2011): 8803–8808. doi:10.1073/pnas.1010042108.

Sweeney, R.Y., Mao, C., Gao, X., Burt, J.L., Belcher, A.M., Georgiou, G., and Iverson, B.L. "Bacterial Biosynthesis of Cadmium Sulfide Nanocrystals". *Chemistry & Biology* 11, no. 11 (2004): 1553–1559. doi:10.1016/j.chembiol.2004.08.022

Tajer Mohammad-Ghazvini, Parisa, Rouha Kasra-Kermanshahi, Ahmad Nozad-Golikand, Majid Sadeghizadeh, Saeid Ghorbanzadeh-Mashkani, and Reza Dabbagh. "Cobalt Separation by *Alphaproteobacterium* MTB-KTN90: Magnetotactic Bacteria in Bioremediation". *Bioprocess and Biosystems Engineering* 39, no. 12 (2016): 1899–1911. doi:10.1007/s00449-016-1664-z.

Tan, Dan, Ying Wang, Yi Tong, and Guo-Qiang Chen. "Grand Challenges for Industrializing Polyhydroxyalkanoates (Phas)". *Trends in Biotechnology* 39, no. 9 (2021): 953–963. doi:10.1016/j.tibtech.2020.11.010.

Tan, Yuanqing, Rong Yao, Rui Wang, Dan Wang, Gejiao Wang, and Shixue Zheng. "Reduction of Selenite to Se(0) Nanoparticles by Filamentous Bacterium *Streptomyces* Sp. ES2-5 Isolated from a Selenium Mining Soil". *Microbial Cell Factories* 15, no. 1 (2016). doi:10.1186/s12934-016-0554-z.

Tang, Cong-Cong, Yu Tian, Heng Liang, Wei Zuo, Zhen-Wei Wang, Jun Zhang, and Zhang-Wei He. "Enhanced Nitrogen and Phosphorus Removal from Domestic Wastewater Via Algae-Assisted Sequencing Batch Biofilm Reactor". *Bioresource Technology* 250 (2018): 185–190. doi:10.1016/j.biortech.2017.11.028.

Taoka, Azuma, Chihiro Umeyama, and Yoshihiro Fukumori. "Identification of Iron Transporters Expressed in the Magnetotactic Bacterium *Magnetospirillum magnetotacticum*". *Current Microbiology* 58, no. 2 (2008): 177–181. doi:10.1007/s00284-008-9305-7.

Thursby, Elizabeth, and Nathalie Juge. "Introduction to the Human Gut Microbiota". *Biochemical Journal* 474, no. 11 (2017): 1823–1836. doi:10.1042/bcj20160510.

Tolson, A. H., and Wang, H. "Regulation of Drug-Metabolizing Enzymes by Xenobiotic Receptors: PXR and CAR". *Advanced Drug Delivery Reviews* 62, no. 13 (2010): 1238–1249. doi: 10.1016/j.addr.2010.08.006

Tombuloglu, Huseyin, Yassine Slimani, Guzin Tombuloglu, Munirah Almessiere, and Abdulhadi Baykal. "Uptake and Translocation of Magnetite (Fe3o4) Nanoparticles and its Impact on Photosynthetic Genes in Barley (*Hordeum Vulgare* L.)". *Chemosphere* 226 (2019): 110–122. doi:10.1016/j.chemosphere.2019.03.075.

Tran, Thao. "Assessment of Iron Oxide Nanoparticle Ecotoxicity on Regeneration and Homeostasis in the Replacement Model System *Schmidtea mediterranea*". *ALTEX* 2019. doi:10.14573/altex.1902061.

Urban, James E. "Adverse Effects of Microgravity on the Magnetotectic Bacterium Magnetospirillum magnetotacticum". *Acta Astronautica* 47, no. 10 (2000): 775–780. doi:10.1016/s0094-5765(00)00120-x.

Urrutia-Ortega, Ismael M., Luis G. Garduño-Balderas, Norma L. Delgado-Buenrostro, Verónica Freyre-Fonseca, José O. Flores-Flores, Arturo González-Robles, and José Pedraza-Chaverri et al. "Food-Grade Titanium Dioxide Exposure Exacerbates Tumor Formation in Colitis Associated Cancer Model". *Food And Chemical Toxicology* 93 (2016): 20–31. doi:10.1016/j.fct.2016.04.014.

Varman, Arul M, Yi Yu, Le You, and Yinjie J Tang. "Photoautotrophic Production of D-Lactic Acid in an Engineered Cyanobacterium". *Microbial Cell Factories* 12, no. 1 (2013): 117. doi:10.1186/1475-2859-12-117.

Vivekanandhan, Singaravelu. Danny Tang, Manjusri Misra, and Amar Kumar Mohanty "Biological Synthesis Of Silver Nanoparticles Using *Glycine max* (Soybean) Leaf Extract: An Investigation on Different Soybean Varieties". *Journal of Nanoscience and Nanotechnology* 9, no. 12 (2009). doi:10.1166/jnn.2009.2201.

Vijayanandan, A.S. and Balakrishnan, R.M. "Biosynthesis of Cobalt Oxide Nanoparticles Using Endophytic Fungus *Aspergillus nidulans*". *Journal of Environmental Management* 218 (2018): 442–450. doi: 10.1016/j.jenvman.2018.04.032

Vorvolakos, Th., S. Arseniou, and M. Samakouri. "There is No Safe Threshold for Lead Exposure: A Literature Review". *Psychiatriki* 27, no. 3 (2016): 204–214. doi:10.22365/jpsych.2016.273.204.

Wainwright, Katlyn E., Melissa A. Miller, Bradd C. Barr, Ian A. Gardner, Ann C. Melli, Tim Essert, Andrea E. Packham, Tin Truong, Manuel Lagunas-Solar, and Patricia A. Conrad. "Chemical Inactivation of Toxoplasma Gondii Oosysts in Water". *Journal of Parasitology* 93, no. 4 (2007): 925–931. doi:10.1645/ge-1063r.1.

Wan, Alex H. L., Robert J. Wilkes, Svenja Heesch, Ricardo Bermejo, Mark P. Johnson, and Liam Morrison. "Assessment and Characterisation of Ireland's Green Tides (*Ulva* Species)". *PLOS ONE* 12, no. 1 (2017): e0169049. doi:10.1371/journal.pone.0169049.

Wang, Dali, Zhifen Lin, Ting Wang, Zhifeng Yao, Mengnan Qin, Shourong Zheng, and Wei Lu. "Where Does the Toxicity of Metal Oxide Nanoparticles Come from: The Nanoparticles, the Ions, Or a Combination of Both?" *Journal of Hazardous Materials* 308 (2016): 328–334. doi:10.1016/j.jhazmat.2016.01.066.

Wang, Fengbin, Yangqi Gu, J. Patrick O'Brien, Sophia M. Yi, Sibel Ebru Yalcin, Vishok Srikanth, and Cong Shen et al. "Structure of Microbial Nanowires Reveals Stacked Hemes That Transport Electrons Over Micrometers". *Cell* 177, no. 2 (2019): 361–369. e10. doi:10.1016/j.cell.2019.03.029.

Williams, Katherine, Jessica Milner, Mary D. Boudreau, Kuppan Gokulan, Carl E. Cerniglia, and Sangeeta Khare. "Effects of Subchronic Exposure of Silver Nanoparticles on Intestinal Microbiota and Gut-Associated Immune Responses in the Ileum of Sprague-Dawley Rats". *Nanotoxicology* 9, no. 3 (2014): 279–289. doi:10.3109/17435390.2014. 921346.

Xia, Tian, Wenqing Lai, Miaomiao Han, Meng Han, Xi Ma, and Liying Zhang. "Dietary Zno Nanoparticles Alters Intestinal Microbiota and Inflammation Response in Weaned Piglets". *Oncotarget* 8, no. 39 (2017): 64878–64891. doi:10.18632/oncotarget.17612.

Xie, Jianping, Jim Yang Lee, Daniel I. C. Wang, and Yen Peng Ting. "Silver Nanoplates: From Biological to Biomimetic Synthesis". *ACS Nano* 1, no. 5 (2007): 429–439. doi:10.1021/nn7000883.

Yan, Lei, Huiyun Da, Shuang Zhang, Viviana Morillo López, and Weidong Wang. "Bacterial Magnetosome and its Potential Application". *Microbiological Research* 203 (2017): 19–28. doi:10.1016/j.micres.2017.06.005.

Yang, Ying, Chunli Li, Shujun Ni, Haifeng Zhang, and Caihong Dong. "Ultrastructure and Development of Acanthocytes, Specialized Cells in *Stropharia Rugosoannulata*, Revealed by Scanning Electron Microscopy (SEM) and Cryo-SEM". *Mycologia* 113, no. 1 (2020): 65–77. doi:10.1080/00275514.2020.1823184.

Yin, Wen, Yiting Wang, Lu Liu, and Jin He. "Biofilms: The Microbial "Protective Clothing" In Extreme Environments". *International Journal of Molecular Sciences* 20, no. 14 (2019): 3423. doi:10.3390/ijms20143423.

Yu, Xiaoniu, and Jianguo Jiang. "Phosphate Microbial Mineralization Removes Nickel Ions from Electroplating Wastewater". *Journal of Environmental Management* 245 (2019): 447–453. doi:10.1016/j.jenvman.2019.05.091.

Yusuf, Ahmed, Samar Al Jitan, Corrado Garlisi, and Giovanni Palmisano. "A Review of Recent and Emerging Antimicrobial Nanomaterials in Wastewater Treatment Applications". *Chemosphere* 278 (2021): 130440. doi:10.1016/j.chemosphere.2021.130440.

Zhang, Z., Ke, M., Qu, Q., Peijnenburg, W. J. G. M., Lu, T., Zhang, Q., Yizhi Ye, Y., Xu, P., Du, B., Sun, L., and Qian, H. "Impact of Copper Nanoparticles and Ionic Copper Exposure on Wheat (*Triticum aestivum* L.) Root Morphology and Antioxidant Response". *Environmental Pollution* 239 (2018): 689–697. doi: 10.1016/j.envpol.2018.04.066

Zhang, Li, and Thien-Fah Mah. "Involvement of a Novel Efflux System in Biofilm-Specific Resistance to Antibiotics". *Journal of Bacteriology* 190, no. 13 (2008): 4447–4452. doi:10.1128/jb.01655-07.

Zheng, Ping, Bei Pu, Bing Yu, Jun He, Jie Yu, Xiangbing Mao, and Yuheng Luo et al. "The Differences Between Copper Sulfate and Tribasic Copper Chloride on Growth Performance, Redox Status, Deposition in Tissues of Pigs, and Excretion in Feces". *Asian-Australasian Journal of Animal Sciences* 31, no. 6 (2018): 873–880. doi:10.5713/ajas.17.0516.

Zielińska, Aleksandra, Filipa Carreiró, Ana M. Oliveira, Andreia Neves, Bárbara Pires, D. Nagasamy Venkatesh, and Alessandra Durazzo et al. "Polymeric Nanoparticles: Production, Characterization, Toxicology and Ecotoxicology". *Molecules* 25, no. 16 (2020): 3731. doi:10.3390/molecules25163731.

Zou, Huibin, Mengxun Shi, Tongtong Zhang, Lei Li, Liangzhi Li, and Mo Xian. "Natural and Engineered Polyhydroxyalkanoate (PHA) Synthase: Key Enzyme in Biopolyester Production". *Applied Microbiology and Biotechnology* 101, no. 20 (2017): 7417–7426. doi:10.1007/s00253-017-8485-0.

Zou, Long, Fei Zhu, Zhong-er Long, and Yunhong Huang. "Bacterial Extracellular Electron Transfer: A Powerful Route to the Green Biosynthesis of Inorganic Nanomaterials for Multifunctional Applications". *Journal of Nanobiotechnology* 19, no. 1 (2021). doi:10.1186/s12951-021-00868-7.

8 Bioinformatics Integration to Biomass Waste Biodegradation and Valorization

Chhavi Thakur and Jata Shankar

Jaypee University of Information Technology, Solan, India

CONTENTS

DOI: 10.1201/9781003187721-8

8.1   INTRODUCTION

Food waste and by-products are associated throughout the food supply chain for example, agricultural production and preliminary preservation, industrial process, dispersion, commerce, and consumption (domestic consumption, restaurants, catering services). Materials produced at the moment of consumption (meal preparation, food leftovers, wasted food) are classified as post-user waste, whereas wastes generated prior, throughout the supply chain, are classified as pre-consumer waste. Wastage of food does not only mean reduced amounts of available food but also means loss of embedded energy in the form of water and fertilizer. In a nutshell, reducing and valorizing food waste and by-products has considerable potential to improve total food system sustainability by addressing all three sustainability dimensions: environment, society, and economy (Gollagher, Campbell, and Bremner 2017).

Wastes and by-products generated at each step of the food supply chain can have a wide range of characteristics, both in terms of quantity and in composition of the material streams. Diverse management techniques and appropriate valorization pathways are required (Kumar et al. 2020). First-generation valorization techniques based on the utilization of whole material streams are the most suitable and frequently utilized in this regard, while others are still being researched. Despite the rising volumes of material, valorization of ensuing wastes and by-products is economically and technically less feasible because of hygiene concerns and health dangers associated with meat and fish processing (Mirabella, Castellani, and Sala 2014). Second-generation valorization aimed at supplying high-value compounds with minimal contamination risk would be beneficial for plant-derived food waste (Pfaltzgraff et al. 2013). Plant-derived agricultural waste products are divided into different categories (Ajila et al. 2012): crop wastes and residues; plant wastes and residues; plant wastes from fruit and vegetable processing industry by-products and legume milling industry; grain and oil business by-products; sugar,starch, and confectionery industry by-products; and distilleries and breweries by-products. In addition to solid effluents, food processing generates a large amount of wastewater with high biological content. F usage in a range of sectors, millions of dangerous compounds have been manufactured (Ellis and Wackett 2012). These compounds are frequently released into the environment as a result of human activities, contaminating soil and water (Arora and Shi 2010). However, many chemicals persist in the environment, posing serious health risks to living species; as a result, it is critical to eliminate these substances from the ecosystem (Arora and Shi 2010). The breakdown of chemicals or xenobiotic substances by bacteria and plants is called biodegradation (Andrady 1998). Toxic substances are degraded by certain bacteria via mineralization or co-metabolism (Arora, Sasikala, and Ramana Ch 2012). Microbes break down harmful substances by using them as carbon and energy sources during the mineralization mechanism, although co-metabolism results in toxic molecules being biotransformed into less toxic compounds (Arora, Srivastava, and Singh 2014). Microbial remediation (Arora and Bae 2014) is a new method for removing hazardous substances first from the environment (Agrawal and Verma 2021).

Several more microbes can utilize toxic chemicals as their exclusive carbon and energy sources due to the presence of enzymes such as monooxygenases, dioxygenases, reductases, deaminases, and dehalogenases. The genes encoding various enzymes have been identified in a broad variety of microorganisms as well as cloned into bacteria that boost bioremediation effectiveness (Arora, Srivastava, and Singh 2010; Chaturvedi et al. 2021). Bioinformatics, which is being integrated into all branches of biological sciences, provides a platform for researchers to create useful computational tools for human and environmental sustainability (Debes and Urrutia 2004; Katara 2013). Bioinformatics has been merged with biodegradation throughout the last few decades, and various bioinformatics tools beneficial in the field of biodegradation are available. Various databases (Ellis, Roe, and Wackett 2006), toxicity prediction tools (Greene 2002), prediction of biodegradation pathway (Gao, Ellis, and Wackett 2011), and next-generation sequencing (Chen et al. 2013; McClymont and Soyer 2013) are examples of these tools.

It is estimated that each year around 1.3 billion metric tons of solid trash are accumulated around the world. By 2025, this number is predicted to rise to 2.2 billion metric tons, with developing countries accounting for nearly all of the growth (Paritosh et al. 2017). Waste management is increasingly being considered as a strategic method to ensure compliance with the recently adopted legislation addressing the disposal of biodegradable leftovers (Gollagher, Campbell, and Bremner 2017). Composting has appeared as a viable option for addressing a wide range of organic waste. Composting is defined as the degradation of biodegradable matter by a complex microbial ecosystem in the existence of gram-positive and gram-negative bacteria and fungi in anaerobic environment (Hansgate et al. 2005; Sykes, Jones, and Wildsmith 2007). Bioaerosols generated during many essential compost processes pose a risk (Persoons et al. 2010). Greenhouse gases are likewise released, and thus the ultimate compost product, depending on its quality, can and will be utilized as fertilizer (Lim, Lee, and Wu 2016). Biomethanization is a green waste valorization technique in which organic waste is biodegraded in anaerobic conditions by microbial colonies. This method of anaerobic digestion produces biogas, which is composed of around 65% methane and 35% carbon dioxide (Mata-Alvarez 2005). Biogas may be utilized as an energy source in the same manner that natural gases can since it contains methane (Amon et al. 2007; Dai et al. 2017). The biogas produced via biomethanization, for example, can be used to heat buildings, power generators, or fuel automobiles. Biomethanization may also examine a variety of biological wastes, such as urban green waste, industry food waste, solid wastes, and livestock manure from cattleand pigs (Bouallagui et al. 2005). When compared to composting, waste management employing anaerobic digestion is observed to be less dangerous to the ecosystem, and a cost-effective approach. Thus, analysis of the relationship between bioinformatics tools and their application in biodegradation becomes important.

8.2 FIRST-GENERATION VALORISATION

8.2.1 Overview of Primary Methods

First-generation valorization techniques utilize entire organic matter fluxes when they eventuate, with just minimal pre-treatment as needed. Food waste as animal feed has been used for millennia in traditional agriculture; the key problems are generic appropriateness of specific substrates, seasonal accessibility and fluctuation in nutritional contents, fast spoiling, and economic impact, particularly when pre-treatment is used. Soil conditioner or fertilizer can be used either by proliferating untreated food waste (e.g., citrus debris, tomato waste, or olive husks) or even after decomposition or anaerobic digestion (Van Dyk et al. 2013). Food wastes can also be used for energy production, usually through anaerobic digestion (AD) with biogas synthesis (particularly for wet feedstocks) or thermochemical transformation. In mushroom culture, lignin-rich materials (e.g., brewery dregs, coffee manufacturing by-products, maize stalk husks, tomato skins) are often used (Liguori, Amore, and Faraco 2013). For wastewater treatment, adsorbents can be used (Kosseva 2011; Agrawal and Verma 2022a). Cascaded procedures, which successively integrate multiple alternatives to enhance overall advantages, are frequently possible with minimal incremental logistical work. The utilization of husks (mill by-products) as debris in animal barns, accompanied by the energetic valorization of the resultant manure via anaerobic digestion with biogas generation, is one example (Kusch-Brandt et al. 2011).

8.2.2 Biogas Production

In recent decades, the adsorption of garbage has become a state-of-the-art technique. Nonetheless, successful implementation necessitates a detailed understanding of the method and associated technologies, as well as an enhanced focus while the facility is in operation. The procedure is vulnerable, both in high and low quantities of volatile fatty acids (VFA) Despite ammonia remaining an important problem (Rajagopal, Massé, and Singh 2013), it is now well established that adding trace elements to a food waste digestion mechanism that generates VFA buildup helps to stabilize the process (Banks et al. 2011). Waste food products and organic components with a high carbon-to-nitrogen ratio can be co-digested in an efficient manner.

8.2.3 Biohydrogen Production

In light of the upcoming "hydrogen economy," which has been widely discussed in recent times, one proposed strategy to solve the problem of energy supply, while considering climate change concerns, includes the generation of hydrogen and its application as clean energy. Some difficulties, such as hydrogen preservation methods with potential usage as a fuel, remain unresolved in this approach nowadays. Moreover, hydrogen's properties necessitate specialized circumstances, including high pressure, the use of specialized materials to decrease diffusion and leakage, and rigorous security measures. Furthermore, the low volumetric power density of liquid hydrogen (about one-third that of compressed natural gas, or CNG) and the need for infrastructure upgrades impede the implementation of this strategy. Hydrogen Fuel Injection (HFI) is currently a viable solution. The objective of HFI is to combine a gaseous fuel with hydrogen to produce a mix with good combustion properties. As hydrogen has a faster flame speed and a lesser ignition energy requirement than other traditional fuels, a tiny quantity of hydrogen added to the fuel improves its utilization and minimizes the number of pollutants (mostly NOx) released into the environment, while preserving CNG's energy efficiency. The use of this combination does not necessitate any adjustments to CNG engines, the storage system, or the infrastructure. HCI patented this combination, and the gasoline became known as hythane. Wide research on this fuel has been conducted (for example, by ENEA in Italy), and the findings have been confirmed. In actuality, both methane and hydrogen are generated utilizing nonrenewable energy sources, through the revamping of fossil fuels and the creation of syngas, a gaseous combination of CO and H_2, which is used as a step in the production of synthetic natural gas (SNG).

8.2.3.1 Basic Concepts of Biohydrogen Generation by Biohythane and Dark Fermentation

Hydrogen is generated biologically through the action of hydrogenase and nitrogenase enzyme complexes, which become involved in three primary biochemical functions: water biophotolysis, photo-fermentation, and dark fermentation (Meher Kotay and Das 2008). Because it may be integrated with anaerobic digestion to generate biohydrogen and biogas, dark fermentation is currently the approach that is most promising for biohythane (Meher Kotay and Das 2008). Some anaerobic bacteria, such as *Thermo Anaerobacterium* spp., *Clostridium* spp., *Bacillus*, and *Enterobacter* produce biohydrogen via dark fermentation on substrates such as carbohydrate-rich substances (Reith, Wijffels, and Barten 2003). One of the most challenging aspects of perpetual biohydrogen generation via dark fermentation (DF) is maintaining an appropriate pH value; in addition, the buildup of organic acids (generated during fermentative metabolism) can lower the pH. Numerous techniques for controlling pH have been proposed, including chemical additions or addition of protein-rich substrates (Valdez-Vazquez 2009). Recently, some authors have suggested an intriguing pH control method that involves connecting the dark fermentation and anaerobic digestion processes in series and employing reuse of effluent from anaerobic digestion, which is extremely useful in buffering agents to regulate the pH during fark fermentation. Furthermore, as previously stated, biohythane may be produced from the overall system. Different substrates might be utilized to produce biohythane, however, the breakdown of carbohydrates to H_2 and organic acids provides a majority of H_2 per mole of the substrate from a thermodynamic standpoint (Reith, Wijffels, and Barten 2003). As a result, biowaste is a highly intriguing substrate for biohydrogen generation through DF since the majority of organic waste is made up of different materials by carbohydrate (simple sugars, starch and cellulose).

8.2.3.2 Biohydrogen and Biohythane Synthesis from Biowaste: Techniques and Applications

Although thermophilic production of biohydrogen is frequently utilized at pilot or laboratory scale, biohydrogen generation by dark fermentation typically occurs both at thermophilic (55°C) and mesophilic (35–37°C) conditions (Lodi). Biowastes are plentiful biological substances collected from municipal solid waste or the food sector, and they are a resource with a high energy content that could be utilized. Parameters like temperature and substrates, as well as other process parameters, can examine the biowaste exploitation mechanism (Hawkes et al. 2007).

8.2.3.2.1 *Method of Hydraulic Retention Time (HRT)*

H_2 production rises as HRT drops; practically, employing a short HRT over two to five days in a continuous stirred tank reactor (CSTR) system provides methanogens by washing, thereby allowing fermentative hydrogen bacteria to proliferate (Hawkes et al. 2007). To avoid wash-out, the HRT must be stronger and larger than the specified growth rate of the hydrogen-producing bacteria (Kongjan and Angelidaki 2010).

8.2.3.2.2 *Method of Organic Loading Rate (OLR)*

The OLRs used vary from a range of 14 to 38.5 kg VS/h (m3rd). Specifically, the more the organic load supplied, the lower the effective H_2 production. This may be owing to a buildup of hydrolyzed compounds that are toxic to the bacteria's cells. High levels of organic acid, as well as pH, may well have a detrimental influence on H_2 production. Undissociate acids act as uncouplers, enabling protons to move inside the cell membrane in large enough numbers, which alters the pH gradient across the membrane. As an outcome, a cell detoxification mechanism has been connected to the transition to solventogenesis to avoid antagonistic effects (Valdez-Vazquez 2009).

8.2.3.2.3 *Hydrogenotrophic Activity Inhibition*

In most research aimed at maximizing the fermenting phase, pH control (Han et al. 2005) and substrate formation, including thermal processing (100 ° C, 10 min), have been used. In this framework, an unconventional method has been developed in recent decades as a substitute for using chemical compounds for external pH control throughout the DF phase (Cecchif et al. 2005), and to provide appropriate nutrients, and use of diluted utilizedfeedstock (Frac and Ziemiński 2012).

8.2.3.2.4 Biohydrogen and Biomethane Yields

Experiments were conducted with reactors larger than 200 liters, with average outputs of hydrogen and methane ranging from 51.0 to 66.7 LH2/kL. $LH_2/kgVS_{feed}$ and from 350 to 483 $LCH_4/kgVS_{feed}$ (Cavinato et al. 2011; Giuliano et al. 2014). Using the best yields, 16 m3 of hydrogen and 110 m3 of methane per tonne of biowaste may be obtained, resulting in about 166 m3 biohythane per tonne of biowaste (0.8 m3/$kgVS_{added}$). With a specific energy need of 20–40 kWh per tonne of trash processed, the production of energy that may be produced is 404 kWh per tonne of waste.

8.2.3.3 The Two-phase Methodology of Automatic Control and Its Successful Application Using Research Developments and Potential Barriers

Full-scale biohydrogen and biohythane process deployment are presently being assessed, both economically and for long-term viability. In reality, prolonged recirculation can cause a rise in alkalinity in the system, which favors the development of hydrogenotrophic bacteria, resulting in poor hydrogen yields and ammonia buildup. Application of variable recirculation flow enables the mechanism control, and prevents ammonia inhibition in the process. The progression of automatic process control and its use of real-time monitoring helps the verification of process conservation using a control logic focused on set-point, automatically maintaining the optimum circumstance for microorganisms and, as a consequence, maximizing biohydrogen and biomethane yields.

8.3 SECOND-GENERATION VALORIZATION

8.3.1 OVERVIEW OF PRIMARY METHODS

Second-generation valorization methods focus on the use of particular compounds with high added value to cover a wide range of applications and provide a diverse range of products to the market. Sugars, starch, lipids, phenols, phytochemicals, amino acids, pectin, and other biomass components are used in applications to transform or improve these compounds to create specific target products. There are many different second-generation valorization paths to choose from, and their applicability will vary depending on the scenario and market. A detailed overview is provided by Kusch-Brandt et al. (2016). Despite the separation of the material sources, the total valorization of feedstock remains a distinct priority. The goal is to improve resource utilization efficiency, primarily through integrated manufacturing of both specialty and commodity items to increase market adaptability and cost-effectiveness (Koutinas et al. 2014). For effluents and by-products from specialized domains such as the dairy and olive oil industries, biorefinery ideas with component extraction and conversion, as well as their diversion into appropriate high-value production systems, are usually achieved (Kosseva 2011). According to research findings, a variety of different sectors, such as vegetable processing by-products, and citrus peels, or other fruit (Kosseva 2011; Pfaltzgraff et al. 2013), have the most potential for elevated applications like biosolvents, flavors, fragrance elements, enzymes, pharmaceutical products, and diverse organic acids.

8.3.2   KEY CHALLENGES

In most cases, recovering particular components necessitates extensive material stream processing. Purification, enrichment, and conversion techniques are all included in this category. Each stage will increase prices and resource consumption, with untapped potential posing an environmental risk (Bhardwaj et al. 2020). This emphasizes the importance of including extra valorization processes in advanced valorization methods to fully utilize the value of resources, for example, in simultaneous or later manufacture of animal feed, energy conversion, and compost generation from no longer usable process wastes. Implementing industrial symbiosis is an essential component of such biorefinery techniques' effectiveness (Kumar and Verma 2022). Due to material streams, seasonal impacts, and general processing, changes in composition (including shifting pH levels) occur. Furthermore, because their major ingredients are organic, and they have a significant water content in aggregate, they are prone to bacterial contamination and deterioration, making logistics a critical issue in defining prospective valorization opportunities. Key challenges include material availability, logistics, and flexible units of operation (Koutinas et al. 2014). In general, highly concentrated quantities are advantageous. Biomass transportation and any pre-treatment, like water content reduction, incur considerable expenses and need the use of extra starting resources such as fuel or power. Overall, specialized biorefinery techniques or incorporating by-product use into existing industrial facilities, can thus be expected to be particularly suited in second-generation valorization techniques.

8.3.3   APPLICATIONS OF FUNCTIONAL FOODS

The modern consumer prefers functional foods that provide health advantages in addition to regular nutritive value (Siró et al. 2008), prompting the food sector to produce functional goods to suit rising customer expectations (Bigliardi and Galati 2013). Functional food components with bioactive characteristics that can benefit public wellness are presently primarily obtained from the primary human diet. At the very same time, by-products created from industrial food processing include important primary or secondary metabolites that may be extracted and reintroduced into the food supply chain, possibly contributing to the emergence of niche markets for novel components in the agri-food economy. This method promotes public health while also assuring effective resource usage, decreased impact on the environment, and minimized waste disposal costs. One utilization of minimum protein-rich food by-products appears to be active peptide-based functional food research (Udenigwe2014), suggesting the enormous potential of agriculture-waste valorization in the creation of high-value goods (Udenigwe 2014). Fruit and vegetable usage, in particular, has also been linked to improved health consequences, such as a lower risk of major chronic diseases, particularly cardiovascular disease (Dauchet et al. 2006). As a result, by-products from these meals may include useful chemicals. Food by-products containing functional components can be valued for nutritional benefits or as sources of bioactive chemicals, which are frequently enriched in the by-products and can be isolated for use in nutraceuticals and pharmaceuticals.

Soybean meal, vegetable and fruit peels, wheat straw, seed husks, and fish processing by-products are currently food by-products that could be valorized for beneficial products. Proteins, polyphenols, carotenoids, polyunsaturated fatty acids (PUFAs), and polysaccharides are among the useful components found in food by-products. Macromolecules can be utilized as nutrients in food, and the secondary metabolites could be used as multifunctional agents in disease states to modulate aberrant biological processes. The above-mentioned food processing by-products are currently mainly underutilized. More effective isolation and recovery will be required for better use. Several technologies are known and are being researched further. For research and development and nutritional reasons, by-product proteins can be extracted via solubilization, and isoelectric precipitation, including membrane technologies (Smithers 2008). Secondary metabolites are extracted using a variety of processes, including solvent, supercritical fluid extraction, and enzyme treatment, as well as membrane processing (Gómez-Guillén et al. 2011).Hydraulic pressing has historically been used to extract oils, although enzymatic and solvent extraction is becoming more common.

Valorization of food through functional food applications is unquestionably an important consideration for a sustainable food system, determining the nutritional and therapeutic advantages of components. Before commercializing by-product-derived functional meals, therefore, there are several possible difficulties to address. Obtaining economies of scale, resolving regulatory limitations, and gaining acceptability for the goods are all major hurdles. The following concerns, among others, must be considered and will necessitate more research. First, food by-product valorization's economic feasibility for the functional food industry. This promotes the development of relatively low technologies capable of efficiently extracting the essential. Second, quantities and quality of functional components, and a trade-off analysis. This involves determining which fractions of the food waste are of interest, determining how much can be recovered, evaluating any additional by-products produced by valorization, and doing cost-benefit assessments (Kusch-Brandt et al. 2014).

8.4 BIOINFORMATICS SYSTEMS: TOOLS AND DATABASES

A growing number of databases have been created in recent years to give information on the biodegradation of chemicals. All such data can be divided into two types of databases: chemical and biodegradable. Biodegradative databases record information about chemical biodegradation, such as xenobiotic-degrading microorganisms, toxic chemical metabolic pathways, enzymes, and genes engaged in biodegradation.

8.4.1 Databases

A growing array of databases has been created in recent years to give data on chemicals and their microbial degradation. These databases can be divided into two types: chemical databases and biodegradable databases. Chemical databases can be used to classify chemicals, assess their risk or characterize their environmental features, toxicity, and distribution (see Table 8.1).

TABLE 8.1
Inventory of Chemical Databases

Database	Description	Reference/Weblink
Databases describing features of chemicals, as well as their toxicity, management, and distribution diseases.		
GENE-TOX: Genetic Toxicology Data Bank	For more than 3,000 chemicals, the National Library of Medicine at the National Institute of Health (NIH) aims at providing genetic toxicity testing results based on the expert peer assessment of accessible scientific literature.	https://toxnet.nlm.nih.gov/newtoxnet/genetox.htm
Toxics Release Inventory (TRI)	Data of specific harmful toxic chemicals and how they have been valorized as waste.	http://toxnet.nlm.nih.gov/cgi-bin/sis/htmlgen?TRI
Integrated Risk Information System (IRIS)	Data on nearly 500 compounds, including hazard identification and dose-response analyses.	http://toxnet.nlm.nih.gov/cgi-bin/sis/htmlgen?IRIS
Toxicology Data Network (TOXNET)	A computerized database covering toxicology and associated disciplines such as harmful chemicals, environmental health, and toxic emissions.	(Wexler 2001)
EnviChem	A scientific database with data on 3,000 compounds and test results. Information on the effects and interaction of chemicals with the environment can be found.	http://www.echemportal.org/echemportal/participant/participantinfo.action?participantID=5&pageID=2
Acutoxbase	This database can be used for *in vitro* acute toxicity research.	(Kinsner-Ovaskainen et al. 2009)
International Uniform Chemical Information Database (IUCLID)	2,600 chemical compounds are assessed for their physicochemical features, environmental origin, toxicity, and ecotoxicity.	http://iuclid.eu/
Comparative ToxicogenomicsDatabase (CTD)	This database delves into the genetic underpinnings of how toxicants affect human health.	http://toxnet.nlm.nih.gov/cgi-bin/sis/htmlgen?CTD
Chemical identification, structure, and classification databases		
ECHA Classification & Labeling Inventory	Manufacturers and importers have reported and recorded information on the categorization and labeling of chemicals.	(Schoning 2011)
ChemIDplus	There are 420,000 chemical records in the database, with 3D structure models for over 335,000 substances and 645,000 variants.	http://toxnet.nlm.nih.gov/cgi-bin/sis/htmlgen?CHEM

8.4.1.1 The University of Minnesota Biocatalysis/Biodegradation Database (UM-BBD)

These databases include biodegradative oxygenases (OxDBase), and the Biodegradation Network-Molecular Biology databases such as Bionemo, MetaCyc, and BioCyc. The biodegradation database (UM-BBD) is accessible free (http://umbbd.ethz.ch). It provides information in a variety of disciplines, such as genes, and biotransformation rules, enzymes, and microbial degradation methods (Ellis, Roe, and Wackett 2006). This database focuses primarily on xenobiotic compound metabolic pathways, that are accessible in both text and graphic formats. Pathways are a set of multistep enzyme processes that begin with a starting chemical and progress through the creation of intermediates. The functional diversity of bacteria species can break down a chemical compound by multiple pathways. The UM-BBD provides all known pathways for a single compound, as well as information on the bacteria and enzymes involved in its degradation. Two bacterial breakdown routes are shown in Figure 8.1 of the UM-BBD pathway map of 2-nitrobenzoic acid. Both pathways began with the production of 2-hydroxylaminobenzoic acid, which was then degraded by two distinct bacteria by two different mechanisms. This database is further categorized into several others domains like NCBI, Enzymes, BRENDA, and ExPASy to give information on xenobiotic breakdown genes and enzymes (Ellis, Roe, and Wackett 2006).

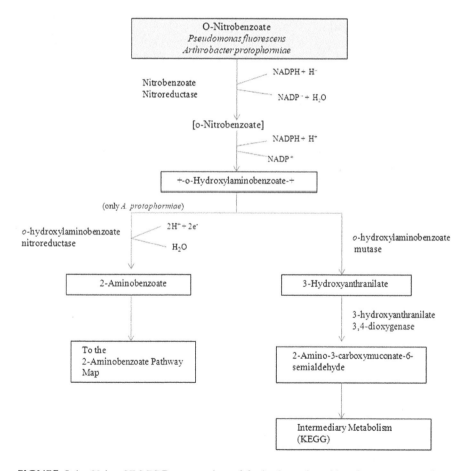

FIGURE 8.1 Using UM-BBD, generation of 2-nitrobenzoic acid pathway map gas been represented (http://umbbd.ethz.ch/onb/onb_map.html).

8.4.1.2   OxDBase

A database established by the IMTECH (CSIR) at Chandigarh, India (http:// www.imtech.res.in/raghava/oxdbase/), incorporates oxygenase data gathered from research publications and databases (Arora et al. 2009). Its most significant enzymes, responsible for the aerobic breakdown of aromatic compounds, are oxygenases (Arora et al. 2009). Monooxygenases and dioxygenases are the two kinds of oxygenases. Monooxygenases catalyze the integration of a single atom of O_2 into the substrate, whereas dioxygenases catalyze the insertion of two atoms (Arora et al. 2009). Aromatic ring hydroxylating dioxygenases (ARHD) and aromatic ring cleavage dioxygenases (ARCD) are two types of dioxygenases. ARHD catalyzes aromatic ring hydroxylation, while ARCD catalyzes aromatic ring cleavage (Arora et al. 2009). There are two types of ARCDs: extradiol and intradiol. Extradiol ARCDs break rings between hydroxylated and non-hydroxylated carbons in the near vicinity, while intradiol ARCDs break rings between hydroxylated carbons and subsequent non-hydroxylated carbons (Arora et al. 2009). 237 distinct oxygenases are included in OxDBase, comprising 118 monooxygenases, 119 dioxygenases, and additional enzymes. All enzyme files entail information about the reaction(s) in which enzymes also become implicated, their common names, and also some related terms, structures and gene linkages, literature study research articles, families and subfamilies, and connections to a variety of external data sources.

8.4.1.3   The Bionemo Database

This was established by the Spanish National Cancer Research Center's structural Computational Biology Group (http://bionemo.bioinfo.cnio.es) (Carbajosa et al. 2008). Bionemo is a manually curated database containing data on genes and proteins engaged in biodegradation metabolism. Protein sequences, domains, and structures are included in protein information, while genomic information includes sequence data, regulatory sequences, and coding regions (Carbajosa et al. 2008). Bionemo is a companion to UM-BBD (Carbajosa et al. 2008), which emphasizes biochemical elements of biodegradation. Bionemo was created by manually connecting sequence database entries with biodegradation reactions using data taken from research papers (Carbajosa et al. 2008). The fundamental biochemical network is linked to information regarding its transcription units and regulating factors for genes involved in biodegradation. Around 1503 chemical reactions, 993 enzymes, 543 microorganisms, 219 microbial degradation pathways, 250 metabolism rules, and 76 naphthalene 1, 2- dioxygenase processes, 50 functional groups, as well as 109 toluene dioxygenase interactions are presently included in the UM-BBD database. This database is also linked to several other databases (Carbajosa et al. 2008).

8.4.1.4   MetaCyc

This is a library of metabolic pathways extracted from scientific experimental literature, with over 2097 experimentally established metabolic processes from over 2460 distinct species. This is the world's biggest curated collection of metabolic processes across all life domains (Caspi et al. 2016). This database supports metabolic engineering, facilitates comparative analysis of biochemical networks, and serves as an encyclopedia for metabolism (Caspi et al. 2016). It contains relevant data about the metabolic pathways associated with primary and secondary metabolic pathways of life forms derived from their genomic sequences. The BioCyc group of SRI International established and administers this database.

8.4.1.5 BioCyc

This is a database (http://biocyc.org/) with over 2988 organism-specific pathways and genomes (PGDBs). Each PGDB comprises a single organism's whole genome and anticipated metabolic process. The pathway tool program uses MetaCyc as a baseline database to estimate pathways (Caspi et al. 2016). Information regarding metabolites, enzymes, and reactions may be found in the predicted metabolic path. BioCyc PGDBs also provide details on anticipated operons, transport systems, and pathway-hole additives (Caspi et al. 2016). Multiple methods for retrieving and analyzing PGDBs are available on BioCyc pathway tool-based web pages, covering gene expression studies, metabolomics as well as other Metabolism with related chemicals, enzymes, and gene transcripts (Caspi et al. 2016). This database can be accessed for free at http://metacyc.org/. MetaCyc has a wide range of scientific applicability. It can, for example, give reference data for metabolic massive datasets computational simulation (Caspi et al. 2016). The Bioinformatics Research Group of SRI International created such a database.

8.4.2 PATHWAY PREDICTION SYSTEMS

Non-biochemical or biochemically based methodologies have been used to construct pathway prediction algorithms (Medema et al. 2012). Non-biochemically based route prediction systems produce reactions between substances using statistical inference procedures (Soh and Hatzimanikatis 2010). Bayesian method (Green and Karp 2004), comparative genomics (Piskur et al. 2007), metabolic network alignment (Cheng, Harrison, and Zelikovsky 2009) and machine learning approaches (Dale, Popescu, and Karp 2010) are examples of these systems. These approaches are quite useful for identifying missing network links (Soh and Hatzimanikatis 2010). The downside of these strategies is that they are dependent solely on statistical inference, which means that many of them may be biochemically unfeasible (Soh and Hatzimanikatis 2010). Biochemically based route prediction systems rely on biotransformation rules that are knowledge and skills. The role of several pathway prediction systems in the realm of biodegradation is summarized in Table 8.2.

8.4.2.1 The UM-BBD-Pathway Prediction System (PPS)

It is accessible at http://umbbd.ethz.ch/predict/, which is a component of UM-BBD. The PPS can also be used to predict microbial metabolic pathways for chemical compound breakdown (Gao, Ellis, and Wackett 2011). Biotransformation rules developed from reactions discovered in the UM-BBD database or the literature are used to make accurate predictions. It provides accurate predictions for substances that are comparable to those that have been documented in the scientific literature as having biodegradation pathways (Gao, Ellis, and Wackett 2011). The processes

TABLE 8.2

Systems of Pathway Prediction

System	Function	Reference
PathPred	Predicts microbial biodegradation of ambient chemicals and plant secondary metabolite production pathways.	(Moriya et al. 2010)
CarbonSearch	An algorithm that monitors the conservation of atoms moving through metabolic networks to identify paths within them.	(Heath, Bennett, and Kavraki 2010)
Metabolic Tinker	All pathways between two chemicals are predicted.	(McClymont and Soyer 2013)
DESHARKY	A Monte Carlo algorithm that uses a database of known enzyme reactions to identify metabolic pathways from target chemicals. Also included are amino acid sequences for homologous enzymes from species that are phylogenetically related.	(Rodrigo et al. 2008)
UM-PPS	Based on biotransformation regulations, predicts microbial breakdown pathways for xenobiotic substances.	(Gao, Ellis, and Wackett 2011)

of 4-nitrophenol degradation, for example, have been fully examined. On the other hand, 2-fluro-4-nitrophenol and 2-bromo-4-nitrophenol have not. Since the structures of 2-fluro-4-nitrophenol and 2-bromo-4-nitrophenol are similar to that of 4-nitophenol, PPS can accurately predict their degradation. Users could either sketch the structure and generate SMILES or directly insert SMILES to enhance a compound's prognosis in the system.

8.4.2.2 PathPred

The KEGG (Kyoto Encyclopedia of Genes and Genomes) REACTION database and the KEGG RPAIR database from the KEGG are knowledge-based pathway prediction systems (http://www.genome.jp/tools/pathpred/) (Moriya et al. 2010). The KEGG REACTION database not only contains all recognized enzymatic reactions according to the IUBMB enzyme nomenclature, but also contains reactions across the KEGG metabolic processes (Moriya et al. 2010). KEGG RPAIR is a library of biologically structural transformation patterns for substrate–product pairings that are mostly found in the KEGG REACTION database. PathPred is a web-based program that predicts probable enzyme-catalyzed reaction pathways given a query molecule using transformation patterns and chemical structure matching of substrate-product pairs. This service includes possible reactions and converted molecules, as well as a tree-shaped graph that shows all predicted reaction routes. Considering chemicals with biological similarities to KEGG compounds, PathPred-based predictions are quite accurate. PathPred includes reference routes for microbial biodegradation of environmental chemicals and plants' secondary metabolite biosynthesis. Users can choose one of the reference pathways based on their goals (Moriya et al. 2010). There are many other user-friendly approaches for finding a query pipeline. A query compound can be entered in one of three ways: the MDL mol-file format, the

SMILES representation, and the KEGG compound identifier. When employing the xenobiotics biodegradation reference pathway, the query shall be the molecule to be biodegraded, whereas the query for the secondary metabolites biosynthesis reference pathway shall be the final product of biosynthesis. The outcomes of the predictions are correlated to genomic data (Moriya et al. 2010). Whether or not enzymes for these reactions are recognized, the PathPred server provides new and alternative reactions. For an enzyme with an unknown EC number, the enzyme tool (http://www. genome.jp/tools/e-zyme/) can help to find the same. Regarding the assignment of EC numbers, a genomic scan for prospective genes with significant sequence homology of known genes corresponding to the similar EC sub-subclass can be performed (Moriya et al. 2010).

8.4.2.3 The Biochemical Network Integrated Computer Explorer (BNICE)

This is s a computer program for creating new routes based on the reaction rules of the Enzyme Commission categorization system BNICE generates all possible routes from a given destination or initial molecule (Finley, Broadbelt, and Hatzimanikatis 2009). Based on the Gibbs free energy of the process, BNICE then screens all alternative pathways for thermodynamic feasibility and picks feasible innovative thermodynamic pathways (Soh and Hatzimanikatis 2010). As per Soh and Hatzimanikatis, the pathways discovered by BNICE can be further evaluated using established pathway methodologies like thermodynamics-based flux balance analysis (FBA) GrowMatch, which allows researchers to investigate the total effect of these novel pathways on metabolic network performance in host organisms. For metabolic engineering, synthetic biology, and xenobiotic biodegradation, FBA could help forecast maximal yield, phenotypic changes, impacts of gene knockouts, modifications in bioenergetics of the system, and modifications in bioenergetics of the system (Soh and Hatzimanikatis 2010). BNICE can be used to discover novel metabolic processes pathways and so forth.

8.4.2.4 From Metabolite to Metabolite (FMM)

From Metabolite to Metabolite (FMM) is a web server publicly easily accessed at http://FMM.mbc.nctu.edu.tw/and which can browse all possible outcome pathways between defined input and output molecules within and between different species, predicated on the KEGG database as well as other interventional biological databases. By merging KEGG maps and KEGG LIGAND data, FMM can build integrated pathway maps (Chou et al. 2009). This website includes information about the associated enzymes, genes, and organisms, as well as a framework called "comparative analysis," which allows metabolic pathways from different species to be evaluated. FMM is a powerful tool for developing drugs, biofuels, and metabolic engineering (Chou et al. 2009). Users can only examine xenobiotic metabolic pathways about which data is available in the KEGG database for biodegradation activities.

8.4.2.5 Metabolic Tinker

A web-based application (http://osslab.ex.ac.uk/tinker.aspx) which can be used to create synthetic metabolic pathways between user-defined target and reference molecules (McClymont and Soyer 2013). Employing a specialized heuristic search approach, Metabolic Tinker explores the whole known metabolic universe for thermodynamically viable routes (McClymont and Soyer 2013). The program includes the Universal Reaction Network (URN), a directed graph that represents the whole collection of defined reactions and compounds from the Rhea database (McClymont and Soyer 2013). This graph's nodes and edges symbolize metabolites and reactions, correspondingly, and the complete network represents the existing metabolic universe (McClymont and Soyer 2013). Metabolic Tinker uses basic search algorithms used in computer science and graph theory to find possible metabolic routes between two molecules via this URN. The source and target compounds' Rhea/ Chemical Entities of Biological Interest (chEBI) identification numbers are necessary to accomplish the search (McClymont and Soyer 2013).

8.4.3 CHEMICAL TOXICITY PREDICTION USING COMPUTATIONAL METHODOLOGIES

Computational methodologies for calculating chemical toxicity are rapidly evolving (Benfenati 2007). Various strategies have been introduced in recent years that employ computer programs to estimate the toxicology of chemical substances (Zheng et al. 2006). Using complex algorithms, quantitative structure-regulatory activity relationship (QSAR) models determine toxicity related to physical properties of chemical structures along with molecular weight or the molecular descriptors of benzene rings (Eriksson et al. 2003). Some foregoing commercially and widely accessible model sources are Sarah Nexus which is a tool (http://www.lhasalimited.org/products/sarah-nexus.htm) for predicting the mutagenicity of substances. VirtualToxLab is a tool for predicting the hazardous effects of medicines, chemicals, and natural products (Vedani et al. 2009). CAESAR is a tool for assessing chemical toxic effects within REACH (Cassano et al. 2010). Services like ToxiPred predict the toxicity of tiny chemical compounds in aqueous solutions in tetrahymena pyriformis (Mishra et al. 2014).

8.4.4 NEXT-GENERATION SEQUENCING: GENOME SEQUENCES OF XENOBIOTIC DEGRADING BACTERIA

With the notion of "from genomics to metabolomics" next-generation sequencing (NGS) has sparked a revolution in biodegradation and bioremediation. First-generation sequencing corresponds to the automated Sanger technique of sequencing, while NGS corresponds to newer technologies discovered for sequencing (Metzker 2010). NGS platforms like Roche/454, Illumina/Solexa, SOLiD/Life, HelicosBioSciences, and the Polonator Instruments are now available (Metzker 2010). The key stages of NGS are the generation of short reads and their subsequent matching to a standard genome sequence. The latter phase is critical for NGS technologies, and a multitude of computer tools, such as SSAKE (Warren et al. 2006), SOAPdenovo (Li et al. 2009), AbySS (Simpson et al. 2009), and Velvet (Zerbino and Birney 2008), were used for genomic sequences assembly. Gene prediction and its functional annotation are the following stages after the sequence reads are integrated into contigs. The most widely used gene prediction method for biochemical pathways is GLIMMER (Gene Locator and Interpolated Markov ModelER), which uses interpolated Markov models to identify the coding region on the microbial genome (Delcher et al. 1999), and to find homologous genes, manually assess and evaluate the predicted coding area sequences, or use automatic annotation software. For bacterial annotation, there are many online and offline programs available, including RAST (Aziz et al. 2008), BASys (Domselaar et al. 2005), and WeGAS (LEE et al. 2009) as well as offline tools like AGeS (Kumar et al. 2011), DIYA (Stewart, Osborne, and Read 2009), and PIPA (Yu et al. 2008). MICheck (Cruveiller et al. 2005) could also be used to detect syntactic problems in annotated sequences. Bacterial genomics is the analysis of bacteria's entire genome to anticipate genes associated with biodegradation as well as other metabolic processes. Employing NGS technology, the entire genomes of numerous xenobiotic-degrading bacteria have been sequenced and several xenobiotic-degrading genes have been discovered by gene predictions and annotations of the bacterial genomes (Lee et al. 2012). In silico study of the bacterial genome predicts metabolic pathways for xenobiotic biodegradation and provides a comprehensive perspective of the metabolic network of specific bacteria (Vilchez-Vargas, Junca, and Pieper 2010). The genome of xenobiotic degrading bacteria can be used to predict many metabolic processes (Romero et al. 2013). Genes and associated activities may be predicted using whole-genome sequences, and novel biocatalysts can be discovered using whole-genome sequences (Vilchez-Vargas, Junca, and Pieper 2010).The integration of genomic and proteomic methods will provide new insights into metabolism at the organism level (Vilchez-Vargas, Junca, and Pieper 2010).

8.5 COMPUTER-AIDED MOLECULAR DESIGN (CAMD)

CAMD is a process for creating chemical engineering paths using a wide variety of organic feedstock molecules to achieve targeted functionality. The molecular structures and their link to attributes that enable functions constitute the CAMD knowledge base. Quantitative Structure-Property Relationships (QSPR) are used to analyze feedstock and product macromolecules. A system of substructures can be found in molecular structure, which each contributes to different parts of the molecule's attributes. The benefit of CAMD is that it allows you to analyze the entire universe of feasible feedstock molecule transformations. CAMD employs three different types of QSPR. Group Contribution, which determines a molecule's features based on the number of sub-structures it includes, is perhaps the most widely used. Butanol, for example, which has the structure depicted in Figure 8.2, is defined by the groups 1xCH3, 3xCH2, and 1xOH). The value of each group to a targeted property is then calculated by multiplying the number of groups by a coefficient for each group's contribution to the target value. The coefficients are obtained from vast datasets spanning many different compounds and relating to the target property.

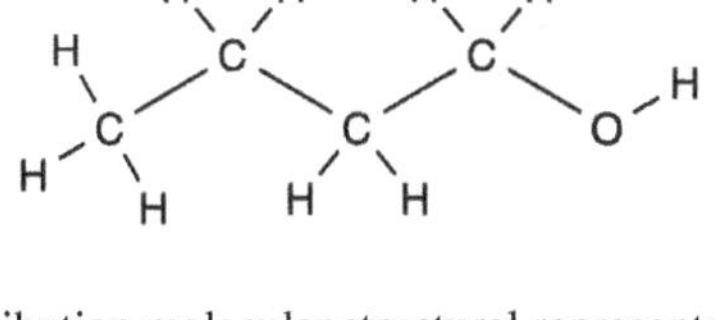

FIGURE 8.2   Group contribution molecular structural representation example.

Topological Indices and Signature Descriptors are the other two relational approaches employed in CAMD. Each of these programs plots geometric bonding and electrical properties of molecular substructures onto 2D and 3D graphs and assign values to each property. The characteristics of the molecules are then connected to functionalities and their permutations using regression coefficients from multiple observations (training set) to properties of the molecule like its anti-inflammatory activity, aqueous solubility, refractive index, vapor pressure, viscosity, heat capacity, and biodegradability. This property-based interpretation of structure allows the creation of molecules with target qualities, as well as the creation of a convert pathway among both the feedstock molecule and the target, all while accounting for the enumeration property quantities. CAMD has largely been utilized to design single-molecule products, with a smaller focus on mixtures, which have higher computational needs. This environmental component could be included in CAMD modeling since QSPR is indeed the cornerstone of QSAR (Quantitative Structure-Activity Relationship), which is used to discover alternatives for hazardous substances (Sheppard et al. 2019).

8.5.1 CAMD Software

If the intended benefit of the research community's outputs is to be recognized, they must be translated into practical tools, particularly widely utilized software. There are two kinds of software in this category:

8.5.1.1 Library-based

Molinspiration Cheminformatics (Molinspiration Cheminformatics 2018) is an established software package that uses a library of empirical data and prior modeling to guide user modification of molecular characteristics and processing.

8.5.1.2 Intelligent

Intelligent packages include libraries and also utilize algorithms to identify patterns and correlations in data, enabling them to predict a conversion path and generate molecules with user-specified properties from feedstock molecules. RXN for Chemistry, a smart, free, cloud-based solution from IBM's research division, has just been introduced. Users use a specially developed interface to identify their originating molecules in structural terms, then include reactants, reagents, and processing parameters from preset libraries and their resources. The methods were developed using two text-mined patent sets of 500k and 350k patents, respectively. Structures are taken from them and stored as SMILES sequences (text-based representations of chemical structures). The algorithms were tested using real-world data and found to be 88% accurate, which is 10% above comparable models (Sheppard et al. 2019).

8.6 TECHNOLOGY COEFFICIENTS

The economic and environmental feasibility of food waste valorization is based on process technology. Process software programs like Aspen and ProSim help with system design in process engineering. Large libraries of data for particular procedures, thermodynamics, and various key aspects of a system computation algorithm are available. The software tools describe the exact technological procedures which can be used to make modifications. Data is assimilated for incumbent technologies, although not at an abstract level to allow for dynamic identification of many more radically optimal technologies. Data is collected to incumbent technologies, but not at an abstraction level to allow for flexible identification of more and more radically optimal technologies. Users do, however, identify and describe 25 separate microbial activities using metrics collected through a literature review. Numerous microbial transformations of sugarcane bagasse for the very same products were assessed (Moncadam, Aristizábal Marulanda, and Cardona 2016) utilizing published microbe-specific kinetic models as the major reference data.

Neither of these features a comparison of technology from various technological fields. Rivas, Castro-Hernández, and Villanueva (Fernandez Rivas et al. 2017) have developed a relatively simple and also the most flexible model yet, which uses prognosis to compare any aspect – such as technical, economic, environmental, or safety – that may affect the accomplishing of a goal. This is done by calculating the "Intensification Factor" (IF) as follows: where the magnitude of a factor before a technology change seems to be the value just after technology change, as well as being a weighting exponent. It includes any sort of value for a domain including temperature, pressure, or flow rate, in the case of the technology domain.

8.7   BIOMETHANIZATION: GREEN WASTE VALORISATION TECHNOLOGY

Biomethanization is a revolutionary green-waste valorization process in which organic waste is decomposed by microorganisms. Efficient management of waste is being increasingly acknowledged as a strategy for fulfilling recently formed requirements regulating the disposal of organic residues; as a result, more facilities are projected to be built.

8.7.1   METHODOLOGY

8.7.1.1   Site Sampling and Biomethanization Facilities

In Quebec (eastern Canada), two biomethanization facilities (BFs)have been established. Both facilities focus on different sorts of garbage in diverse ways and under various conditions. Primary and secondary sludge from wastewater treatment facilities (Agrawal and Verma 2022b), as well as organic industrial food waste, are treated in the first BF. Trash is then processed in mesophilic conditions at the plant, which has a capacity of 40,000 metric tons of waste per year. The second facility, with a capacity of 27,000 metric tons per year, processes household garbage under thermophilic conditions. Two sites were sampled: the reception site, which featured shredding, and the mixing site, where organic waste is combined with a buffer before even being placed in the digesters. This process was followed by air sampling, Fungal spore concentration, DNA extraction, and real-time Polymerase Chain Reaction (PCR) quantification (Mbareche et al. 2018).

8.7.1.2   Next-generation Sequencing

Amplicon amplification, equimolar pooling, and sequencing are techniques performed in next-generating sequencing. The sequence-specific sections described by Tedersoo (Tedersoo et al. 2015) and the relevant data therein were used to amplify the areas utilizing a two-step dual-indexed PCR method specifically built for Illumina instruments. Firstly, the gene-specific sequence was fused to Illumina TruSeq sequencing primers, and PCR was conducted. The second cycle of PCR was performed to include barcodes (dual-indexed) and missing sequences necessary for Illumina sequencing using a 50- to 100-fold dilution of this purified product as a template. The second PCR used the same cycling parameters as the first, but with 12 cycles. The PCR products were purified and verified for purity using a DNA7500 Bioanalyzer chip, and then quantified using a Nanodrop 1000 spectrophotometer (Shankar et al.2018).

8.7.1.3   Sequencing Data Processing

The bioinformatics process created was utilized in this investigation during a composting study (Mbareche et al. 2017). The paired-end reads produced from the sequencing were merged (MOTHUR 1.35.1) after demultiplexing the FASTQ files (Schloss et al. 2009). MOTHUR was used to do qualitative filtering by removing reads with unclear sequences. Reads with a length of fewer than 100 bytes and more than 450 bytes were also deleted. To decrease the computational load, similar sequences were merged, and the numbers of copies of the very same sequence were shown. USEARCH was used to complete the deduplication process (version 7.0.1090; (Edgar 2010). ITSx, which utilizes HMMER3 (Mistry et al. 2013) to match input sequences against a collection of models created from several different internal transcribed spacer (ITS) region sequences found in diverse species, was then used to extract the chosen area of fungal origin from its sequences. Only fungi-specific sequences are used for further analysis. UPARSE was used to cluster operational taxonomic units (OTUs) using a 97% similarity threshold (version 7.1; (Edgar 2013)). UCHIME was used to find chimeric sequences and delete them (Edgar et al. 2011).

8.8   CONCLUSION

Significant valorization processes nowadays involve developing t land, animal feed, effluent treatment of compost, and anaerobic digestion. While the aim of the first-generation valorization method is making the use of material streams as they emerge to be now regarded state-of-the-art, the key issue is broad adoption and enhancing overall efficiency, for instance through cascaded usage. Second-generation valorization pathways rely on the sequential production of several types of goods, such as fine chemicals, commodity products, and biofuels, by combining appropriate recovery and conversion methods for particular components' bio-tools: These are also used to sequence and annotate the whole genomes of several putative bacteria with potential for degrading biomass waste. There have been systems developed for targeting food waste. It entails the use of computer-aided molecular design (CAMD) and technical coefficients to detect and quantify food waste flow at various scales, analyze them, encapsulate relevant conversion technologies, and allow for the evaluation of the economic, environmental, and social consequences. Several databases have been established to provide data on chemicals and their microbial degradation. These databases allow users to obtain data based on their research interests. Chemical databases, for example, can be used to collect information on toxicology, risk analysis, and the environmental factors of chemicals. In addition, multiple bioinformatics techniques for predicting chemical toxicity have been constituted. Such tools can be used to forecast the toxicity of substances. Furthermore, numerous pathway prediction algorithms are available for anticipating the degradation channels of compounds whose degradation mechanisms have not been identified through literature. Experimental studies should be conducted in the future to evaluate the anticipated mechanisms. Moreover, next-generation data sequencing can be used to sequence the genomes of various xenobiotic-degrading bacteria, and gene-annotation has been used to identify the genes and enzymes participating in biodegradation. Molecular methods, in combination with bioinformatics technologies, may give a unique insight into the genetics of biodegradation in the near future. Biomethanization is a significant green-waste valorization technique that involves the method of revalorizing residual organic material to produce biogas. The composition mix of this biogas varies, but it is mostly made up of methane (i.e., 60– 85%) and carbon dioxide (approximately 35%), as well as water vapor, ammonia, hydrogen gas, and hydrogen sulfide. After being dried and cleansed, biogas is utilized as a fuel for powerplants, fuel cells, gas boilers, and many other energy applications. This gas is created by the anaerobic fermentation of organic waste in a digester.

REFERENCES

Agrawal, K., and P. Verma. 2021. Phytoremediation for the treatment of various types of pollutants: A multi-dimensional approach. In: Shah, M.P. (Ed) *Removal of Refractory Pollutants from Wastewater Treatment Plants*, 1–16, CRC Press, Boca Raton.

Agrawal, K., and P. Verma. 2022a. An overview of wastewater treatment facilities in Asian and European countries In: Shah, M.P. (Ed) *Wastewater Treatment*, 1–12, CRC Press, Raton.

Agrawal, K., and P. Verma. 2022b. An overview of various algal biomolecules and its applications. In: Shah, M., Rodriguez-Couto, S., De La Cruz, C.B.V., Biswas, J. (Eds) *An Integration of Phycoremediation Processes in Wastewater Treatment*, 1–15, Elsevier, USA.

Ajila, C. M., S. K. Brar, M. Verma, R. D. Tyagi, S. Godbout, and J. R. Valero. 2012. Bioprocessing of agro-byproducts to animal feed. *Critical Reviews in Biotechnology* 32 (4): 382–400.

Amon, Thomas, Barbara Amon, Vitaliy Kryvoruchko, Werner Zollitsch, Karl Mayer, and Leonhard Gruber. 2007. Biogas production from maize and dairy cattle manure— Influence of biomass composition on the methane yield. *Agriculture, Ecosystems & Environment* 118 (1): 173–182.

Andrady, Anthony L. 1998. Biodegradation of plastics: Monitoring what happens. In *Plastics Additives: An A-Z reference*, edited by G. Pritchard. Dordrecht: Springer Netherlands.

Arora, Pankaj Kumar, and Hanhong Bae. 2014. Bacterial degradation of chlorophenols and their derivatives. *Microbial Cell Factories* 13 (1): 31.

Arora, Pankaj K., Manish Kumar, Archana Chauhan, Gajendra P. S. Raghava, and Rakesh K. Jain. 2009. OxDBase: A database of oxygenases involved in biodegradation. *BMC Research Notes* 2 (1): 67.

Arora, P. K., C.H. Sasikala, and V. Ramana Ch. 2012. Degradation of chlorinated nitroaromatic compounds. *Appl Microbiol Biotechnol* 93 (6): 2265–2277.

Arora, Pankaj, and Wenxin Shi. 2010. Tools of bioinformatics in biodegradation. *Reviews in Environmental Science and Biotechnology* 9: 211–213.

Arora, Pankaj, Alok Srivastava, and Vijay Singh. 2010. Application of monooxygenases in dehalogenation, desulphurization, denitrification and hydroxylation of aromatic compounds. *Journal of Bioremediation & Biodegradation* 1: 112.

Arora, Pankaj Kumar, Alok Srivastava, and Vijay Pal Singh. 2014. Bacterial degradation of nitrophenols and their derivatives. *Journal of Hazardous Materials* 266: 42–59.

Aziz, Ramy, Daniela Bartels, Aaron Best, et al. 2008. The RAST server: Rapid annotations using subsystems technology. *BMC Genomics* 9: 75.

Banks, Charles, Yue Zhang, Ying Jiang, and Sonia Heaven. 2011. Trace element requirements for stable food waste digestion at elevated ammonia concentrations. *Bioresource Technology* 104: 127–135.

Benfenati, Emilio. 2007. Predicting toxicity through computers: A changing world. *Chemistry Central Journal* 1: 32.

Bhardwaj, N., K. Agrawal, and P. Verma. 2020. Algal biofuels: An economic and effective alternative of fossil fuels. In: Srivastava, N., Srivastava, M., Mishra, P.K., Gupta, V.K. *Microbial Strategies for Techno-economic Biofuel Production*, 59–83. Springer, Singapore.

Bigliardi, Barbara, and Francesco Galati. 2013. Innovation trends in the food industry: The case of functional foods. *Trends in Food Science & Technology* 31 (2): 118–129.

Bouallagui, H., Y. Touhami, R. Ben Cheikh, and M. Hamdi. 2005. Bioreactor performance in anaerobic digestion of fruit and vegetable wastes. *Process Biochemistry* 40 (3): 989–995.

Carbajosa, Guillermo, Almudena Trigo, Alfonso Valencia, and Ildefonso Cases. 2008. Bionemo: Molecular information on biodegradation metabolism. *Nucleic Acids Research* 37: D598–D602.

Caspi, Ron, Richard Billington, Luciana Ferrer, et al.2016. The MetaCyc database of metabolic pathways and enzymes and the BioCyc collection of pathway/genome databases. *Nucleic Acids Research* 44 (D1): D471–D480.

Cassano, Antonio, Alberto Manganaro, Todd Martin, et al.2010. CAESAR models for developmental toxicity. *Chemistry Central Journal* 4 (Suppl.1): S4.

Cavinato, Cristina, David Bolzonella, Francesco Fatone, F. Cecchi, and Paolo Pavan. 2011. Optimization of two-phase thermophilic anaerobic digestion of biowaste for hydrogen and methane production through reject water recirculation. *Bioresource Technology* 102: 8605–8611.

Cecchif, P. Battistoni, P. Pavan, D. Bolzonella, and L. Innocenti. 2005. Digestione anaerobica della frazione organica dei rifiuti solidi. Aspetti fondamentali, progettuali, gestionali, di impatto ambientale ed integrazione con la depurazione delle acque reflue. APAT, Manual and guideline, APAT – Agenzia per la Protezione dell'Ambiente e per i Servizi Tecnici, Rome Pg 1–178.

Chaturvedi, V., R.K. Goswami, and P. Verma. 2021. Genetic engineering for enhancement of biofuel production in microalgae. In: Verma, P. (Ed) *Biorefineries: A Step Towards Renewable and Clean Energy*, 539–559. Springer, Singapore.

Chen, Lei, Jing Lu, Jian Zhang, Kai-Rui Feng, Ming-Yue Zheng, and Yu-Dong Cai. 2013. Predicting chemical toxicity effects based on chemical-chemical interactions. *PloS One* 8 (2): e56517–e56517.

Cheng, Qiong, Robert Harrison, and A. Zelikovsky. 2009. MetNetAligner: A web service tool for metabolic network alignments. *Bioinformatics (Oxford, England)* 25: 1989–1990.

Chou, Chih-Hung, Wen-Chi Chang, Chih-Min Chiu, Chih-Chang Huang, and Hsien-Da Huang. 2009. FMM: A web server for metabolic pathway reconstruction and comparative analysis. *Nucleic Acids Research* 37: W129–W134.

Cruveiller, Stéphane, Jérôme Saux, David Vallenet, Aurélie Lajus, Stéphanie Bocs, and Claudine Medigue. 2005. MICheck: A web tool for fast checking of syntactic annotations of bacterial genomes. *Nucleic Acids Research* 33: W471–W479.

Dai, X., C. Hu, D. Zhang, and Y. Chen. 2017. A new method for the simultaneous enhancement of methane yield and reduction of hydrogen sulfide production in the anaerobic digestion of waste activated sludge. *Bioresource Technology* 243: 914–921.

Dale, Joseph M., Liviu Popescu, and Peter D. Karp. 2010. Machine learning methods for metabolic pathway prediction. *BMC Bioinformatics* 11 (1): 15.

Dauchet, Luc, Philippe Amouyel, Serge Hercberg, and Jean Dallongeville. 2006. Fruit and vegetable consumption and risk of coronary heart disease: A meta-analysis of cohort studies. *The Journal of Nutrition* 136: 2588–2593.

Debes, Jose, and Raul Urrutia. 2004. Bioinformatics tools to understand human diseases. *Surgery* 135: 579–585.

Delcher, A. L., D. Harmon, S. Kasif, O. White, and S. L. Salzberg. 1999. Improved microbial gene identification with GLIMMER. *Nucleic Acids Research* 27 (23): 4636–4641.

Domselaar, Gary, Paul Stothard, Savita Shrivastava, et al.2005. BASys: A web server for automated bacterial genome annotation. *Nucleic Acids Research* 33: W455–W459.

Edgar, Robert. 2010. Search and clustering orders of magnitude faster than BLAST. *Bioinformatics (Oxford, England)* 27 (16): 2194–2200.

Edgar, Robert 2013. UPARSE: Highly accurate OTU sequences from microbial amplicon reads. *Nature Methods* 10: 996–998.

Edgar, Robert, Brian Haas, Jose Clemente, Christopher Quince, and Rob Knight. 2011. UCHIIME improves sensitivity and speed of chimera detection. *Bioinformatics (Oxford, England)* 27: 2194–2200.

Ellis, Lynda B. M., and Lawrence P. Wackett. 2012. Use of the University of Minnesota biocatalysis/biodegradation database for study of microbial degradation. *Microbial Informatics and Experimentation* 2 (1): 1.

Ellis, L. B., D. Roe, and L. P. Wackett. 2006. The University of Minnesota biocatalysis/biodegradation database: The first decade. *Nucleic Acids Research* 34 (Database issue): D517–D521.

Eriksson, Lennart, Joanna Jaworska, Andrew Worth, Mark Cronin, Robert McDowell, and Paola Gramatica. 2003. Methods for reliability and uncertainty assessment and for applicability evaluations of classification- and regression-based QSARs. *Environmental Health Perspectives* 111: 1361–1375.

Fernandez Rivas, Elena Castro-Hernández David, A. Villanueva Perales, and Walter Meer. 2017. Evaluation method for process intensification alternatives. *Chemical Engineering and Processing: Process Intensification* 123: 221–232.

Finley, Stacey D., Linda J. Broadbelt, and Vassily Hatzimanikatis. 2009. Computational framework for predictive biodegradation. *Biotechnology and bioengineering* 104 (6): 1086–1097.

Frac, M., and K. Ziemiński. 2012. Methane fermentation process for utilization of organic waste. *International Agrophysics* 26: 317–330.

Gao, Junfeng, Lynda B. M. Ellis, and Lawrence P. Wackett. 2011. The University of Minnesota pathway prediction system: Multi-level prediction and visualization. *Nucleic Acids Research* 39 (Web Server issue): W406–W411.

Giuliano, A., L. Zanetti, F. Micolucci, and C. Cavinato. 2014. Thermophilic two-phase anaerobic digestion of source-sorted organic fraction of municipal solid waste for bio-hythane production: Effect of recirculation sludge on process stability and microbiology over a long-term pilot-scale experience. *Water Science and Technology* 69 (11): 2200–2209.

Gollagher, Margaret, Jenny Campbell, and Anne-Marie Bremner. 2017. Collaboration achieves effective waste management design at brookfield place perth, Western Australia. *Procedia Engineering* 180: 1763–1772.

Gómez-Guillén, M. C., B. Giménez, M. E. López-Caballero, and M. P. Montero. 2011. Functional and bioactive properties of collagen and gelatin from alternative sources: A review. *Food Hydrocolloids* 25 (8): 1813–1827.

Green, Michelle L., and Peter D. Karp. 2004. A Bayesian method for identifying missing enzymes in predicted metabolic pathway databases. *BMC Bioinformatics* 5 (1): 76.

Greene, N. 2002. Computer systems for the prediction of toxicity: An update. *Advanced Drug Delivery Reviews* 54 (3): 417–431.

Han, S., Sang-Hyoun Kim, Hyun-Woo Kim, and H. Shin. 2005. Pilot-scale two-stage process: A combination of acidogenic hydrogenesis and methanogenesis. *Water Science and Technology: A Journal of the International Association on Water Pollution Research* 52: 131–138.

Hansgate, Ann M., Patrick D. Schloss, Anthony G. Hay, and Larry P. Walker. 2005. Molecular characterization of fungal community dynamics in the initial stages of composting. *FEMS Microbiology Ecology* 51 (2): 209–214.

Hawkes, Freda R., Ines Hussy, Godfrey Kyazze, Richard Dinsdale, and Dennis L. Hawkes. 2007. Continuous dark fermentative hydrogen production by mesophilic microflora: Principles and progress. *International Journal of Hydrogen Energy* 32 (2): 172–184.

Heath, Allison P., George N. Bennett, and Lydia E. Kavraki. 2010. Finding metabolic pathways using atom tracking. *Bioinformatics (Oxford, England)* 26 (12): 1548–1555.

Katara, Pramod. 2013. Role of bioinformatics and pharmacogenomics in drug discovery and development process. *Network Modeling Analysis in Health Informatics and Bioinformatics* 2 (4): 225–230.

Kinsner-Ovaskainen, Agnieszka, Radosław Rzepka, Robert Rudowski, Sandra Coecke, Thomas Cole, and Pilar Prieto. 2009. Acutoxbase, an innovative database for in vitro acute toxicity studies. *Toxicology in Vitro: An International Journal Published in Association with BIBRA* 23: 476–485.

Kongjan, Prawit, and Irini Angelidaki. 2010. Extreme thermophilic biohydrogen production from wheat straw hydrolysate using mixed culture fermentation: Effect of reactor configuration. *Bioresource Technology* 101 (20): 7789–7796.

Kosseva, Maria. 2011. Management and processing of food wastes.

Koutinas, Apostolis, Anestis Vlysidis, Daniel Pleissner, et al. 2014. ChemInform abstract: Valorization of industrial waste and by-product streams via fermentation for the production of chemicals and biopolymers. *Chemical Society Reviews* 43: 2587–2627.

Kumar, Kamal, Valmik Desai, Li Cheng, et al. 2011. AGeS: A software system for microbial genome sequence annotation. *PloS One* 6 (3): e17469–e17469.

Kumar, B., N. Bhardwaj, K. Agrawal, and P. Verma. 2020. Bioethanol production: Generation-based comparative status measurements. In: Srivastava, N., Srivastava, M., Mishra, P.K., Gupta, V.K. (Eds) *Biofuel Production Technologies: Critical Analysis for Sustainability*, 155–201, Springer, Singapore.

Kumar, B., and P. Verma. 2022. Pichia pastoris: Multifaced fungal cell factory of biochemicals for biorefinery applications. In: Deshmukh, S.K., Sridhar, K.R., Badalyan, S.M. (Eds) *Fungal Biotechnology Prospects, and Avenues*, 1–28: CRC Press, Boca Raton.

Kusch-Brandt, Sigrid, Britt Schumacher, Hans Oechsner, and Winfried Schäfer. 2011. Methane yield of oat husks. *Biomass and Bioenergy* 35: 2627–2633.

Kusch-Brandt, Sigrid, Chibuike Udenigwe, Cristina Cavinato, Marco Gottardo, and Federico Micolucci. 2016. Value-added utilization of agro-industrial residues.

Kusch-Brandt, Sigrid, Chibuike Udenigwe, Marco Gottardo, F. Micolucci, and Cristina Cavinato. 2014. First- and second-generation valorisation of wastes and residues occurring in the food supply chain.

Lee, Seung Hyeon, Hyun Mi Jin, Hyo Jung Lee, Jeong Myeong Kim, and Che Ok Jeon. 2012. Complete genome sequence of the BTEX-degrading bacterium Pseudoxanthomonas spadix BD-a59. *Journal of Bacteriology* 194 (2): 544–544.

Lee, Daesang, Hwajung Seo, Chankyu Park, and Kiejung Park. 2009. WeGAS: A web-based microbial genome annotation system. *Bioscience, Biotechnology, and Biochemistry* 73 (1): 213–216.

Li, Ruiqiang, Hongmei Zhu, Jue Ruan, et al. 2009. De novo assembly of human genomes with massively parallel short read sequencing. *Genome Research* 20: 265–272.

Liguori, R., A. Amore, and V. Faraco. 2013. Waste valorization by biotechnological conversion into added value products. *Applied Microbiology and Biotechnology* 97 (14): 6129–6147.

Lim, Su Lin, Leong Hwee Lee, and Ta Yeong Wu. 2016. Sustainability of using composting and vermicomposting technologies for organic solid waste biotransformation: Recent overview, greenhouse gases emissions and economic analysis. *Journal of Cleaner Production* 111: 262–278.

Mata-Alvarez, J. 2005. IWA Publishing. https://doi.org/10.2166/9781780402994

Mbareche, Hamza, Marc Veillette, Laetitia Bonifait, et al. 2017. A next generation sequencing approach with a suitable bioinformatics workflow to study fungal diversity in bioaerosols released from two different types of composting plants. *Science of The Total Environment* 601–602: 1306–1314.

Mbareche, Hamza, Marc Veillette, Marie-Ève Dubuis, et al. 2018. Fungal bioaerosols in biomethanization facilities. *Journal of the Air & Waste Management Association* 68 (11): 1198–1210.

McClymont, Kent, and Orkun S. Soyer. 2013. Metabolic tinker: An online tool for guiding the design of synthetic metabolic pathways. *Nucleic Acids Research* 41 (11): e113–e113.

Medema, Marnix, Renske Raaphorst, Eriko Takano, and Rainer Breitling. 2012. Computational tools for the synthetic design of biochemical pathways. *Nature Reviews. Microbiology* 10: 191–202.

Meher Kotay, Shireen, and Debabrata Das. 2008. Biohydrogen as a renewable energy resource—Prospects and potentials. *International Journal of Hydrogen Energy* 33 (1): 258–263.

Metzker, Michael L. 2010. Sequencing technologies – the next generation. *Nature Reviews Genetics* 11 (1): 31–46.

Mirabella, Nadia, Valentina Castellani, and Serenella Sala. 2014. Current options for the valorization of food manufacturing waste: A review. *Journal of Cleaner Production* 65: 28–41.

Mishra, Nitish, Deepak Singla, Sandhya Agarwal, and Gajendra Raghava. 2014. ToxiPred: A server for prediction of aqueous toxicity of small chemical molecules in T. Pyriformis. *Journal of Translational Toxicology* 1 (1): 21–27.

Mistry, Jaina, Robert D. Finn, Sean R. Eddy, Alex Bateman, and Marco Punta. 2013. Challenges in homology search: HMMER3 and convergent evolution of coiled-coil regions. *Nucleic Acids Research* 41 (12): e121–e121.

Moncadam, Jonathan, Valentina Aristizábal Marulanda, and Carlos Ariel Cardona. 2016. Design strategies for sustainable biorefineries. *Biochemical Engineering Journal* 116 122–134.

Moriya, Yuki, Daichi Shigemizu, Masahiro Hattori, et al. 2010. PathPred: An enzyme-catalyzed metabolic pathway prediction server. *Nucleic Acids Research* 38: W138–W143.

Paritosh, Kunwar, Sandeep K. Kushwaha, Monika Yadav, Nidhi Pareek, Aakash Chawade, and Vivekanand Vivekanand. 2017. Food waste to energy: An overview of sustainable approaches for food waste management and nutrient recycling. *BioMed Research International* 2017: 2370927.

Persoons, Renaud, Sylvie Parat, Muriel Stoklov, Alain Perdrix, and Anne Maitre. 2010. Critical working tasks and determinants of exposure to bioaerosols and MVOC at composting facilities. *International Journal of Hygiene and Environmental Health* 213 (5): 338–347.

Pfaltzgraff, Lucie, Mario De Bruyn, Emma Cooper, Vitaliy Budarin, and James Clark. 2013. Food waste biomass: A resource for high-value chemicals. *Green Chemistry* 15: 307–314.

Piskur, J., K. D. Schnackerz, G. Andersen, and O. Bjornberg. 2007. Comparative genomics reveals novel biochemical pathways. *Trends in Genetics* 23 (8): 369–372.

Rajagopal, Rajinikanth, Daniel I. Massé, and Gursharan Singh. 2013. A critical review on inhibition of anaerobic digestion process by excess ammonia. *Bioresource Technology* 143: 632–641.

Reith, J. H., René Wijffels, and H. Barten. 2003. Bio-methane and bio-hydrogen: Status and perspectives of biological methane and hydrogen production. *Dutch Biol Hydrogen* 32: 9–13.

Rodrigo, Guillermo, Javier Carrera, Kristala Prather, and Alfonso Jaramillo. 2008. DESHARKY: Automatic design of metabolic pathways for optimal cell growth. *Bioinformatics (Oxford, England)* 24: 2554–2556.

Romero, María Jose, Valentina Méndez, Loreine Agulló, and Michael Seeger. 2013. Genomic and functional analyses of the gentisate and protocatechuate ring-cleavage pathways and related 3-hydroxybenzoate and 4-hydroxybenzoate peripheral pathways in *Burkholderia xenovorans* LB400. *PloS One* 8: e56038.

Schloss, Patrick D., Sarah L. Westcott, Thomas Ryabin, et al. 2009. Introducing mothur: Open-source, platform-independent, community-supported software for describing and comparing microbial communities. *Applied and Environmental Microbiology* 75 (23): 7537–7541.

Schoning, G. 2011. Classification & labelling inventory: Role of ECHA and notification requirements. *Annali dell'Istituto Superiore di Sanita* 47 (2): 140–145.

Shankar, Jata, Gustavo C. Cerqueira, Jennifer R. Wortman, Karl V. Clemons, and David A. Stevens. 2018. RNA-Seq profile reveals Th-1 and Th-17-type of immune responses in mice infected systemically with *Aspergillus fumigatus*. *Mycopathologia* 183 (4): 645–658.

Sheppard, Phil, Guillermo Garcia-Garcia, Jamie Stone, and Shahin Rahimifard. 2019. A complete decision-support infrastructure for food waste valorisation. *Journal of Cleaner Production* 247: 119608.

Simpson, Jared, Kim Wong, Shaun Jackman, Jacqueline Schein, Steven Jones, and Inanç Birol. 2009. ABySS: A parallel assembler for short read sequence data. *Genome Research* 19: 1117–1123.

Siró, István, Emese Kápolna, Beáta Kápolna, and Andrea Lugasi. 2008. Functional food. Product development, marketing and consumer acceptance – A review. *Appetite* 51 (4): 456–467.

Smithers, Geoffrey W. 2008. Whey and whey proteins—From 'gutter-to-gold'. *International Dairy Journal* 18 (7): 695–704.

Soh, K. C., and V. Hatzimanikatis. 2010. DREAMS of metabolism. *Trends Biotechnol* 28 (10): 501–508.

Stewart, Andrew, Brian Osborne, and Timothy Read. 2009. DIYA: A bacterial annotation pipeline for any genomics lab. *Bioinformatics (Oxford, England)* 25: 962–963.

Sykes, Peter, Ken Jones, and John D. Wildsmith. 2007. Managing the potential public health risks from bioaerosol liberation at commercial composting sites in the UK: An analysis of the evidence base. *Resources, Conservation and Recycling* 52 (2): 410–424.

Tedersoo, Leho, Sten Anslan, Mohammad Bahram, et al. 2015. Shotgun metagenomes and multiple primer pair-barcode combinations of amplicons reveal biases in metabarcoding analyses of fungi. *MycoKeys* 10: 1–43.

Udenigwe, Chibuike C. 2014. Bioinformatics approaches, prospects and challenges of food bioactive peptide research. *Trends in Food Science & Technology* 36 (2): 137–143.

Valdez-Vazquez, Idania. 2009. Hydrogen production by fermentative consortia. *Renewable and Sustainable Energy Reviews* 13: 1000–1013.

Van Dyk, J. S., R. Gama, D. Morrison, S. Swart, and B. I. Pletschke. 2013. Food processing waste: Problems, current management and prospects for utilisation of the lignocellulose component through enzyme synergistic degradation. *Renewable and Sustainable Energy Reviews* 26: 521–531.

Vedani, A., M. Smiesko, M. Spreafico, O. Peristera, and M. Dobler. 2009. VirtualToxLab - in silico prediction of the toxic (endocrine-disrupting) potential of drugs, chemicals and natural products. Two years and 2,000 compounds of experience: A progress report. *ALTEX* 26 (3): 167–176.

Vilchez-Vargas, Ramiro, Howard Junca, and Dietmar Pieper. 2010. Metabolic networks, microbial ecology and 'omics' technologies: Towards understanding in situ biodegradation processes. *Environmental Microbiology* 12: 3089–3104.

Warren, René L., Granger G. Sutton, Steven J. M. Jones, and Robert A. Holt. 2006. Assembling millions of short DNA sequences using SSAKE. *Bioinformatics* 23 (4): 500–501.

Wexler, Philip. 2001. TOXNET: An evolving web resource for toxicology and environmental health information. *Toxicology* 157: 3–10.

Yu, Chenggang, Nela Zavaljevski, Valmik Desai, Seth Johnson, Fred Stevens, and Jaques Reifman. 2008. The development of PIPA: An integrated and automated pipeline for genome-wide protein function annotation. *BMC Bioinformatics* 9: 52.

Zerbino, Daniel R., and Ewan Birney. 2008. Velvet: Algorithms for de novo short read assembly using de Bruijn graphs. *Genome Research* 18 (5): 821–829.

Zheng, Mingyue, Zhiguo Liu, Chunxia Xue, et al. 2006. Mutagenic probability estimation of chemical compounds by a novel molecular electrophilicity vector and support vector machine. *Bioinformatics* 22 (17): 2099–2106.

9 Cell Surface Engineering
A Fabrication Approach Toward Effective Valorization of Waste

Anupama Binoy
Indian Institute of Technology Palakkad, Palakkad, India

Satyam Singh
Indian Institute of Technology Indore, Indore, India

Sushabhan Sadhukhan
Indian Institute of Technology Palakkad, Palakkad, India

CONTENTS

9.1   INTRODUCTION

Depletion of fossil fuels and the release of greenhouse gases are admonitory for finding an alternative solution to the increasing energy demands of the growing population. Simultaneously, unchecked accumulation of organic waste is a rising environmental issue globally. The World Bank has estimated that by 2025, production of solid wastes from urban areas across the world may reach 2.2 billion metric tons per year, and waste release rates could plausibly double in the coming 20 years (Hoornweg & Bhada-Tata, 2012). This waste, which was previously considered futile, is now receiving remarkable attention as a sustainable resource for manufacturing several useful substances and fuels (Arancon et al., 2013). Bigger government initiatives and incentives have been structured for the utilization of these sustainable and renewable sources for the production of bioenergy through waste valorization and biomineralization (Hill, 2007). The objective of valorization is the recycling of waste into valuable resources, successful disposal, and the process of reusing. Waste valorization also aims for the utilization of renewable energy, removal of toxic chemicals, and development of eco-friendly products. Many valorization techniques are showing effective responses according to industrial demands such as the flow chemical technique to convert waste into a useful resource (Serrano-Ruiz et al., 2012), solid-state fermentation, use of pyrolysis in the synthesis of fuels (Heo et al., 2010; Luque et al., 2012), and utilization of microbes to dispose of complex wastes and generate fuel (Klein-Marcuschamer et al., 2012; Wulff et al., 2006).

Microorganisms have been employed for the production of an array of bio-ingredients such as drugs, dyes, chemicals, as well as fuels (Bhardwaj et al., 2020a). The organisms known to be at the forefront of utilizing organic waste for bioethanol fermentation are *Saccharomyces cerevisiae*, as they are efficient at hydrolyzing complex sugars as well as fermenting them to produce ethanol. In other words, yeast can biosynthesize enzymes that are necessary for hydrolyzing and fermenting sugars. However, one of the key stumbling blocks in controlling the cost involved in biofuel production is enzymes and their efficient use (Kumar and Verma, 2020a). Several attempts have been made to develop genetically engineered microbial strains which are more efficient in depolymerizing organic masses, thereby making the process easy and economical.

Yet another biotechnological tool is known as "cell surface engineering" or "arming technology" where surface modification of microbial cells is achieved via either functional integration or introduction of biomimetics such as nanostructure carrier, supramolecular, nano-scale, peptide-based, micro-fluids, or biosensor molecules (Budihardjo et al., 2021; Chen et al., 2014). To date, many techniques have been investigated for the surface engineering of microbial cells, for example, layer by layer modification (Li et al., 2021), mineralization (Jonathan & Steinmetz, 2011), encapsulation (Yang et al., 2011), and genetic engineering (Falade, 2021). Bacterial cells have been modified with different polymers and magnetic nanoparticles (Jia et al., 2017; Konnova et al., 2016). The generation of the whole-cell biocatalyst is possible by displaying proteins and/or peptides on the cell surface of yeast or bacteria by genetic engineering. Although these processes require a lot of optimization for the development of appropriate microorganisms, they present a promising platform in decidedly increasing the valorization of waste as well as reducing the price as compared to traditional methods such as separate hydrolysis and fermentation (SHF) and simultaneous saccharification and fermentation (SSF) processes. In this chapter, therefore, we will focus on shedding light on different strategies and methods used in the development of cell surface engineering for construction of bioengineered microorganisms.

9.2 CONVENTIONAL STRATEGIES USED IN THE PRODUCTION OF BIOETHANOL

Very recently, biodiesel and bioethanol have gained ground in the form of bioenergy produced from feedstock obtained from biomass waste. Although biofuels are produced prominently by the hydrolysis of starch-rich biomass, lignocellulosic biomass such as sugarcane bagasse, corn stover, rice, wheat straw, and switchgrass presents an abundant and less expensive option for the same (Huang & Clubb. 2017). This abundant polymer is available as forest and agricultural waste, solid municipal waste, and also industrial waste such as from pulp and paper industries and certain food-packaging industries.

The main components in lignocellulosic biomass are cellulose, hemicellulose, and lignin. Cellulose, the central component in lignocellulosic biomass is a polymer of glucose with β-1, 4 glycosidic bonds. Hemicellulose which encloses the cellulosic part is again complex in nature and composed of a variety of carbohydrates such as xylose, mannose, arabinose, glucuronic acid, galactose, etc. The outermost rigid and complex component is lignin composed of phenylpropane units. These feedstocks, on one hand, are cheap; economical as well as serving as a potential carbon source (Kumar & Verma, 2020b), but on the other hand pose a major drawback as raw material for biofuel production (Chaturvedi &Verma, 2013) because of their recalcitrant and stubborn structure which leads to reduced hydrolytic enzyme exposure and poor product yield. This requires additional methods such as chemical, thermal, or biochemical methods (Wiselogel et al., 2018).

The chemical pretreatment methods include the use of harsh, corrosive chemicals, and later on, employ several steps for the final recovery of the products. The unwanted by-products formed during chemical pretreatment methods inhibit the growth of microbes during the fermentation process making the removal of these inhibitors essential before the addition of the microbial inoculum. Moreover, in chemical methods, the enormous carbon source present in the lignin and hemicellulose portion of the biomass is not well utilized (Manisha & Yadav, 2017). The next important step is the production of cellulase enzyme for hydrolysis (Bhardwaj et al., 2020b). There are two different cellulase enzyme complexes, called cellulosomes, produced by aerobic and anaerobic cellulolytic microorganisms. The aerobic cellulose-degrading microorganism's non-complexed cellulosomes are extracellularly secreted in the culture. While the anaerobic organism, for example, *Clostridium thermocellum*, produces a complex cellulosome with the various catalytic domain and enzymatic functions assembled in a scaffoldin. The specific activity of the enzyme complex produced by anaerobic microorganisms is high but the titers produced are very low as compared to the complex produced by the aerobic microorganisms, and that is the reason why the non-complexed cellulosomes are recommended in industries. The aerobic fungi *Trichoderma reesei* are commonly used organisms producing high titers of two major component enzymes that is, Cellulose1,4-β cellobiosidase (*CBHI and CBHII*) (Schwarz, 2001). For the optimum hydrolysis of the whole lignocellulosic components, it is essential to acquire synergistic activity between the different enzymes such as ligninolytic, hemicellulases, and cellulolytic enzymes. There are other factors like crystallinity index of cellulose, microbial source of enzymes, structural features, composition, source of lignocellulosic component, and product inhibition as well.

Microbial fermentation of sugar or starch-based feedstock is a well-established protocol for the production of ethanol (Kumar et al., 2020). But when it comes to making a more complex but sustainable choice for the feedstock in the form of cellulosic biomass, it demands the development of potential microorganisms. The chosen microorganism should be able to ferment not only hexose, but also pentose sugars, which are present in lignocellulosic components, and also should be able to withstand the effect of various inhibitors produced therein. In this context, microorganisms that possess multiple substrate specificities are selected and transformed into better ethanol fermenters through advanced technologies like recombinant/metabolic engineering (Dien et al., 2003; Hahn-Hägerdal et al., 2007; Ho et al., 1998; Zaldivar et al., 2001). Finally, product recovery is required from the batch reactors which is mostly done through the distillation process. The recovery of ethanol is much lower in the case of lignocellulosic biomass as compared to sugar-based feedstocks.

Economic investigation reveals that conversion of lignocellulosic biomass to ethanol makes up for more cost despite the zero cost inferred on the sustainable feedstocks. The higher cost is inferred from the pretreatment step followed by enzyme hydrolysis and enzyme production. To overcome mainly the cost associated with using lignocellulosic biomass as the feedstock, it is amenable to develop more effective strategies to transform from the separate hydrolysis and fermentation process (SHF) to the simultaneous saccharification and fermentation (SSF) process, which is a more consolidated approach. The simultaneous saccharification and co-fermentation processes SSCF) gives more advantage over the SSF process as it can co-ferment hexose and pentose sugars (Ho et al., 1998; Wyman, 2007). This is achieved through several advanced bioengineering techniques. The consolidated bioprocessing is even more enhanced and beneficial as it combines all the processes involved from enzyme production to glucose fermentation (Figure 9.1).

9.3 CONSOLIDATED BIOPROCESSING (CBP)

Consolidated bioprocessing (CBP) is conceptualized as extreme consolidation of four processes: enzyme production (mostly enzymes from the family glycoside hydrolase (GH)), hydrolysis of lignocellulose components, and finally simultaneous fermentation of hexose as well as pentose sugars followed by recovery of products through distillation. All these processes are carried out by the addition of monoculture which is capable of producing hydrolytic enzymes, hydrolyzed polysaccharides, as well as

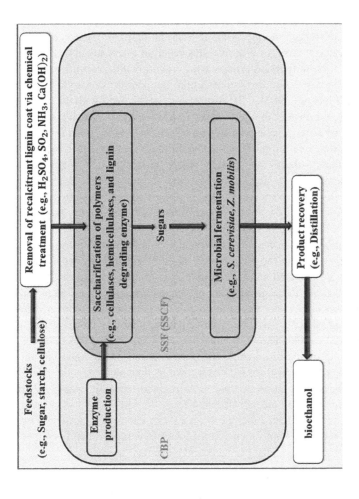

FIGURE 9.1 A pictorial presentation of bioethanol production differentiating between simultaneous saccharification and fermentation (SSF), simultaneous saccharification and co-fermentation processes (SSCF), and consolidated bioprocessing (CBP).

utilizing the hydrolyzed sugars as the sole carbon source producing bioethanol and other by-products. Therefore, CBP is also termed as "direct microbial conversions" (Parisutham et al., 2014). The conjunction of all these processes in a single reactor results in cost-effectiveness. Several fungi and bacterias have been employed for this purpose including *Clostridium thermocellum, Saccharomyces cerevisiae, Fusarium oxysporum, Paecilomyces* sp., and *Neurospora crassa.* CBP utilizes anaerobic bacteria with a tertiary cellulase microbes system, or cellulosomes along with a cellulase enzyme system. This is known as the microbes enzyme synergy scenario. Due to the increased reaction temperature in reactors thermophiles are preferred in CBP, such as *Clostridium thermocellum* that has an optimal function temperature of 65°C. The elevated temperature accelerates reaction rates, avoids the requirement of cooling reactors, reduces interference of contaminants, and also aids in direct recovery of ethanol through distillation at high temperatures (Argyros et al., 2011), (Sonnleitner & Fiechter, 1983). CBP ensures microorganisms with the consortium of all enzymes are not present in nature and the development of such microorganisms is taken care of through genetic engineering (Chaturvedi et al., 2021). Two strategies are adapted for the generation of consolidated microorganisms (Mbaneme-Smith & Chinn, 2015). One is the native strategy where naturally occurring cellulolytic or cellulosomes possessing organisms are selected and engineered to ferment sugars. Some of the cellulolytic bacteria that are preferred for native strategy are cellulolytic microorganisms like *Clostridium phytofermentans* and *Thermanaerobacterium* sp. Naturally, cellulosome-possessing organisms, such as *Clostridium thermocellum* and *Clostridium cellulolyticum* are suitable for genetic engineering (Lynd et al., 2005). Another method is the recombinant cellulolytic strategy where noncellulolytic organisms are selected and recombinantly engineered to saccharify as well as ferment sugars. Here, the ethanalogens or sugar-fermenting microorganisms such as *Saccharomyces cerevisiae, Escherichia coli* are genetically engineered to ensure CBP (Olson et al., 2012). These bioengineered organisms are seen performing better in comparison to their natural counterparts in CBP. The major disadvantage of CBP is the longer period required for the fermentation process and biofuel yields. Another method cultivated from CBP is known as consolidated biosaccharification (CBS) where the fermentation process is segregated from the whole process (Liu et al., 2020). This strategy ensures efficient fermentation without interference from the hydrolytic process. Ultimately, the economic production of sufficient quantities of biofuels will require advancements in feedstock development and better strategies in its conversion to value-added products and biofuels (Table 9.2).

9.3.1 Structural Composition of Natural Cellulosomes of *Clostridium thermocellum* and Recombinant *Minicellulosomes*

The crystalline structure of cellulose microfibril along with tightly conjunct hemicellulose and lignin makes it tough for degradation by any single enzyme or many enzymes separately. It requires a consortium of enzymes bridged together with noncatalytic binding modules making a complex enzymatic system. This structure is naturally secreted mainly by the anaerobic bacteria inhabiting the places rich in lignocellulosic materials (Bayer et al., 2013; Doi & Kosugi, 2004). This complexed enzyme scaffold containing a consortium of cellulolytic enzymes exhibited by anaerobic organisms such as *Clostridium thermocellum* is known as cellulosomes (Bayer et al., 2004). Cellulosomes contain a fibrillar scaffolding protein with enzymes positioned on them. The intricate interaction between the dockrine present at the noncatalytic scaffoldin module and the cohesin protein on the bacterial cell envelope helps in the anchoring of catalytic enzymes of the cellulosome on the bacterial cell wall (Fontes & Gilbert, 2010). In addition, the complex also accommodates a carbohydrate-binding domain apart from the active substrate binding site which aids in enzyme-substrate interaction. The dockrine module's main job is to assemble the cellulosomes and cohesion mediates cellulosome cell-surface interaction. Figure 9.2 displays the structure of natural cellulosome exhibited on the cell surface of *Clostridium sp* (Brás et al., 2016). The cellulosomes are heterogeneous in nature because of the species-specific differences in the scaffoldin module. Most of the scaffoldin module is known to have 6–9 cohesin molecules that can accommodate 26 different enzymes. This fact opens up wide opportunities for modulating the structural composition of cellulosomes artificially.

With the help of advanced technologies in enzyme architectonic, the composition and structure of cellulosome can be controlled. This can be achieved by engineering artificial scaffoldins which can interact with distinct cohesin modules obtained from natural cellulosome-containing species (Moraïs et al., 2012). Simultaneously, the cellulolytic enzymes can be modified and assembled on genetic grounds to integrate into the specific scaffoldin (Kahn et al., 2018). Cell surface display of cellulosome on the bacterial or fungal cell wall is an attractive strategy for the hydrolysis of complex substrates. Many organisms like *Saccharomyces cerevisiae*, lactic acid bacteria, *Bacillus subtilis*, and *Escherichia coli*, among others, have been employed for the same purpose (Schwarz, 2001). In these organisms, the assembly of cellulosomes can be achieved by an *ex-vivo* technique where the component of cellulosomes exhibiting the scaffoldin are externally displayed on the surface of microorganisms, *in-vivo* technique where the different elements are co-expressed in the microorganisms with genetic modification, as well as through applying cell consortium strategy (Doi & Kosugi, 2004).

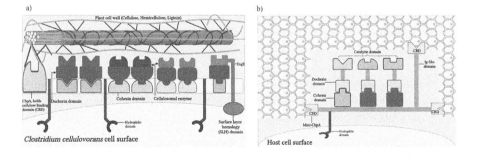

FIGURE 9.2 A model of cellulosome explaining the architecture: (a) schematic representation of the structure of cellulosomes from *Clostridium sp.* (b) modular diagram of the recombinant cellulosome.

9.4 PROTEIN CELL SURFACE DISPLAY

Cell surface display is a powerful technique through which desired proteins or enzymes can be displayed on the surface of the host cells by anchoring proteins. It has a plethora of applications such as vaccine development, antibody production, peptide library screening, removal of harmful agents through bioadsorption, and formation of whole-cell biocatalyst. The cell surface technique enables the endowment of novel functions from different organisms and displays them in close proximity to each other on the surface of host organisms.

9.4.1 APPROACHES FOR CELL SURFACE DISPLAY

There are various approaches adopted for the display of target/passenger proteins on the surface of the host cells. One of the approaches to display these proteins on the host cells is to fuse the N-terminal or C-terminal end of the anchor protein to the passenger protein using recombinant technology and expressed in the host cells. The choice of N or C terminal fusion site is critical to the activity of the protein displayed on the surface. Passenger proteins are also fused to carrier proteins through the sandwich fusion approach, where target proteins are inserted into permissive regions apart from the N/C-terminal region on carrier proteins. The target protein fused to the anchor protein is linked with a signal sequence in order to translocate them to the membrane. The anchor protein tethered to the target protein links to the cell surface of the bacterial cell using covalent or noncovalent binding (Tanaka & Kondo, 2015; Tanaka et al., 2012). While displaying the passenger protein the optimum expression and activity of the enzyme should be carefully controlled. In this context, it is necessary to choose an appropriate signal sequence as well as a promoter for the regulation of the enzyme. In addition to this, introduction of required enhancing elements can prove useful in increasing the efficiency of the enzymes. Various system biology approaches can be useful in doing so. Strategies for modifying the signal sequences are also a promising approach for increasing the effective display of the target protein (Rakestraw et al., 2009). The presence of proper protein folding mechanisms in the host organisms can be enhanced by overexpression or deletion of host cell genes. The stability of the anchor protein should also be improved so that it can withstand the degrading enzymes present in the periplasmic region of the bacterial cell surface (Tanaka & Kondo, 2015).

9.4.2 YEAST – ARMING YEAST

Saccharomyces cerevisiae have historically been used as ethanalogens in fermenting sugars to produce alcohol. An enormous amount of research has also taken place to establish yeast as a cellulose-degrading whole-cell biocatalyst. The mannoproteins present on the yeast cell surface are attached to the cell surface with β-1,6 glucans and are also rich in serine/threonine amino acids. These proteins are known to possess a glycosylphosphatidylinositol (GPI) anchor attachment sequence at their C-terminal. Thus, by targeting the signal sequence for mannoproteins, passenger proteins are anchored on GPI anchors on the surface of yeast cells. In addition to GPI anchors, agglutinin, and flocculin anchoring systems are widely practiced. Several enzymes are displayed on the surface of the yeast cell surface. Here are some the examples like β-glucosidase-displaying yeast strain which can ferment directly xylose and cellobiose, yeast codisplaying xylanase and β-xylosidase, and so on (Lee et al., 2003).

Cell surface engineered yeast is termed "arming yeast". The novel yeast is armed with a consortium of enzymes such as: glucoamylase, α-amylase, carboxymethylcellulase, glucosidase, and lipase. The gene-encoding glucoamylase with its secretion signal peptide, the truncated-gene fragment of the α-amylase with its prepro peptide secretion signal, the cellulose-degrading enzyme (carboxymethylcellulase, one of the endo-type cellulases, and β-glucosidase), and lipase were obtained from *Rhizopus oryzae, Bacillus stearothermophilus, Aspergillus aculeatus*, and *Rhizopus oryzae* respectively. All these enzymes were displayed on the yeast cell surface using the C-terminal half of yeast α-agglutinin (Ueda & Tanaka, 2000).

Researchers have explored yeast in displaying complex cellulose-degrading cellulosomes as well as noncomplexed cellulose-degrading systems on their surface. Wen *et al* reported the first recombinant yeast display system able to produce minicellulosomes which consisted of a mini scaffoldin attached to the cell surface through an α-agglutinin anchor (Wen et al., 2010). This strain accommodated three different cellulose-degrading enzymes. Very recently, Marimuthu et al. were successfully able to architect the largest cellulolytic complex, accommodating 63 different enzymes, and express it in the yeast called *Kluyveromyces marxianus* to display them on its cell surface (Anandharaj et al., 2020).

9.5 CELL SURFACE ENGINEERING IN THE VALORIZATION OF WASTE

9.5.1 Construction of Whole-cell Biocatalyst for Biofuel Production

The concept of the whole-cell biocatalyst is advantageous over a single biocatalyst or microbes used as biocatalyst due to its effective, simple, and environment-friendly approach. In the transformation of lignocellulosic materials into biofuels, where there is an involvement of many enzymes for the complete utilization of available carbon sources present in them, the introduction of the whole-cell biocatalyst is a boon. Cell surface engineering can pose newer approaches in overcoming the recalcitrant nature of lignocellulosic mass. Overexpressed enzymes can be displayed on the surface of host microbes to convert them as whole-cell catalysts. Genes encoding enzymes like α-amylase, β-glucanase, and β-glucosidase can be displayed on the surface of microbes to yield positive valorization of lignocellulosic biomass (Anteneh & Franco, 2019).

Among all the organisms, the use of yeast cells for the production of the whole-cell biocatalyst is the most compelling owing to: 1) a larger surface area enabling display of heterogeneous molecules; 2) excellent protein folding machinery along with post-translational modification of secreted proteins, which allows properly folded proteins to display on the surface; 3) yeast like *Saccharomyces cerevisiae, Pichia pastoris*, among others, which are included in the GRAS (Generally regarded as safe) category allowing them to be used in food and brewery industries; and lastly

TABLE 9.1

Studies on Biofuel Production *via* Cell-surface Engineering of *Saccharomyces cerevisiae*

Displayed Proteins/Enzymes	Species	Substrates	Applications	References
Glucoamylase	*Rhizopus oryzae*	Starch	Bioethanol production	(Murai et al., 1997)
α-amylase	*Bacillus stearothermophilus*	Starch	Bioethanol production	(Murai et al., 1999)
α-amylase	*Streptococcus bovis*	Starch	Bioethanol production	(Shigechi et al., 2004)
Carboxymethylce-llulase (CMCase) and β-glucosidases (BGLs)	*Aspergillus aculeatus*	Cellulose	Cellobiose or oligosaccharide production	(Murai et al., 1998)
β-glucosidases (BGLs) Endoglucanase (EG) Cellobiohydrolase II (CBH II)	*Aspergillus aculeatus, Trichoderma reesei*	Cellulose	Bioethanol production	(Tsai et al., 2009)
Endoglucanase II (EG II) and cellobiohydrolase II (CBH II)	*Trichoderma reesei*	Phosphoric acid-swollen cellulose	Bioethanol production	(Fujita et al., 2004)
Endoglucanases (EGs) and β-glucosidases (BGLs)	*Aspergillus oryzae*	β-glucan	Bioethanol production	(Kotaka et al., 2008)
Xylanase II (XYN II) and β-xylosidase (XylA)	*Trichoderma reesei, Aspergillus oryzae*	Xylan	Xylose production	(Katahira et al., 2004)
Xylose reductase (XR) and Xylitol dehydrogenase (XDH), Xylulokinase (XK)	*Pichia stipitis Saccharomyces cerevisiae*	Xylan	Saccharification and fermentation of xylan	(Katahira et al., 2006)
Alginate lyases	*Saccharophagus degradans*	Alginate	Monosaccharides	(Takagi et al., 2016)

4) yeast is easily cultivable in large density due to its well-known fermentation characteristics (Ye et al., 2021). Table 9.1 represents the different enzymes from various organisms used to surface display lignocellulolytic enzymes for the construction of whole-cell biocatalyst using *Saccharomyces cerevisiae*. Construction of minicellulosomes and genetically structured cellulosomes can also be used in cell surface display for the formation of the whole-cell biocatalyst (Bayer et al., 2004). Despite being an efficacious technique, cell surface engineering is not being utilized at its full potential. To make use of this technique, a grasp all of its prospects is necessary.

9.5.2 FOR BIOADSORPTION OF HEAVY AND TOXIC METAL AND BIOREMEDIATION

Owing to rapid industrialization, environmental pollution is at its peak. More than 80% of Indian water resources are being contaminated posing a high risk for human health, and an increase in the concentration of heavy metals in the aquatic ecosystem could be deleterious (Kashyap et al., 2022). Heavy metals have a high affinity for water which makes their physical separation very difficult (Kumar et al., 2021). The main source of heavy metal contamination in the environment is the mining and metal plating industries' effluents (Mahmood et al., 2012). They can enter our food chain via aquatic flora and fauna and are detrimental to human health while causing terminal diseases like cancer and neurological disorders (Alengebawy et al., 2021).

Various chemical, physical, and biological methods like precipitation, filtration, ion exchange, solvent extraction, coagulation, flocculation, and so on, have been devised in past years for the removal of heavy metals (Ahmed & Ahmaruzzaman, 2016; Bhattacharjee et al., 2020; Lesmana et al., 2009; Goswami et al., 2021). However, all the above-mentioned methods have some drawbacks such as the use of toxic metals, being ineffective at a lower concentration, high sludge volume production, high cost, pH dependency, and more. To overcome these drawbacks, the development of new methods using sustainable strategies is vital (Table 9.2).

Adsorption is a physical process in which solid particles are accumulated on a given surface. When this process is coupled with a biological surface, it is called bioadsorption. Cells have different proteins, peptides, or polysaccharides on their surface which can be utilized to trap heavy metals using adsorption. It is a noninvasive technique that allows us to use the same biomass again. Reusability of cells for adsorption of heavy metals and easy removal of metals from the surface reduces the recovery cost.

S. cerevisiae, apart from having many industrial and fermentation applications, can also be applied in cell surface engineering for bioadsorption. To develop a strain

TABLE 9.2
Cell Surface Engineering in the Production of Value-added Products and Industrial Commodities

Displayed Proteins/ Enzymes	Species	Product	References
BGL-1, EG-II	*Saccharomyces cerevisiae*	β-Glycosidase	(Inokuma et al., 2016)
Xylose reductase (XR), β-D-glucosidase (BGL), Xylosidase (XYL), and Xylanase (XYN)	*Saccharomyces cerevisiae*	Xylitol	(Guirimand et al., 2019)
Endoglucanase, Xylanase	*Escherichia coli*	Butanol, Pinene	(Bokinsky et al., 2011)
Hemicellulases	*Escherichia coli*	Succinate	(Zheng, Chen, Zhao, Wang, & Zhao, 2012a)
α-amylase	*Lactobacillus casei*	Lactic acid	(Narita et al., 2006)
α-amylase	*Corynebacterium glutamicum*	Lysine	(Tateno et al., 2009)
α-amylase	*Corynebacterium glutamicum*	Poly-hydroxy butyrate (PHB)	(Song et al., 2013)
PhoA	*Gluconobacter oxydans*	Alkaline phosphatase	(Blank & Schweiger, 2018)

with high adsorbent efficiency toward divalent heavy metals, a chelating peptide, Hexa-His was cloned and expressed onto the cell surface of yeast resulting in adsorption of Cu^{2+} and Ni^{2+} (Kuroda et al., 2001). Intragenus cloning of *E.coli* ModE protein was successfully done by Nishitani et al., to adsorb molybdenum, one of the rare earth metals, using α-agglutinin expression system in *Saccharomyces cerevisiae*. Yeast cells have shown higher adsorption of molybdenum at low pH and easy recovery using papain with phosphate buffer saline (PBS) at pH 7.4 (Nishitani et al., 2010). Recently Wei and coworkers developed an engineered strain of *S. cerevisiae*, by cloning MerR gene with α-agglutinin expression system. The engineered strain showed a higher affinity for mercury in a wide pH range making it a highly efficient mercury adsorbing system (Wei et al., 2018).

Adaptive modification of the cell surface of yeast has proven useful in adsorption and the stress tolerance of cells has also been increased by creating a barrier on the surface to limit the entrance of pollutants inside the cells. These strategies can be further developed for the engineering of strains for generalized and selective bioadsorption of heavy metals.

9.6 EVOLUTION OF ENZYMES BY CELL SURFACE ENGINEERING

Numerous enzymes have shown remarkable potential in waste valorization due to their increased specificity, product selectivity, lesser energy requirements, and enhanced environmental sustainability compared to synthesized catalysts (Jemli et al., 2016). However, due to a decrease in stability happening during recovery, the use of enzymes in their native structure is minimal (Andler & Goddard, 2018). Cell surface engineering curtails this limitation by protecting the enzymes from unstable and inhibitory elements. Cell surface engineering is delivered through a genetic engineering technique for immobilizing one or many enzymes on the cell membrane. Therefore, it allows the enzyme to interact with low as well as high molecular weight substrate more effectively (Kuroda & Ueda, 2011). Cell surface engineered microbes promise a broad variety of functions such as protein identification (Gera et al., 2013), antibody production, waste biorefineries (Liu et al., 2016), and biodegradation of waste products (Tanaka & Kondo, 2015), contributing towards a sustainable environment. Compared to other microbes, the response of yeast in the cell surface display is enthralling owing to the large size of yeasts, which facilitate numerous types of enzymes/proteins displayed on the surface (Shibasaki et al., 2001). The performance of yeasts in fermentation is promising as they can be easily grown to higher density in a cost-effective medium. Furthermore, most of the yeast varieties such as *Saccharomyces cerevisiae, Yarrowia lipolytica, Pichia pastoris are* also considered as GRAS (Generally regarded as safe), which compels their utilization in food industries (Boder & Wittrup, 1997). Yeast is the most suitable candidate for cell surface engineering because of its structural properties such as a rigid cell wall (200 nm thickness), and a bilayer structure that consists of an outer fibrillar layer with inner β-linked glucans (Kondo & Ueda, 2004). Thus, protein or peptide display on yeast is comparatively easy and flexible. In the following section, we will discuss several methods utilized or experimented to improve the enzyme evolution by yeast cell surface engineering.

9.6.1 TAILORING GPI ANCHORS

Glycosylphosphatidylinositol (GPI) anchored proteins are a type of glucanase-removable membrane mannoproteins connected with β-1,6-glucan through the GPI anchor (e.g., Cwp2p and Sed1p) (Figure 9.3). They maintain the structural rigidity of cell walls along with the cell functions (Orlean, 2012). A GPI anchoring approach led to the engineering of several types of enzymes on the cell membrane. Many studies on single-cell eukaryotic microorganisms have revealed various surface proteins anchored on the outer membrane through GPIs. GPI-anchored proteins such as agglutinins (a-agglutinin and α-agglutinin), flocculins have been broadly used in yeast cell surface engineering (Ye et al., 2021). The yeast cell surface engineering enables binding of the enzymes or proteins to the GPI protein anchoring domain through covalent anchoring on the cell membrane (Kondo & Ueda, 2004; Liu et al., 2014). To establish consolidated bioprocessing (CBP) efficient strains in generating enzymes as well as converting sugars into valuable products, saccharolytic enzymes *via* GPI-based cell surface engineering were installed on the membrane of *Saccharomyces cerevisiae* (van Zyl et al., 2007).

9.6.2 INHIBITOR TOLERANCE

During pretreatment of the lignocellulosic component using harsh chemicals, a lot of inhibitors are generated (Taherzadeh et al., 1997). These elements act as inhibitors for the growth and metabolism of fermenting microbes; eventually, it has become the bottleneck problem for biofuel manufacturing industries. These inhibitors are generally categorized into three types; furans, weak acids, and phenolic acids (Palmqvist & Hahn-Hägerdal, 2000) (Table 9.3).

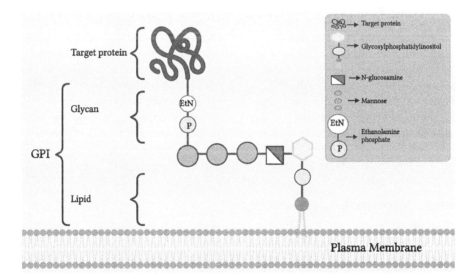

FIGURE 9.3 Structure of glycosylphosphatidylinositol (GPI) anchors. The structure explains the arrangement of molecular moieties which constitute GPI anchors.

TABLE 9.3

Role of Furan, Weak Acids, and Phenolic Acids on the Growth and Fermentation Performance of *Saccharomyces Cerevisiae*

No.	Inhibitor Category	Inhibitory Products	Mechanism	Target Species	References
1	Furans	2-furaldehyde (furfural) 5-hydroxymethyl-2-furaldehyde (HMF)	Inhibits the oxidative metabolism and fermentation of yeasts	*Saccharomyces cerevisiae*	(Li et al., 2017; Sanchez & Bautista, 1988)
2	Weak acids	Acetic acid Formic acid Levulinic acid	Decreased ethanol yield and productivity in *Saccharomyces cerevisiae* fermentations	*Saccharomyces cerevisiae*	(Alvira et al., 2013; Larsson et al., 1999)
3	Phenolic acids	Catechol Coniferyl alcohol Vanillin Hydroquinone Cinnamic acid *p*-coumeric acid	Limit the growth of *S. cerevisiae* and ethanol formation	*Saccharomyces cerevisiae*	(Gu et al., 2019; Larsson et al., 2000)

Different strategies have been identified to enhance the fermentation capability and circumvent the inhibitory effects such as the alkali treatment, sulfite treatment, laccase treatment, evaporation, and anion exchange (Larsson et al., 1999). However, these approaches cause a reduction in fermentable sugars. Successfully developing a new strain of yeast to combat these inhibitors is the need of the hour. The plausible solution to overcoming the inhibitors present in lignocellulose hydrolysates is to enhance the adaptability of fermenting microbes by developing the acquired resistance against the inhibitor formed during the pretreatment step. In this context, Koppram et al. reported continuous batch culture in the presence of twelve inhibitors along with long chemostat exposure by availing the spruce hydrolysates. The resulting strains enhanced the utilization of sugars by 25–38%, however, it also enhanced ethanol production by up to 32–50% compared to *Saccharomyces cerevisiae* strain TMB3400 (Koppram et al., 2012). Furthermore, ^{13}C metabolic flux analysis on two *Saccharomyces cerevisiae* strains was revealed; the absence of ATP and NADP is the probable reason for decreased fermentability in the presence of inhibitory products. This indicates better ATP and NADP synthesis can be a plausible approach to enhance yeast resistance to fermentation blockers (Guo et al., 2016).

9.6.3 METABOLIC ENGINEERING APPROACH

Metabolic engineering is a systemic approach to target in their entirety biochemical or more specifically metabolic pathways, which directly or indirectly influence the activity or production of a molecule (Tyo et al., 2007). Bioethanol (biofuel) production *via* yeast (the ideal host for CBP) can be significantly enhanced by applying systemic metabolic engineering (Choi et al., 2019). Metabolically engineered microbes exhibit improved biofuel production, resistance against toxic materials, increased utilization of a wide range of carbon sources or increased substrate choice, survival in unfavorable environmental conditions, and so forth. Certain metabolic pathways could be improved through genetic modification of the enzymes involved in the regulation and production of biofuel. Recent advancements in the system biology approach provide integrated information about the different pathways in an organism in a comprehensive manner. Together with system biology, genetic and metabolic engineering exhibited certain novel biological applications *via* constructing the unique genetically modified microbial system. Current progress in metabolic engineering allows the tailoring of novel biochemical pathways from the parent microbe into a superior microbe. This approach has revolutionized the production of biofuels from waste in biorefineries. *Saccharomyces cerevisiae* strain houses enzymes that can readily ferment glucose, however, cannot solubilize or hydrolyze sugar polymers present in lignocellulosic biomass, a commonly used feedstock in bioethanol production (Young, Lee, & Alper, 2010). Using metabolic engineering, scientists were able to screen enzymes that can readily hydrolyze even the most complex polysaccharides and engineer *Saccharomyces cerevisiae* by expressing those hydrolytic enzymes on its surface (Buschke et al., 2013). For example, *Saccharomyces cerevisiae* metabolically engineered to display minicellulosome on the cell surface can saccharify cellulose, thereby increasing utilization of more carbon sources leading to improved bioethanol titer of 1.4 g/L/h (Fan et al., 2012). In a separate study Ha et al. reported a novel metabolically engineered *Saccharomyces cerevisiae* strain for the cofermentation of the mixture of cellobiose and xylose for enhanced bioethanol production, approximately up to 0.65 g/L/h (Ha et al., 2011). During bioethanol production, glycerol is released as a by-product which is necessary for the regulation of osmotic balance in the cells of yeast. However, it has been published that deletion of gpd1 and gpd2 gene, coding cytosolic nicotinamide adenine dinucleotide hydrogen (NADH-reduced) dependent glycerol-3-phosphate dehydrogenases (GlpD), has decreased/abolished glycerol production and led to incrementation in the carbon supply for bioethanol production (Kurylenko et al., 2016). Continuous overactivation of PDC1 and ILV2 genes in *Saccharomyces cerevisiae via* an optogenetic pathway to regulate cellular metabolism yielded 8.49 g/L iso-butanol and 41.9 g/L bioethanol production (Zhao et al., 2018). To increase the protein folding tendency, Bip protein (endoplasmic reticulum chaperon protein) along with Pdi1p (disulfide isomerase) were successfully overexpressed in *Saccharomyces cerevisiae*. Furthermore, Golgi membrane protein Ca^{2+}/Mn^{2+}ATPase Pmr1p was downregulated to achieve glycosylation of cellulose. The resulting strain SK12-50 produced 4.7 g/L bioethanol (Song et al., 2018). Taken together, metabolic engineering along with cell-surface display could be a major approach to solve the majority of aforementioned issues.

9.6.4 Genetic Engineering Approach

When enzymes are selected for the production of biofuel, the efficacy of enzyme activity, selectivity, substrate specificity, and stability of the enzyme could be enhanced *via* applying protein engineering approaches (Choi & Chan, 2015; Mutanda et al., 2021). The efficacy of enzymatic activity relies on many key factors such as pH, type of substrate, temperature, co-factors, enzymatic regulation, among others. (Bhan et al., 2013; Biggs et al., 2014). The protein engineering technique is directly linked with the evolution for the improvement of enzyme efficiency. Recent advancements in prior knowledge of amino acid sequences, function, protein structure, and dynamics significantly minimize the time required for protein engineering of the enzymes (Wang et al., 2016; Wu et al., 2010). It has been reported that a higher concentration of arginine is known to maintain the cell membrane integrity, thus increasing the bioethanol tolerance. In this context, an enzyme involved in the catalysis of arginine was rationally selected for the mutation which led to increased bioethanol production (Shima et al., 2003). In the continuation of the previous study, a study was conducted on *S. cerevisiae*, and obtained results suggesting that these properties can be attained by the mTN3 transposon dependent random mutation approach (Kim et al., 2011). Five thermotolerant (grown at 42°C) and enhanced bioethanol-tolerant mutant strains (up to 15%) were isolated and their increased growth rate was compared with the control strain. Several other genes were also selected for increased bioethanol tolerance by protein engineering approaches, resulting in enhanced production of bioethanol. These genes are cAMP receptor, URA7/CTP synthetase, aminopeptidase, calcineurin. Tn1, Tn4, Tn5, arginine permeases from *Clostridium. thermocellum, Zymomonas mobilis, Escherichia coli*, and *Saccharomyces cerevisiae* (Ulaganathan et al., 2017). Substitution of an amino acid in the hexose transporter gene leads to increased xylose utilization in the *Saccharomyces cerevisiae* strain (Apel et al., 2016). Using cost-effective and less time-consuming strategies like that of the advanced protein engineering approaches in bioethanol production provides the immense possibility of enhancing the success of the process.

9.6.5 Immobilization of Enzymes on the Cell Surface

Enzyme immobilization states for confinement of enzymes over a physically confined area (onto a support or matrix) without compromising their original catalytic activity, thereby leads to the use of enzymes multiple times or continuously. Enzyme immobilization through cell surface display allows the repeated utilization of various enzymes and cells in multiple batch fermentation, reducing the cost of yeast culture. Enzyme immobilization can be divided into four groups: carrier binding, cross-linking, entrapping, and a combination of these three approaches (Sanda et al., 2011). Due to low toxicity and cost-effectiveness, enzyme immobilization is feasible in biofuel production as well as the food industry. Many natural support materials including alginates, κ-carrageenan, agar, and agarose are robust gel materials and used for enzyme entrapment (Oztop et al., 2003). In another report, endoglucanase II (EGII) from *Trichoderma reesei* and β-glucosidase (BGL1) from *Aspergillus aculeatus* were co-immobilized on the yeast membrane resulting in a novel strain which enhanced saccharification as well as the fermentation of cellulose to approximately 17 g/L/48 h of bioethanol production from 45 g/L β-glucan with an ethanol substrate conversion yield (ESCY) of ~74%. Further, this strain was genetically engineered by the addition of *Trichoderma reesei* cellobiohydrolase II (CBH II) and this novel strain resulted in 3 g/L/40 h of bioethanol, with an ESCY of ~61% from 10 g/L phosphoric acid cellulose (Fujita et al., 2004).

Zheng et al. utilized *Saccharomyces cerevisiae* for the fermentation of sugar molasses and acquired increased bioethanol production up to 6.55 g/L/h. Moreover, enzyme immobilization on the yeast cell surface increases enzymatic activities and stability. Immobilization of enzymes provides feasibility in increasing the bioethanol concentration as well as incrementation in the cell density. Hence, immobilized yeast exhibits great potential for cost-effective commercialization of bioethanol production (Zheng et al., 2012b).

9.7 CONCLUSION

This chapter demonstrates the use of the display of proteins or enzymes on the cell surface of microorganisms for the efficient and economical conversion of lignocellulosic biomass to bioethanol and other value-added commodities. Utilizing a consortium of enzymes instead of single and separate enzymes has helped in deciphering consolidated bioprocessing, which was a success. The display of proteins on the microbial cell surface helped to reduce the drawbacks incurred in conventional methods of bioethanol production as well as improving the CBP strategy. Further, the discovery of cellulosomes initiated the elaboration of the idea of displaying the enzymes using anchors on the surface of host organisms. Cell surface engineering was the key to developing armed organisms. One of the best examples was "arming yeast". Taking a reductionist approach, more knowledge, starting with genes and progressing to the coded enzymes for different organisms along with their regulation, can be gathered through system biology approaches. Different advanced technologies like genetic engineering, protein engineering, and metabolic engineering, among others. have the potential to bring advancement in cell surface engineering and develop more potent whole-cell biocatalysts.

ACKNOWLEDGMENTS

The authors gratefully acknowledge the financial support from the Indian Institute of Technology Palakkad and Department of Science and Technology (DST)-SERB, Govt. of India for providing all the infrastructure for writing this book chapter. Anupama Binoy is supported by the National Post-doctoral Fellowship granted by DST-SERB (PDF/2020/1950), Govt. of India.

REFERENCES

Ahmed, M. J. K., & Ahmaruzzaman, M. (2016). A review on potential usage of industrial waste materials for binding heavy metal ions from aqueous solutions. *Journal of Water Process Engineering*, 10, 39–47. https://doi.org/10.1016/J.JWPE.2016.01.014

Alengebawy, A., Abdelkhalek, S. T., Qureshi, S. R., & Wang, M.-Q. (2021). Heavy metals and pesticides toxicity in agricultural soil and plants: Ecological risks and human health implications. *Toxics*, 9(3), 1–34. https://doi.org/10.3390/TOXICS9030042

Alvira, P., Moreno, A. D., Ibarra, D., Sáez, F., & Ballesteros, M. (2013). Improving the fermentation performance of *Saccharomyces cerevisiae* by laccase during ethanol production from steam-exploded wheat straw at high-substrate loadings. *Biotechnology Progress*, 29(1), 74–82. https://doi.org/10.1002/btpr.1666

Anandharaj, M., Lin, Y.-J., Rani, R. P., Nadendla, E. K., Ho, M.-C., Huang, C.-C., ... Li, W.-H. (2020). Constructing a yeast to express the largest cellulosome complex on the cell surface. *Proceedings of the National Academy of Sciences*, 117(5), 2385–2394. https://doi.org/10.1073/PNAS.1916529117

Andler, S. M., & Goddard, J. M. (2018). Transforming food waste: How immobilized enzymes can valorize waste streams into revenue streams. *Npj Science of Food*, 2(1), 1–11. https://doi.org/10.1038/s41538-018-0028-2

Anteneh, Y. S., & Franco, C. M. M. (2019). Whole cell actinobacteria as biocatalysts. *Frontiers in Microbiology*, (FEB), 77. https://doi.org/10.3389/FMICB.2019.00077

Apel, A. R., Ouellet, M., Szmidt-Middleton, H., Keasling, J. D., & Mukhopadhyay, A. (2016). Evolved hexose transporter enhances xylose uptake and glucose/xylose co-utilization in *Saccharomyces cerevisiae*. *Scientific Reports*, 6(1), 1–10. https://doi.org/10.1038/srep19512

Arancon, R. A. D., Lin, C. S. K., Chan, K. M., Kwan, T. H., & Luque, R. (2013). Advances on waste valorization: New horizons for a more sustainable society. *Energy Science & Engineering*, 1(2), 53–71. https://doi.org/10.1002/ESE3.9

Argyros, D. A., Tripathi, S. A., Barrett, T. F., Rogers, S. R., Feinberg, L. F., Olson, D. G., ... Caiazza, N. C. (2011). High ethanol titers from cellulose by using metabolically engineered thermophilic, anaerobic microbes. *Applied and Environmental Microbiology*, 77(23), 8288–8294. https://doi.org/10.1128/AEM.00646-11

Bayer, E. A., Belaich, J. P., Shoham, Y., & Lamed, R. (2004). The cellulosomes: Multienzyme machines for degradation of plant cell wall polysaccharides. *Annual Review of Microbiology*, 58, 521–554. https://doi.org/10.1146/ANNUREV.MICRO.57.030502.091022

Bayer, E. A., Shoham, Y., & Lamed, R. (2013). Lignocellulose-decomposing bacteria and their enzyme systems. *The Prokaryotes: Prokaryotic Physiology and Biochemistry*, 215–266. https://doi.org/10.1007/978-3-642-30141-4_67

Bhan, N., Xu, P., & Koffas, M. A. (2013). Pathway and protein engineering approaches to produce novel and commodity small molecules. *Current Opinion in Biotechnology*, 24(6), 1137–1143. https://doi.org/10.1016/J.COPBIO.2013.02.019

Bhardwaj, N., Agrawal, K., & Verma, P. (2020a). Algal biofuels: An economic and effective alternative of fossil fuels. In: *Microbial Strategies for Techno-economic Biofuel Production*, 59–83. Springer, Singapore.

Bhardwaj, N., Kumar, B., Agrawal, K., & Verma, P., (2020b). Bioconversion of rice straw by synergistic effect of in-house produced ligno-hemicellulolytic enzymes for enhanced bioethanol production. *Bioresource Technology Reports*, 10, 100352.

Bhattacharjee, C., Dutta, S., & Saxena, V. K. (2020). A review on biosorptive removal of dyes and heavy metals from wastewater using watermelon rind as biosorbent. *Environmental Advances*, 2, 100007. https://doi.org/10.1016/J.ENVADV.2020.100007

Biggs, B. W., De Paepe, B., Santos, C. N., De Mey, M., & Kumaran Ajikumar, P. (2014). Multivariate modular metabolic engineering for pathway and strain optimization. *Current Opinion in Biotechnology*, 29(1), 156–162. https://doi.org/10.1016/J.COPBIO.2014.05.005

Blank, M., & Schweiger, P. (2018). Surface display for metabolic engineering of industrially important acetic acid bacteria. PeerJ, 6, e4626. https://doi.org/10.7717/peerj.4626

Boder, E. T., & Wittrup, K. D. (1997). Yeast surface display for screening combinatorial polypeptide libraries. *Nature Biotechnology*, 15(6), 553–557. https://doi.org/10.1038/NBT0697-553

Bokinsky, G., Peralta-Yahya, P. P., George, A., Holmes, B. M., Steen, E. J., Dietrich, J., ... Simmons, B. A. (2011). Synthesis of three advanced biofuels from ionic liquid-pretreated switchgrass using engineered *Escherichia coli*. *Proceedings of the National Academy of Sciences*, 108(50), 19949–19954. https://doi.org/10.1073/pnas.1106958108

Brás, J. L. A., Pinheiro, B. A., Cameron, K., Cuskin, F., Viegas, A., Najmudin, S., ... Fontes, C. M. G. A. (2016). Diverse specificity of cellulosome attachment to the bacterial cell surface. *Scientific Reports*, 6(1), 1–12. https://doi.org/10.1038/srep38292

Budihardjo, M. A., Effendi, A. J., Hidayat, S., Purnawan, C., Lantasi, A. I. D., Muhammad, F. I., & Ramadan, B. S. (2021). Waste valorization using solid-phase microbial fuel cells (SMFCs): Recent trends and status. *Journal of Environmental Management*, 277, 111417. https://doi.org/10.1016/J.JENVMAN.2020.111417

Buschke, N., Schäfer, R., Becker, J., & Wittmann, C. (2013). Metabolic engineering of industrial platform microorganisms for biorefinery applications--optimization of substrate spectrum and process robustness by rational and evolutive strategies. *Bioresource Technology*, 135, 544–554. https://doi.org/10.1016/J.BIORTECH.2012.11.047

Chaturvedi, V., Goswami, R.K., & Verma, P. (2021). Genetic engineering for enhancement of biofuel production in microalgae. In: *Biorefineries: A Step Towards Renewable and Clean Energy*. 539–559. Springer. Singapore.

Chaturvedi, V., & Verma, P. (2013). An overview of key pretreatment processes employed for bioconversion of lignocellulosic biomass into biofuels and value-added products. *Biotech*, 3(5), 415–431.

Chen, W., Wang, G., & Tang, R. (2014). Nanomodification of living organisms by biomimetic mineralization. *Nano Research*, 7(10), 1404–1428. https://doi.org/10.1007/S12274-014-0509-9

Choi, Y., & Chan, A. P. (2015). PROVEAN web server: A tool to predict the functional effect of amino acid substitutions and indels. *Bioinformatics*, 31(16), 2745–2747. https://doi.org/10.1093/BIOINFORMATICS/BTV195

Choi, K. R., Jang, W. D., Yang, D., Cho, J. S., Park, D., & Lee, S. Y. (2019). Systems metabolic engineering strategies: Integrating systems and synthetic biology with metabolic engineering. *Trends in Biotechnology*, 37(8), 817–837. https://doi.org/10.1016/J.TIBTECH.2019.01.003

Dien, B. S., Cotta, M. A., & Jeffries, T. W. (2003). Bacteria engineered for fuel ethanol production: Current status. *Applied Microbiology and Biotechnology*, 63(3), 258–266. https://doi.org/10.1007/S00253-003-1444-Y

Doi, R. H., & Kosugi, A. (2004). Cellulosomes: Plant-cell-wall-degrading enzyme complexes. *Nature Reviews Microbiology*, 2(7), 541–551. https://doi.org/10.1038/nrmicro925

Falade, A. O. (2021). Valorization of agricultural wastes for production of biocatalysts of environmental significance: Towards a sustainable environment. *Environmental Sustainability*, 4(2), 317–328. https://doi.org/10.1007/S42398-021-00183-9

Fan, L.-H., Zhang, Z.-J., Yu, X.-Y., Xue, Y.-X., & Tan, T.-W. (2012). Self-surface assembly of cellulosomes with two miniscaffoldins on *Saccharomyces cerevisiae* for cellulosic ethanol production. *Proceedings of the National Academy of Sciences*, 109(33), 13260–13265. https://doi.org/10.1073/PNAS.1209856109

Fontes, C. M., & Gilbert, H. J. (2010). Cellulosomes: Highly efficient nanomachines designed to deconstruct plant cell wall complex carbohydrates. *Annual Review of Biochemistry*, 79, 655–681. https://doi.org/10.1146/ANNUREV-BIOCHEM-091208-085603

Fujita, Y., Ito, J., Ueda, M., Fukuda, H., & Kondo, A. (2004). Synergistic saccharification, and direct fermentation to ethanol, of amorphous cellulose by use of an engineered yeast strain codisplaying Three types of cellulolytic enzyme. *Applied and Environmental Microbiology*, 70(2), 1207. https://doi.org/10.1128/AEM.70.2.1207-1212.2004

Gera, N., Hussain, M., & Rao, B. M. (2013). Protein selection using yeast surface display. *Methods (San Diego, Calif.)*, 60(1), 15–26. https://doi.org/10.1016/J.YMETH.2012.03.014

Goswami, R.K., Agrawal, K., & Verma, P. (2021). Bioremediation of heavy metals from wastewater: a current perspective on microalgae-based future. *Letters in Applied Microbiology*, 1–21.

Gu, H., Zhu, Y., Peng, Y., Liang, X., Liu, X., Shao, L., ... Li, J. (2019). Physiological mechanism of improved tolerance of *Saccharomyces cerevisiae* to lignin-derived phenolic acids in lignocellulosic ethanol fermentation by short-term adaptation. *Biotechnology for Biofuels*, 12(1), 1–14. https://doi.org/10.1186/s13068-019-1610-9

Guirimand, G., Inokuma, K., Bamba, T., Matsuda, M., Morita, K., Sasaki, K., ... Kondo, A. (2019). Cell-surface display technology and metabolic engineering of *Saccharomyces cerevisiae* for enhancing xylitol production from woody biomass. *Green Chemistry*, 21(7), 1795–1808. https://doi.org/10.1039/C8GC03864C

Guo, W., Chen, Y., Wei, N., & Feng, X. (2016). Investigate the metabolic reprogramming of *Saccharomyces cerevisiae* for enhanced resistance to mixed fermentation inhibitors via 13C metabolic flux analysis. *PLOS ONE*, 11(8), e0161448. https://doi.org/10.1371/JOURNAL.PONE.0161448

Ha, S.-J., Galazka, J. M., Kim, S. R., Choi, J.-H., Yang, X., Seo, J.-H., ... Jin, Y.-S. (2011). Engineered *Saccharomyces cerevisiae* capable of simultaneous cellobiose and xylose fermentation. *Proceedings of the National Academy of Sciences*, 108(2), 504–509. https://doi.org/10.1073/PNAS.1010456108

Hahn-Hägerdal, B., Karhumaa, K., Fonseca, C., Spencer-Martins, I., & Gorwa-Grauslund, M. F. (2007). Towards industrial pentose-fermenting yeast strains. *Applied Microbiology and Biotechnology*, 74(5), 937–953. https://doi.org/10.1007/S00253-006-0827-2

Heo, H. S., Park, H. J., Park, Y. K., Ryu, C., Suh, D. J., Suh, Y. W., ... Kim, S. S. (2010). Bio-oil production from fast pyrolysis of waste furniture sawdust in a fluidized bed. *Bioresource Technology*, 101(1), S91–S96. https://doi.org/10.1016/J.BIORTECH.2009.06.003

Hill, J. (2007). Environmental costs and benefits of transportation biofuel production from food- and lignocellulose-based energy crops. A review. *Agronomy for Sustainable Development*, 27(1), 1–12. https://doi.org/10.1051/AGRO:2007006

Ho, N. W., Chen, Z., & Brainard, A. P. (1998). Genetically engineered *Saccharomyces* yeast capable of effective cofermentation of glucose and xylose. *Applied and Environmental Microbiology*, 64(5), 1852–1859. https://doi.org/10.1128/AEM.64.5.1852-1859.1998

Hoornweg, D., & Bhada-Tata, P. (2012). What a waste: A global review of solid waste management, Urban development series; knowledge papers no. 15. World Bank, Washington, DC https://openknowledge.worldbank.org/handle/10986/17388

Huang, G. L., & Clubb, R. T. (2017). Progress towards engineering microbial surfaces to degrade biomass. *Biomass Volume Estimation and Valorization for Energy*. https://doi.org/10.5772/65509

Inokuma, K., Bamba, T., Ishii, J., Ito, Y., Hasunuma, T., & Kondo, A. (2016). Enhanced cell-surface display and secretory production of cellulolytic enzymes with *Saccharomyces cerevisiae* Sed1 signal peptide. *Biotechnology and Bioengineering*, 113(11), 2358–2366. https://doi.org/10.1002/bit.26008

Jemli, S., Ayadi-Zouari, D., Hlima, H. B., & Bejar, S. (2016). Biocatalysts: Application and engineering for industrial purposes. *Critical Reviews in Biotechnology*, 36(2), 246–258. https://doi.org/10.3109/07388551.2014.950550

Jia, H.-R., Zhu, Y.-X., Chen, Z., Wu, F.-G. (2017). Cholesterol-assisted bacterial cell surface engineering for photodynamic inactivation of gram-positive and gram-negative bacteria. *ACS Applied Materials and Interfaces*, 9(19), 15943–15951. https://doi.org/10.1021/ACSAMI.7B02562

Jonathan, K Pokorski, & Nicole F. Steinmetz (2011). The art of engineering viral nanoparticles. *Molecular Pharmaceutics*, 8(1), 29–43. https://doi.org/10.1021/MP100225Y

Kahn, A., Bayer, E. A., & Moraïs, S. (2018). Advanced cloning tools for construction of designer cellulosomes. *Methods in Molecular Biology (Clifton, N.J.)*, 1796, 135–151. https://doi.org/10.1007/978-1-4939-7877-9_11

Kashyap, S., Chandra, R., Kumar, B., & Verma, P. (2022). Biosorption efficiency of nickel by various endophytic bacterial strains for removal of nickel from electroplating industry effluents: an operational study. *Ecotoxicology*, 31, 565–580.

Katahira, S., Fujita, Y., Mizuike, A., Fukuda, H., & Kondo, A. (2004). Construction of a xylan-fermenting yeast strain through codisplay of xylanolytic enzymes on the surface of xylose-utilizing *Saccharomyces cerevisiae* cells. *Applied and Environmental Microbiology*, 70(9), 5407–5414. https://doi.org/10.1128/AEM.70.9.5407-5414.2004

Katahira, S., Mizuike, A., Fukuda, H., & Kondo, A. (2006). Ethanol fermentation from lignocellulosic hydrolysate by a recombinant xylose-and cellooligosaccharide-assimilating yeast strain. *Applied Microbiology and Biotechnology*, 72(6), 1136–1143. https://doi.org/10.1007/s00253-006-0402-x

Ulaganathan, K., Goud, S., Reddy, M., & Kayalvili, U. (2017). Genome engineering for breaking barriers in lignocellulosic bioethanol production. *Renewable and Sustainable Energy Reviews*, 74, 1080–1107. https://doi.org/10.1016/J.RSER.2017.01.028

Kim, H. S., Kim, N. R., Yang, J, & Choi, W. (2011). Identification of novel genes responsible for ethanol and/or thermotolerance by transposon mutagenesis in *Saccharomyces cerevisiae*. *Applied Microbiology and Biotechnology*, 91(4), 1159–1172. https://doi.org/10.1007/S00253-011-3298-Z

Klein-Marcuschamer, D., Oleskowicz-Popiel, P., Simmons, B. A., & Blanch, H. W. (2012). The challenge of enzyme cost in the production of lignocellulosic biofuels. *Biotechnology and Bioengineering*, 109(4), 1083–1087. https://doi.org/10.1002/BIT.24370

Kondo, A., & Ueda, M. (2004). Yeast cell-surface display—applications of molecular display. *Applied Microbiology and Biotechnology*, 64(1), 28–40. https://doi.org/10.1007/S00253-003-1492-3

Konnova, S. A., Lvov, Y. M., & Fakhrullin, R. F. (2016). Nanoshell assembly for magnet-responsive oil-degrading bacteria. *Langmuir*, 32(47), 12552–12558. https://doi.org/10.1021/ACS.LANGMUIR.6B01743

Koppram, R., Albers, E., & Olsson, L. (2012). Evolutionary engineering strategies to enhance tolerance of xylose utilizing recombinant yeast to inhibitors derived from spruce biomass. *Biotechnology for Biofuels*, 5, 32. https://doi.org/10.1186/1754-6834-5-32

Kotaka, A., Bando, H., Kaya, M., Kato-Murai, M., Kuroda, K., Sahara, H., … Ueda, M. (2008). Direct ethanol production from barley β-glucan by sake yeast displaying *Aspergillus oryzae* β-glucosidase and endoglucanase. *Journal of Bioscience and Bioengineering*, 105(6), 622–627. https://doi.org/10.1263/jbb.105.622

Kumar, B., Agrawal, K., & Verma, P. (2021). Current perspective and advances of microbe assisted electrochemical system as a sustainable approach for mitigating toxic dyes and heavy metals from wastewater. *ASCE's Journal of Hazardous, Toxic, and Radioactive Waste*, 25(2), 04020082.

Kumar, B., Bhardwaj, N., Agrawal, K., & Verma, P. (2020). Bioethanol production: Generation-based comparative status measurements. In: *Biofuel Production Technologies: Critical Analysis for Sustainability*, 155–201. Springer, Singapore.

Kumar, B., & Verma, P. (2020a). Application of hydrolytic enzymes in biorefinery and its future prospects. In: *Microbial Strategies for Techno-economic Biofuel Production*, 59–83. Springer.

Kumar, B., & Verma, P. (2020b). Biomass-based biorefineries: An important architype towards a circular economy. *Fuel*, 288, 119622. Elsevier. https://doi.org/10.1016/j.fuel.2020.119622

Kuroda, K., & Ueda, M. (2011). Cell surface engineering of yeast for applications in white biotechnology. *Biotechnology Letters*, 33(1), 1–9. https://doi.org/10.1007/S10529-010-0403-9

Kuroda, K., Shibasaki, S., Ueda, M., & Tanaka, A. (2001). Cell surface-engineered yeast displaying a histidine oligopeptide (hexa-His) has enhanced adsorption of and tolerance to heavy metal ions. *Applied Microbiology and Biotechnology*, 57(5–6), 697–701. https://doi.org/10.1007/s002530100813

Kurylenko, O., Semkiv, M,. Ruchala, J., Hryniv, O., Kshanovska, B., Abbas, C., … Sibirny, A. (2016). New approaches for improving the production of the 1st and 2nd generation ethanol by yeast. *Acta Biochimica Polonica*, 63(1), 31–38. https://doi.org/10.18388/ABP.2015_1156

Larsson, S., Quintana-Sáinz, A., Reimann, A., Nilvebrant, N.-O., & Jönsson, L. J. (2000). Influence of lignocellulose-derived aromatic compounds on oxygen-limited growth and ethanolic fermentation by *Saccharomyces cerevisiae*. *Twenty-First Symposium on Biotechnology for Fuels and Chemicals*, 617–632. Springer. https://doi.org/10.1007/978-1-4612-1392-5_47

Larsson, S., Reimann, A., Nilvebrant, N.-O., & Jönsson, L. J. (1999). Comparison of different methods for the detoxification of lignocellulose hydrolyzates of spruce. *Applied Biochemistry and Biotechnology*, 77(1), 91–103. https://doi.org/10.1385/ABAB:77:1-3:91

Lee, S. Y., Choi, J. H., & Xu, Z. (2003). Microbial cell-surface display. *Trends in Biotechnology*, 21(1), 45–52. https://doi.org/10.1016/S0167-7799(02)00006-9

Lesmana, S. O., Febriana, N., Soetaredjo, F. E., Sunarso, J., & Ismadji, S. (2009). Studies on potential applications of biomass for the separation of heavy metals from water and wastewater. *Biochemical Engineering Journal*, 44(1), 19–41. https://doi.org/10.1016/J.BEJ.2008.12.009

Li, Y.-C., Gou, Z.-X., Zhang, Y., Xia, Z.-Y., Tang, Y.-Q., & Kida, K. (2017). Inhibitor tolerance of a recombinant flocculating industrial *Saccharomyces cerevisiae* strain during glucose and xylose co-fermentation. *Brazilian Journal of Microbiology*, 48, 791–800. https://doi.org/10.1016/j.bjm.2016.11.011

Li, P., Jiang, Y., Song, R.-B., Zhang, J.-R., & Zhu, J.-J. (2021). Layer-by-layer assembly of Au and CdS nanoparticles on the surface of bacterial cells for photo-assisted bioanodes in microbial fuel cells. *Journal of Materials Chemistry B*, 9(6), 1638–1646. https://doi.org/10.1039/D0TB02642E

Liu, Z., Ho, S. H., Hasunuma, T., Chang, J. S., Ren, N. Q., & Kondo, A. (2016). Recent advances in yeast cell-surface display technologies for waste biorefineries. *Bioresource Technology*, 215, 324–333. https://doi.org/10.1016/J.BIORTECH.2016.03.132

Liu, Y. J., Li, B., Feng, Y., & Cui, Q. (2020). Consolidated bio-saccharification: Leading lignocellulose bioconversion into the real world. *Biotechnology Advances*, 40, 107535. https://doi.org/10.1016/J.BIOTECHADV.2020.107535

Liu, Y., Zhang, R., Lian, Z., Wang, S., & Wright, A. T. (2014). Yeast cell surface display for lipase whole cell catalyst and its applications. *Journal of Molecular Catalysis B: Enzymatic*, 106, 17–25. https://doi.org/10.1016/J.MOLCATB.2014.04.011

Luque, R., Menéndez, J. A., Arenillas, A., & Cot, J. (2012). Microwave-assisted pyrolysis of biomass feedstocks: The way forward? *Energy & Environmental Science*, 5(2), 5481–5488. https://doi.org/10.1039/C1EE02450G

Lynd, L. R., Van Zyl, W. H., McBride, J. E., & Laser, M. (2005). Consolidated bioprocessing of cellulosic biomass: An update. *Current Opinion in Biotechnology*, 16(5), 577–583. https://doi.org/10.1016/J.COPBIO.2005.08.009

Mahmood, Q., Rashid, A., Ahmad, S. S., Azim, M. R., & Bilal, M. (2012). Current status of toxic metals addition to environment and its consequences. *Environmental Pollution*, 21, 35–69. https://doi.org/10.1007/978-94-007-3913-0_2

Manisha, & Yadav, S. K. (2017). Technological advances and applications of hydrolytic enzymes for valorization of lignocellulosic biomass. *Bioresource Technology*, 245, 1727–1739. https://doi.org/10.1016/J.BIORTECH.2017.05.066

Mbaneme-Smith, V., & Chinn, M. S. (2015). Consolidated bioprocessing for biofuel production: Recent advances. *Energy and Emission Control Technologies*, 3, 23–44. https://doi.org/10.2147/EECT.S63000

Moraïs, S., Morag, E., Barak, Y., Goldman, D., Hadar, Y., Lamed, R., … Bayera, E. A. (2012). Deconstruction of lignocellulose into soluble sugars by native and designer cellulosomes. *MBio*, 3(6). https://doi.org/10.1128/MBIO.00508-12

Murai, Toshiyuki, Ueda, M., Kawaguchi, T., Arai, M., & Tanaka, A. (1998). Assimilation of cellooligosaccharides by a cell surface-engineered yeast expressing β-glucosidase and carboxymethylcellulase from *Aspergillus aculeatus*. *Applied and Environmental Microbiology*, 64(12), 4857–4861. https://doi.org/10.1128/AEM.64.12.4857-4861.1998

Murai, T, Ueda, M., Shibasaki, Y., Kamasawa, N., Osumi, M., Imanaka, T., & Tanaka, A. (1999). Development of an arming yeast strain for efficient utilization of starch by co-display of sequential amylolytic enzymes on the cell surface. *Applied Microbiology and Biotechnology*, 51(1), 65–70. https://doi.org/10.1007/s002530051364

Murai, Toshiyuki, Ueda, M., Yamamura, M., Atomi, H., Shibasaki, Y., Kamasawa, N., ... Tanaka, A. (1997). Construction of a starch-utilizing yeast by cell surface engineering. *Applied and Environmental Microbiology*, 63(4), 1362–1366. https://doi.org/10.1128/aem.63.4.1362-1366.1997

Mutanda, I., Li, J., Xu, F., & Wang, Y. (2021). Recent advances in metabolic engineering, protein engineering, and transcriptome-guided insights toward synthetic production of taxol. *Frontiers in Bioengineering and Biotechnology*, 34. https://doi.org/10.3389/FBIOE.2021.632269

Narita, J., Okano, K., Kitao, T., Ishida, S., Sewaki, T., Sung, M.-H., ... Kondo, A. (2006). Display of α-amylase on the surface of *Lactobacillus casei* cells by use of the PgsA anchor protein, and production of lactic acid from starch. *Applied and Environmental Microbiology*, 72(1), 269–275. https://doi.org/10.1128/AEM.72.1.269-275.2006

Nishitani, T., Shimada, M., Kuroda, K., & Ueda, M. (2010). Molecular design of yeast cell surface for adsorption and recovery of molybdenum, one of rare metals. *Applied Microbiology and Biotechnology*, 86(2), 641–648. https://doi.org/10.1007/s00253-009-2304-1

Olson, D. G., McBride, J. E., Joe Shaw, A., & Lynd, L. R. (2012). Recent progress in consolidated bioprocessing. *Current Opinion in Biotechnology*, 23(3), 396–405. https://doi.org/10.1016/J.COPBIO.2011.11.026

Orlean, P. (2012). Architecture and Biosynthesis of the Saccharomyces cerevisiae cell wall. *Genetics*, 192(3), 775–818. https://doi.org/10.1534/GENETICS.112.144485

Öztop, H. N., Öztop, A. Y., Karadağ, E., Işikver, Y., & Saraydin, D. (2003). Immobilization of *Saccharomyces cerevisiae* on to acrylamide–sodium acrylate hydrogels for production of ethyl alcohol. *Enzyme and Microbial Technology*, 32(1), 114–119. https://doi.org/10.1016/S0141-0229(02)00244-2

Palmqvist, E., & Hahn-Hägerdal, B. (2000). Fermentation of lignocellulosic hydrolysates. II: Inhibitors and mechanisms of inhibition. *Bioresource Technology*, 74(1), 25–33. https://doi.org/10.1016/S0960-8524(99)00161-3

Parisutham, V., Kim, T. H., & Lee, S. K. (2014). Feasibilities of consolidated bioprocessing microbes: From pretreatment to biofuel production. *Bioresource Technology*, 161, 431–440. https://doi.org/10.1016/J.BIORTECH.2014.03.114

Rakestraw, J. A., Sazinsky, S. L., Piatesi, A., Antipov, E., & Wittrup, K. D. (2009). Directed evolution of a secretory leader for the improved expression of heterologous proteins and full-length antibodies in S. cerevisiae. *Biotechnology and Bioengineering*, 103(6), 1192. https://doi.org/10.1002/BIT.22338

Sanchez, B., & Bautista, J. (1988). Effects of furfural and 5-hydroxymethylfurfural on the fermentation of *Saccharomyces cerevisiae* and biomass production from *Candida guilliermondii*. *Enzyme and Microbial Technology*, 10(5), 315–318. https://doi.org/10.1016/0141-0229(88)90135-4

Sanda, T., Hasunuma, T., Matsuda, F., & Kondo, A. (2011). Repeated-batch fermentation of lignocellulosic hydrolysate to ethanol using a hybrid *Saccharomyces cerevisiae* strain metabolically engineered for tolerance to acetic and formic acids. *Bioresource Technology*, 102(17), 7917–7924. https://doi.org/10.1016/J.BIORTECH.2011.06.028

Schwarz, W. (2001). The cellulosome and cellulose degradation by anaerobic bacteria. *Applied Microbiology and Biotechnology*, 56(5), 634–649. https://doi.org/10.1007/S002530100710

Serrano-Ruiz, J. C., Luque, R., Campelo, J. M., & Romero, A. A. (2012). Continuous-flow processes in heterogeneously catalyzed transformations of biomass derivatives into fuels and chemicals. *Challenges*, 3(2), 114–132. https://doi.org/10.3390/CHALLE3020114

Shibasaki, S., Ninomiya, Y., Ueda, M., Iwahashi, M., Katsuragi, T, Tani, Y …. Tanaka, A. (2001). Intelligent yeast strains with the ability to self-monitor the concentrations of intra- and extracellular phosphate or ammonium ion by emission of fluorescence from the cell surface. *Applied Microbiology and Biotechnology*, 57(5–6), 702–707. https://doi.org/10.1007/S00253-001-0849-8

Shigechi, H., Koh, J., Fujita, Y., Matsumoto, T., Bito, Y., Ueda, M., … Kondo, A. (2004). Direct production of ethanol from raw corn starch via fermentation by use of a novel surface-engineered yeast strain codisplaying glucoamylase and α-amylase. *Applied and Environmental Microbiology*, 70(8), 5037–5040. https://doi.org/10.1128/AEM.70.8.5037-5040.2004

Shima, J., Sakata-Tsuda, Y., Suzuki, Y., Nakajima, R., Watanabe, H., Kawamoto, S., Takano, H. (2003). Disruption of the CAR1 gene encoding arginase enhances freeze tolerance of the commercial baker's Yeast *Saccharomyces cerevisiae*. *Applied and Environmental Microbiology*, 69(1), 715. https://doi.org/10.1128/AEM.69.1.715-718.2003

Song, X., Li, Y., Wu, Y., Cai, M., Liu, Q., Gao, K., … Qiao, M. (2018). Metabolic engineering strategies for improvement of ethanol production in cellulolytic *Saccharomyces cerevisiae*. FEMS Yeast Research, 18(8), 90. https://doi.org/10.1093/FEMSYR/FOY090

Song, Y., Matsumoto, K., Tanaka, T., Kondo, A., Taguchi, S. (2013). Single-step production of polyhydroxybutyrate from starch by using α-amylase cell-surface displaying system of *Corynebacterium glutamicum*. *Journal of Bioscience and Bioengineering*, 115(1), 12–14. https://doi.org/10.1016/j.jbiosc.2012.08.004

Sonnleitner, B., & Fiechter, A. (1983). Advantages of using thermophiles in biotechnological processes: Expectations and reality. *Trends in Biotechnology*, 1(3), 74–80. https://doi.org/10.1016/0167-7799(83)90056-2

Taherzadeh, Mohammad J., Eklund, Robert, Gustafsson, Lena, Niklasson, Claes, Lidén, G. (1997). Characterization and fermentation of dilute-acid hydrolyzates from wood. *Industrial and Engineering Chemistry Research*, 36(11), 4659–4665. https://doi.org/10.1021/IE9700831

Takagi, T., Morisaka, H., Aburaya, S., Tatsukami, Y., Kuroda, K., & Ueda, M. (2016). Putative alginate assimilation process of the marine bacterium *Saccharophagus degradans* 2–40 based on quantitative proteomic analysis. *Marine Biotechnology*, 18(1), 15–23. https://doi.org/10.1007/s10126-015-9667-3

Tanaka, T., & Kondo, A. (2015). Cell surface engineering of industrial microorganisms for biorefining applications. B*iotechnology Advances*, 33(7), 1403–1411. https://doi.org/10.1016/J.BIOTECHADV.2015.06.002

Tanaka, T., Yamada, R., Ogino, C., & Kondo, A. (2012). Recent developments in yeast cell surface display toward extended applications in biotechnology. *Applied Microbiology and Biotechnology*, 95(3), 577–591. https://doi.org/10.1007/S00253-012-4175-0

Tateno, T., Okada, Y., Tsuchidate, T., Tanaka, T., Fukuda, H., & Kondo, A. (2009). Direct production of cadaverine from soluble starch using *Corynebacterium glutamicum* coexpressing α-amylase and lysine decarboxylase. *Applied Microbiology and Biotechnology*, 82(1), 115–121. https://doi.org/10.1007/s00253-008-1751-4

Tsai, S.-L., Oh, J., Singh, S., Chen, R., & Chen, W. (2009). Functional assembly of minicellulosomes on the *Saccharomyces cerevisiae* cell surface for cellulose hydrolysis and ethanol production. *Applied and Environmental Microbiology*, 75(19), 6087–6093. https://doi.org/10.1128/AEM.01538-09

Tyo, K. E., Alper, H. S., & Stephanopoulos, G. N. (2007). Expanding the metabolic engineering toolbox: More options to engineer cells. *Trends in Biotechnology*, 25(3), 132–137. https://doi.org/10.1016/J.TIBTECH.2007.01.003

Ueda, M., & Tanaka, A. (2000). Cell surface engineering of yeast: Construction of arming yeast with biocatalyst. *Journal of Bioscience and Bioengineering*, 90(2), 125–136. https://doi.org/10.1016/S1389-1723(00)80099-7

van Zyl, W. H., Lynd, L. R., den Haan, R., & McBride, J. E. (2007). Consolidated bioprocessing for bioethanol production using *Saccharomyces cerevisiae*. *Advances in Biochemical
Engineering/Biotechnology*, 108, 205–235. https://doi.org/10.1007/10_2007_061

Wang, H. C., Ho, C. H., Chou, C. C., Ko, T. P., Huang, M. F., ... Wang, A. H. (2016). Using
structural-based protein engineering to modulate the differential inhibition effects of
SAUGI on human and HSV uracil DNA glycosylase. *Nucleic Acids Research*, 44(9),
4440–4449. https://doi.org/10.1093/NAR/GKW185

Wei, Q., Yan, J., Chen, Y., Zhang, L., Wu, X., Shang, S., ... Zhang, H. (2018). Cell surface
display of MerR on *Saccharomyces cerevisiae* for biosorption of mercury. *Molecular
Biotechnology*, 60(1), 12–20. https://doi.org/10.1007/s12033-017-0039-2

Wen, F., Sun, J., & Zhao, H. (2010). Yeast surface display of trifunctional minicellulo-
somes for simultaneous saccharification and fermentation of cellulose to ethanol.
Applied and Environmental Microbiology, 76(4), 1251–1260. https://doi.org/10.1128/
AEM.01687-09

Wiselogel, A., Tyson, S., & Johnson, D. (2018). Biomass feedstock resources and composition. *Handbook on Bioethanol*, 105–118. https://doi.org/10.1201/9780203752456-6

Wu, S. J., Luo, J, O'Neil, K. T., Kang, J., Lacy, E. R., Canziani, G., ... Feng, Y. (2010).
Structure-based engineering of a monoclonal antibody for improved solubility. *Protein
Engineering, Design & Selection: PEDS*, 23(8), 643–651. https://doi.org/10.1093/
PROTEIN/GZQ037

Wulff, N. A., Carrer, H., & Pascholati, S. F. (2006). Expression and purification of cellulase
Xf818 from *Xylella fastidiosa* in *Escherichia coli*. *Current Microbiology*, 53(3), 198–
203. https://doi.org/10.1007/S00284-005-0475-2

Wyman, C. E. (2007). What is (and is not) vital to advancing cellulosic ethanol. Trends in
Biotechnology, 25(4), 153–157. https://doi.org/10.1016/J.TIBTECH.2007.02.009

Yang, S. H., Kang, S. M., Lee, K.-B., Chung, T. D., Lee, H., & Choi, I. S. (2011). Mussel-
inspired encapsulation and functionalization of individual yeast cells. *Journal of the
American Chemical Society*, 133(9), 2795–2797. https://doi.org/10.1021/JA1100189

Ye, M., Ye, Y., Du, Z., & Chen, G. (2021). Cell-surface engineering of yeasts for whole-cell
biocatalysts. *Bioprocess and Biosystems Engineering*, 44(6), 1003–1019. https://doi.
org/10.1007/S00449-020-02484-5

Young, E., Lee, S.-M., & Alper, H. (2010). Optimizing pentose utilization in yeast: The need
for novel tools and approaches. *Biotechnology for Biofuels*, 3(1), 1–12. https://doi.
org/10.1186/1754-6834-3-24

Zaldivar, J., Nielsen, J., & Olsson, L. (2001). Fuel ethanol production from lignocellulose: A
challenge for metabolic engineering and process integration. *Applied Microbiology and
Biotechnology*, 56(1–2), 17–34. https://doi.org/10.1007/S002530100624

Zhao, E. M., Zhang, Y., Mehl, J., Park, H., Lalwani, M. A., Toettcher, J. E., & Avalos, J. L.
(2018). Optogenetic regulation of engineered cellular metabolism for microbial chemical production. *Nature*, 555(7698), 683–687. https://doi.org/10.1038/nature26141

Zheng, Z., Chen, T., Zhao, M., Wang, Z., & Zhao, X. (2012a). Engineering *Escherichia coli*
for succinate production from hemicellulose via consolidated bioprocessing. *Microbial
Cell Factories*, 11(1), 1–11. https://doi.org/10.1186/1475-2859-11-3

Zheng, C., Sun, X., Li, L., & Guan, N. (2012b). Scaling up of ethanol production from sugar
molasses using yeast immobilized with alginate-based MCM-41 mesoporous zeolite
composite carrier. *Bioresource Technology*, 115, 208–214. https://doi.org/10.1016/J.
BIORTECH.2011.11.056

10 Economics of the Biochemical Conversion-based Biorefinery Concept for the Valorization of Lignocellulosic Biomass

Praveen Kumar Ghodke
National Institute of Technology Calicut, Kozhikode, India

Sumit Dhawane
National Institute of Technology (MANIT), Bhopal, India

Cecil Antony
National Institute of Technology Calicut, Kozhikode, India

CONTENTS

DOI: 10.1201/9781003187721-10

10.1   INTRODUCTION

The generation of municipal solid waste (MSW) is one of humanity's biggest concerns today. Increasing MSW is due to a rising urban population and a linear economy. In recent years, the concept of a circular economy has gained popularity (Vea et al., 2018). In a closed-loop system, materials are either recycled endlessly without degradation or returned to the natural ecosystem without harming the environment (Peeters et al., 2014). Recognizing the urgency of urban waste management and the scarcity of resources, we must rethink our economic restructuring. Resources will become scarcer in the future, making recovery and management skills vital for a sustainable global economy (Dahiya et al., 2018).

Biorefinery is defined as "the sustainable processing of biomass into a spectrum of bio-based products including food, feed, chemicals, and/or materials and bioenergy includes biofuels, power, and/or heat" (Bioenergy, 2014). Almost any type of biomass can be used as a viable feedstock in a biorefinery as long as adequate conversion and valorization processes are available (Kumar et al., 2020a). Various feedstocks include biomass from agriculture residue, forestry waste, macro/microalgae, industrial residual waste, household waste, and other organic residues (Pandey et al., 2015). A near-zero-waste biorefinery process would use sequential extraction, fractionation, and conversion operations, and be further subjected to the conversion process of biochemical and thermochemical, with continuous wastestream recycling and energy recovery. (Mishra et al., 2019). As a result, a biorefinery necessitates a multidisciplinary approach that includes chemical sciences, chemical technology, and engineering, along with biochemical engineering, chemistry, biochemistry, biology, and materials science. Although historically biomass was used for heating and building materials, modern biorefineries produce liquid transportation fuels and commercial chemicals from biomass. However, despite the economic and environmental benefits of bio-based chemicals and products, technological challenges remain in the way of widespread industrial biorefinery use. The ideal biorefinery concept focuses on sustainable, catalytic processing of renewable feedstocks while maximizing energy efficiency and minimizing environmental impact (Huber et al., 2006; Agrawal and Verma, 2020). The lignocellulosic biorefinery requires significant research and development to be sustainable and economically successful.

Enzymes are critical in the industry because of their substrate and product specificity, their ability to operate under mild reaction conditions, their low by-product generation, and their high yields of products. They play a significant role in a variety of products and manufacturing processes (Kumar et al., 2020a). All living systems contain enzymes, which act as biological catalysts (Hinnemann et al., 2005). Enzymes are proteinaceous and catalyze several processes. Enzymes have been employed involuntarily for millennia to make wine, cheese, bread, beer, and vinegar.

Enzymes are used in the production of leather and linen (Ravindran and Jaiswal, 2016). Purified enzymes have just recently become widely used in manufacturing processes (Sharma et al., 2007). The expense of using enzymes in industrial processes is a big issue. Enzyme production is a capital-intensive process, and the use of enzymes in many manufacturing processes indirectly influences product pricing (Kumar and Verma, 2021). Plant equipment and installation account for a large portion of annual operating costs. However, raw materials constitute 28% of operating costs (Sharma et al., 2007).

In 2018, India generated 3.3 million metric tons (Central Pollution Control Board (CPCB) report)) of municipal solid waste ((CPCB) and Ministry of Environment, 2015), of which 40–60% was organic waste. High moisture and salt content in organic waste lead to fast breakdown and odor (Abdel-Shafy and Mansour, 2018). If not handled effectively, it can cause greenhouse gas emissions (mainly methane emissions), leachate, and hygiene issues. But it can also be used to produce renewable energy and value-added products like enzymes, organic fertilizers, biopesticides, and bioplastics (Raheem et al., 2018; Sadh et al., 2018; Tyagi and Lo, 2013). Biorefineries use biomass as a feedstock to produce biobased products. The use of wastestreams from agricultural and food production (e.g., bakeries and breweries) has recently gained popularity. Research on the biorefinery of agricultural and food processing wastes have been published, but none on lignocellulose biomass with low-lignin content as feedstock (Hollins et al., 2017; Maria et al., 2012; Zaher et al., 2007). One study also examines the best lignocellulose biorefinery output products and processes.

This study, on the other hand, explores low-lignin lignocellulose biomass as a possible biobased circular economy feedstock. As shown in Figure 10.1, it identifies relevant biorefinery concepts to provide an overview of potential valorization strategies for lignocellulose biomass with low-lignin content. Table 10.1 reports a list of some existing technologies, which differ based on the heterogeneity of the feedstocks, the feedstock structure and chemistry, and the streams of products. To achieve this study's objectives, numerous identified studies that use low-lignin biomass as feedstock to produce commercial chemical and value-added products (beyond

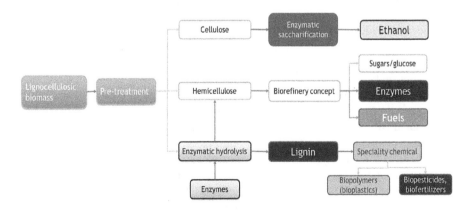

FIGURE 10.1 Biorefinery concept of lignocellulosic biomass.

TABLE 10.1

Feedstocks Used in the Biorefinery Concept (Amore et al., 2016)

Feedstock	Product Valorization	Organization
Wheat and sugar beet	Paper and pulp, ethanol, and succinic acid.	Les Sohettes' complex, France.
Corncobs and corn	Ethanol, corn oil, white sugar, organic acids.	Cargill Inc., USA
Agricultural residues and fiber crops including hemp, miscanthus, and Provence cane	Paper and pulp, sugar syrup, and bioethanol.	CIMV built, France
Municipal solid waste	Butadiene, acetic acid, isobutylene.	Lanzatech, India
Sugarcane, straw, and bagasse	Ethanol.	GranBio, Brazil
Corn starch	Biopolymers, and starch-derived thermoplastics.	Novamont, Italy
Forestry wood	Energy, newsprint, and wood chips.	West Fraser, North America

bioenergy) are reviewed, Additionally, biorefinery conversion efficiencies are assessed and future biorefinery bioeconomyis estimated.

10.2 MARKET POTENTIAL

The worldwide enzyme industry is rapidly expanding. It was valued at over USD 9.9 billion in 2019 and is expected to be worth a whopping USD 15.92 billion by 2027 (Cripwell et al., 2021). Enzymes are primarily employed in the manufacture of a variety of products that individuals use on a daily basis. Furthermore, the renewed interest in bioenergy has resulted in an increase in the demand for enzymes that may be used in the biofuel industry (Mittal and Decker, 2013; Bhardwaj and Verma, 2021). With the aid of recombinant DNA technology and protein engineering, enzymes may be produced for specific applications in a variety of industrial processes. Cellulosic ethanol costs are heavily reliant on the cost of enzymes, and reducing enzyme costs can boost the market potential for biofuels (Yukesh Kannah et al., 2020; Kumar et al., 2020b).

The worldwide bioplastics market was estimated at USD 4.6 billion in 2019, with a compound annual growth rate (CAGR) of 13.8% projected from 2020 to 2027 (Ummalyma et al., 2020). Growing population and urbanization, as well as increased awareness of health-related concerns in Asia Pacific's rising economies, are projected to benefit end-use industries, resulting in increased demand for bioplastics throughout the forecast period. Furthermore, the packaging industry's growing product demand will propel the market even higher. Agriculture, consumer products, textile, automotive and transportation, and building and construction are just a few of the industries that employ bioplastics. This is also expected to aid market expansion. Bioplastics have a unique potential and may thus effectively assist the reduction of greenhouse gas (GHG) emissions (Shrestha et al., 2020).

From an anticipated value of USD 4.3 billion in 2020 to USD 8.5 billion in 2025, the worldwide biopesticides market is expected to develop at a CAGR of 14.7%.

Synthetic chemicals may pollute the environment and contaminate the soil, as well as have negative consequences for the food chain. As a result of this concern, there has been a rise in awareness of residue-free food, with biological products receiving special attention (Miranda et al., 2021).

There is a large scope and need for biopesticide products due to causes such as the prohibition on chemical pesticides in key nations due to worsening soil conditions, extended agriculture practices, and rising worry about residue levels in food items. Fruits and vegetables, cereals and grains, and oilseeds and pulses are the most common biopesticide crops. In recent years, there has been a rise in demand for biopesticides to manage pests in recreational parks, amusement parks, and golf courses. Along with biopesticides, the organic fertilizers market is expected to be worth USD 6.30 billion in 2017 and USD 11.16 billion by 2022, growing at a CAGR of 12.08% during that time period (Lin et al., 2019).

10.3 LIGNOCELLULOSIC BIOMASS WITH LOW-LIGNIN COMPOSITION

Several variables contribute to the variability of lignocellulosic biomass with low lignin content, often known as "green waste". To begin with, the types of plants that grow in green waste differ based on where the waste was collected. Dead eucalyptus leaves, for example, are common in Singapore (Langsdorf et al., 2021), although they are rarely found in other parts of the world. Differences in composition and physical shapes, as well as seasonal fluctuations, provide transportation and storage issues which makes the green waste vary in composition (Tahmoorian and Khabbaz, 2020). Pretreatment of green waste at the point of the collection could save time and money in the long run. From public parks, organic wastes are reasonably easy for authorities to collect separately, while private garden wastes are rarely separated (Malinauskaite et al., 2017). It is difficult to estimate the amount of green waste generated by private households because kitchen wet waste and garden dry waste are frequently merged into organic waste. Nonetheless, some research on the composition of MSW from specific regions, and cities can be determined. The amount of garden-related green waste and dry wood waste in MSW varies, owing to the presence or absence of private gardens in residential buildings (Chojnacka et al., 2020). Hanc et al. reported disparities in biomass collection from the urban settlement, where individual families have small private gardens in megacities, such as Prague, Czech Republic (Hanc et al., 2011). Boldrin and Christensen have also demonstrated the large seasonal fluctuations in green waste. They discovered in Aarhus, Denmark, the amount of private garden waste ranges from 20.4 kg per person per month in the summer to 5.5 kilograms per person per month in the winter (Boldrin and Christensen, 2010). The study found that summer garden waste was dominated by flowers, grass clippings, leaves, hedge trimmings, and sand/soil, whereas winter garden waste was dominated by woody items, dry leaves, and sticks.

Aside from regional and seasonal variations in green waste content, various factors influence the physicochemical features of specific plants. Carbohydrate polymers like lignin (phenolic compounds), hemicellulose and cellulose, and other biological compounds including proteins, fats, fatty esters, organic acids, minerals, and salts make up the composition of herbaceous green waste materials (Mohapatra et al., 2017). Herbaceous green waste's chemical composition varies based on the type's species (hardwood, softwood), developmental stage, soil type, minerals and nutrients provided, and growth environmental conditions. Herrmann et al. found that the amount of dry volatile matter, cellulose, hemicellulose, lignin, fats, fiber, carbohydrates, protein, and sugar in freshly harvested samples from three different grassland biotopes in north Germany varied depending on the harvested season (Herrmann et al., 2014; Mashingo et al., 2008). The effect of growth conditions on prairie/freshwater cordgrass (*Spartina pectinata* L.) was demonstrated by Kim et al., who found considerable variations in the plant's composition over two years (Kim, 2018). Typical herbaceous protein content for different cordgrass was between 2.5 and 5.5 wt.% for grasslands in Oregon, USA (Juneja et al., 2011), 5.2 to 6.3 wt.% for elephant grass (Menegol et al., 2016), 12.3 wt.% for timothy cordgrass, and 18.5 wt.% for alfalfa. The amount of extractives determined in miscanthus and switchgrass are 6.9 wt.% and 13.6 wt.% respectively (Hitzl et al., 2015). The proportion of fats/lipids in yard waste, grass, and leaves were found to be 2.5 wt.% (Lee et al., 2010). The composition of cellulose, hemicellulose, and lignin in biomass resources is perhaps one of the most essential factors. These three polymers combine to generate lignocellulose, the most abundant component of plant biomass. Unlike cellulose and hemicellulose, which are sugar-based macromolecules, lignin is an aromatic polymer made from phenylpropanoid precursors. Cellulose is a linear polymer made up of D-glucose subunits linked by ß-1,4-glycosidic linkages. Cellobiose molecules combine to create fibrils that are held together by hydrogen bonds and the Van der Waals force.

The composition of lignocellulosic components differs significantly amongst green waste. Table 10.2 presents a summary of the composition of various lignocellulosic green waste with lignin concentrations less than 25 wt.% and greater the 25 wt.%. The composition of lignocellulose varies depending on the species, provenance of the green waste, and other factors.

10.4 BIOTECHNOLOGICAL CONVERSION OF GREEN WASTE

In biotechnological processes, typically microorganisms and enzymes are involved in the conversion of green waste to various value-added compounds. The production of liquid fuels, especially bioethanol and gaseous fuels including biogas are two of the most popular microbial conversion processes (Kumar et al., 2020c, 2020b). The composition of low-lignin green waste varies depending on the origin of the waste and has a significant impact on the yield of subsequent value-added products. Food and vegetable wastes are more interesting since they are quickly degradable and often produce higher yields of bioethanol and biogas than other wastes such as papers and cardboard. The rate-limiting aspect of the process is commonly thought to be the effective degradation of complex polymeric molecules (such as cellulose and proteins) to simple molecules (such as sugars and amino acids). Different research groups have studied a variety of pretreatment procedures such as thermochemical (Sarkar and Praveen, 2017), hydrothermal, and enzymatic hydrolysis, as well as

TABLE 10.2

Lignocellulosic Green Waste Composition (Components; Cellulose, Hemi-cellulose, and Lignin) (Anwar et al., 2014; Huang et al., 2010; Keshav et al., 2021)

Lignocellulosic Green Waste	Cellulose	Hemicellulose	Lignin	Other Materials
Rice straw	42.4	25.3	13.4	10
Cotton stalk	35.7	16.5	23.6	N.A.
Corn Stover	19.3	26.5	38.7	9.2
Aspen	50.2	5.9	21.2	8.5
Pine	45.2	23.4	26.2	10.3
Switchgrass	11–31	26–51	26–41	14.6
Groundnut shell	31–39	26–29	26–29	17.2
Sugarcane bagasse	20.4	25.6	42.3	8.2
Wheat straw	15–22	25–33	28–36	6.8
Forestry waste	21–37	22–38	42–51	N.A.

N.A. = Not Available

combinations of chemical, thermal, and enzymatic treatments to increase the solubilization of organic components in low-lignin green waste (Ghodke et al., 2021). The various conversion processes of lignocellulosic biomass and the products of the processes are simplified and are shown in Figure 10.2.

Cellulases and hemicellulases enzymes such as α-amylases, β-amylases, and glucoamylases are involved in starch hydrolysis, lipases enzymes are used fats and oils, and proteases enzymes are used for protein targeting (Bhardwaj et al., 2021). Thus, a wide range of enzymes are involved in enzymatic hydrolysis (or saccharification) of lignocellulosic green waste. The enzymes or enzyme mixes involved are determined by the nature of the raw material, the substance to be hydrolyzed, and the ability of the microorganisms. It is essential not to overprocess green waste because recycling should be techno-economically feasible. Numerous researchers demonstrate the production of succinic acid, levulinic acid, and xylitol from various green waste low-lignin biomass (Buschke et al., 2013; Huber et al., 2006; Langsdorf et al., 2021). Furfuraldehyde and 5-hydroxymethylfurfural (HMF) components produced from lignocellulosic biomass are characterized as inhibitors during the fermentation process, although these are valuable basic compounds (Hu and Gholizadeh, 2019; Kim, 2018). The next sections explain the biotechnological conversion of low-lignin green waste to produce these promising compounds, as well as other prospective candidates.

10.4.1 Production of Enzymes

Several industrial enzymes have been produced using solid-state fermentation. On solid substrates, fermentation promotes the growth of ascomycetes, basidiomycetes, and deuteromycetes (Ravindran and Jaiswal, 2016). The majority of industrial enzymes come from fungi especially conidiospores. In solid-state fermentation and

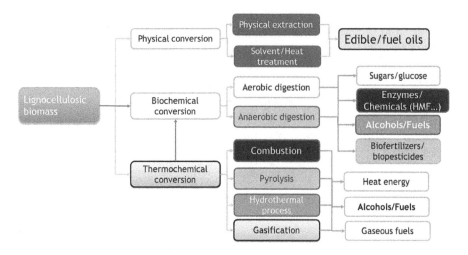

FIGURE 10.2   Various conversion processes of lignocellulosic biomass.

submerged fermentation process, the engineered gene expressions in fungi are different (Sharma et al., 2020). However, solid-state fermentation has not been widely embraced due to its inability to standardize methods and the poor reproducibility of the production process. Additionally, temperatures in the process are less controllable which results in potentially denaturing the enzymes formed in the fermenter. Aeration of the fermentation process can avoid the denaturing of enzymes. However, aeration results in water loss in the process due to evaporation (Mamo and Alemu, 2012). Table 10.3, summarizes the production of different enzymes from low-lignin green waste used as feedstock raw material and the microbial strains involved.

Recent advancements of solid-state fermentation technology have resulted in success, creating strategies to avoid the denaturing of enzymes. A specially designed permeable membrane (based on floropermeable) non-air flow box, allows water vapor to escape and maintain the substrate dry. As a result, enhanced culture development and enzyme synthesis with great reproducibility has been possible (Ito et al., 2013). Recent developments in solid-state fermentation promise excellent volumetric output along with activity product concentrations, while producing less effluent and requiring less downstream processing. However, while employing the solid-state fermentation regime, researchers have documented the emission of volatile organic compounds (VOCs) such as CH_4, N_2O, and NH_3 (Du et al., 2014).

For large-scale enzyme manufacturing, the submerged fermentation technique is the most popular option. In this process, the water-based medium controls pH and temperature while allowing agitation and aeration within the reactor vessel. Additionally, a sterile condition can be maintained in a reactor vessel in order to reduce the risk of contamination. In the submerged fermentation process, process parameters are better controlled and maintain homogeneity in the reactor conditions with even distribution of nutrients and oxygen to the growing microbe. However, excessive metal ions, butylated hydroxytoluene, and hydrogen peroxide presence in the process can cause oxidative stress to grow microbes and lower the process efficiency (Vo et al., 2020).

TABLE 10.3
Summary of Low-lignin Green Waste Sources Used as Raw Material for Enzyme Production

Enzyme	Feedstock	Microbial Strain	Applications	Reference
Amylase	Brewers spent	Catabolite-repressed. *Bacillus subtilis* KCC103	In bleaching & laundry detergents, aquaculture processing, animal nutrition process, and baking industries.	(Cunha et al., 2012)
Xylanase	Coffee by-products	*Penicillium* sp. CFR 303	In bleaching & detergents, aquaculture processing, animal nutrition process, and baking industries.	(Murthy and Naidu, 2012)
Inulinase	Banana peel, wheat bran, rice bran, orange peel, bagasse, and soybean cake	*Aspergillus kawachii, Saccharomyces sp., Penicillium rugulosum* (MTCC-3487)	Fructose corn syrup	(Dilipkumar et al., 2014)
Cellulase	Apple pomace, and banana peel	*Aspergillus niger* NRRL-567 and *Trichoderma viride* GIM 3.0010	Bioethanol production, decolorization of the dyeing process, bleaching, de-inking & detergent processes, and refining industry.	(Dilipkumar et al., 2014)
Protease	Brewer's spent grain, and corn steep liquor	*Streptomyces malaysiensis* AMT-3,	Wet waste processing, food & pharmaceutical processes, and leather industry.	(Nascimento et al., 2011)
Invertase	Bagasse, Orange peel, and pineapple peel.	*S. cerevisiae* NRRL Y-12632, *Aspergillus niger* GH1, and *Cladosporium cladosporioides*	Sucrose	(Veana et al., 2014)
Lipase	Banana peel, potato peel, and cassava peel.	*Aspergillus niger*	Food industries like meat processing, and refining processes such as degreasing agents.	(Ravindran and Jaiswal, 2016)
Transglutaminase	Industrial fibrous and soy residue	*Bacillus circulans* BL32	Food industries like meat processing, leather & cosmetics technologies, and baking industries.	(Songulashvili et al., 2015)

For the manufacturing of different enzymes and bioactive substances, various filamentous fungus species are used in the submerged fermentation process. Although submerged fermentation technology is a proven technology in the industry for the production of enzymes, various studies have shown it is not the best approach for the production of enzymes using filamentous fungus species (Mamo and Alemu, 2012). Filamentous fungus species require a large surface area and specialized design of submerged fermentation reactor vessels, which increases the capital cost. A comparative study of biodiesel production using filamentous fungus enzymes results in higher capital investment for the submerged fermentation process than for the solid-state fermentation process. Submerged fermentation technology generates 80% higher costs than solid-state fermentation technology (Usman et al., 2021).

10.4.1.1 Technical Problems in Isolation and Recovery of Enzymes

Technical challenges are observed for microorganisms to survive on lignocellulosic green waste hydrolysate, which causes low enzyme titer, low enzyme yields, and low enzyme productivity. Some technical issues reported in the literature include maintaining the C: N ratio, low carbohydrate (sugar) levels, and the presence of monosaccharides 5C/6C sugars which cause diauxic development in the culture. However, the existences of microorganisms' inhibitors such as furfuraldehyde, acetic acid, 5-Hydroxymethylfurfural, and phenolic compounds, are a serious concern when using lignocellulosic green waste in the fermentation process. The degradation of lignin during rigorous pretreatment produces microbial inhibitors as a by-product. The exact mechanism of microbial growth inhibition by phenolic compounds is still unknown. Other chemical compounds produced during pretreatment of lignocellulosic green waste which inhibit enzyme activity are: 4-Hydroxybenzoic acid, vanillin, cinnamic acid, and syringaldehyde (Escobar et al., 2020).

The main approach for removing inhibition effects in the fermentation process includes (i) preventing the production of inhibitor compounds during pretreatment and the hydrolysis process, (ii) hydrolysate detoxification, (iii) developing bacteria that can tolerate inhibitor compounds, and (iv) neutralizing the inhibitor chemical compounds. The development of inhibitor chemical compounds can be avoided by carefully selecting lignocellulosic material and using mild pretreatment techniques. Most common physical treatments include vacuum evaporation of inhibitors such as furfuraldehyde, acetic acid, and vanillin. Polyethylene glycol surfactant, ethyl acetate, and activated charcoal are used to remove the phenolic compounds inhibitors. To detoxify the hydrolysate, the $Ca(OH)_2$ compound was used in excess during pretreatment of lignocellulose biomass (Jönsson and Martín, 2016). NH_4OH was employed by Alriksson et al. as a nitrogen source as well as a neutralizing chemical for inhibitors (Alriksson et al., n.d.). The addition of NH_4OH at pH 10.0 resulted in a significant reduction in furfuraldehyde and 5-hydroxymethylfurfural concentrations. This process improved the fermentation efficiency, immobilization of enzymes and reduced the overall inhibition effect (Lyu and Ishida, 2019).

10.4.2 Production of Levulinic acid and Succinic Acid

Popular building block compounds produced from low-lignin green waste include levulinic acid and succinic acid. An acid-catalyzed process produces levulinic acid. Girisuta et al. demonstrated that sulphuric acid accelerated the optimal hydrolysis of water hyacinth to levulinic acid (Girisuta et al., 2008). The water hyacinth leaves were cleaned, cut, and dried before being incubated with an aqueous sulphuric acid solution at a steady temperature for various reaction periods. Water hyacinth input, sulfuric acid, and temperature were explored as process reaction parameters. Levulinic acid was produced at around 55 mol% during pretreatment processes, based on the C6 sugar and high acid concentration > 0.5 M. In the presence of homogeneous acid catalysts, several authors demonstrate the hydrothermal process of the olive tree, giant reed (*Arundo donax* L.) green waste-producing levulinic acid (Galletti et al., 2012). Optimum process parameters for the maximum yield of levulinic acid were observed are 200°C temperature, HCl catalyst, and incubation media used are niobium phosphate and water along with inert atmosphere (Galletti et al., 2012). A few other studies report the synthesis of levulinic acid from cellulose and hemicellulose composition of giant reed yields up to 90% in the presence of acid catalyst HCl (Antonetti et al., 2015). Similarly, two-stage acidic hydrolysis of cellulose and hemicellulose composition of *Miscanthus* × *giganteus* produces levulinic acid (60–75 mol %) and furfural (30 mol %), in the presence of an acid catalyst (H_2SO_4) (Dussan et al., 2013).

Low-lignin green waste was fermented to yield succinic acid. Several reports show that succinic acid may be produced via microbial fermentation of *Miscanthus* × *giganteus* (Dąbkowska et al., 2019). The *Miscanthus* × *giganteus* green waste biomass was pretreated with glycerol and hydrolyzed using an enzyme to break the lignin content. Further, it was fermented in the presence of a species of bacteria known as *Actinobacillus succinogenes* 130Z to produce 82 wt.% of succinic acid (Dąbkowska et al., 2019). The synthesis of succinic acid from arundo donax hydrolysate was demonstrated in the presence of *Basfia succiniciproducens* BPP7 and created yields of 6.5 g L⁻¹ of succinic acid. The most optimal succinic acid concentration of 9.5 g L⁻¹ could be obtained in a 2.5 L batch lab-scale experiment. It was observed that maximum yields of succinic acid were due to the presence of furfuraldehyde, HMF, *p*-hydroxybenzoic aldehyde, or vanillin in the culture obtained from biomass enzymatic hydrolysate. Ventorino et al. reported that a BPM1 strain isolated from a cow's rumen produces succinic acid, lactic acid, and acetic acid from *Arundo donax* enzymatic hydrolysate (Ventorino et al., 2017). Kuglarz et al. observed that bacteria called *Actinobacillus succinogenes* may convert industrial hemp seed oil to succinic acid to the highest yields of 83 wt.% (21.9 g L⁻¹). Enzymatic hydrolysis of hemp seeds produces 74 wt.% of sugar in the presence of 3 wt.% H2O2 at 120°C (Kuglarz et al., 2014).

10.4.3 Production of Furfuraldehyde and 5-Hydroxymethylfurfural

Furfuraldehyde and 5-hydroxymethylfurfural value-added building block compounds can perhaps be produced from low-lignin green waste. A biorefinery technology has been demonstrated by many researchers that utilize *Miscanthus* × *giganteus*, giant red, hyacinth, hempseed, olive green waste, among others, to produce furfural-dehyde and HMF (Rivas et al., 2019). The enzymatic hydrothermal pretreatment of miscanthus produces hemicellulose, cellulose, and lignin-rich extracts used to syn-thesize furfuraldehyde and 5-hydroxymethylfurfural. Solid extract of enzymatic pre-treatment of miscanthus yields 50 wt.% of HMF, while a soluble portion of extract yields 80 wt.% of furfuraldehyde in an acidic medium. A hydrolyzed portion of mis-canthus and switchgrass with steam yields over 90 wt.% of furfural in the presence of sulfuric acid (used as dehydration agent) (Mandalika and Runge, 2012). Switchgrass as feedstock has been utilized to produce furfuraldehyde and HMF in the presence of $AlCl_3 \cdot 6H_2O$ as a catalyst. Improved yields were obtained when catalysts were used in a biphasic mixture of water/tetrahydrofuran. Yields of 65 wt.% furfuraldehyde were derived from the pentose content, while 25 wt.% HMF was derived from the hexose component of the extract portion of hydrolyzed switchgrass.

Recently, conceptual design for the production of furfuraldehyde, HMF, and dimethyl furfural has been established by the scientific community (Sajid et al., 2018). In the new concept, HMF and 5-Ethoxymethyl-2-furfural are proposed as promising future fuels and replacements for traditional fuel. The concept was achieved by designing a novel solid acid catalyst and ionic liquid catalyst. A few authors have demonstrated the synthesis of HMF using weeds: a special catalyst such as Brønsted acidic ionic liquid catalysts was used and observed yields of HMF varied from 10–60 wt.%. While 10–30 wt.% yield of HMF was achieved via the direct conversion of the weeds in the presence of a solid acid catalyst. Amongst the different weeds tested, foxtail weed utilization produced the maximum yield of HMF, while giant reed hydrothermal conversion yielded 70 wt.% in the presence of an acid catalyst.

10.4.4   PRODUCTION OF OTHER VALUABLE SUBSTANCES

Xylitol is one of the commercial value-added products that can be produced from the xylose portion of hydrolyzed low-lignin green waste. Biotechnologically cultured yeast *candida* was capable of producing xylitol from the hydrolysate part of big bluestem perennial prairie grass. Prairie grass hydrolysate was prepared by treating with 1 wt.% sulfuric acid and xylanase enzyme in the pretreatment process. Acid hydrolyzed switchgrass produces xylitol in the presence of *Pichia stipitis* CBS 5773 (Langsdorf et al., 2021). The maximum yield of 50 wt.% xylitol was achieved using low-lignin grassy green waste.

Other value-added chemicals such as polyhydroxyalkanoates (PHAs) (polyhy-droxy butyrate (PHB) and polyhydroxy valerate (PHV)) are also possibly produced from low-lignin grassy green waste (Jiang et al., 2011). The medium-chain lengths of polyhydroxyalkanoates are obtained from perennial ryegrass using a range of bacte-ria *Pseudomonas* strains. A pretreatment of the ryegrass was carried out using 2 wt.% NaOH in water prior to enzymatic hydrolysis. Additional treatment was performed to remove the extra lignin present in ryegrass using sodium chlorite/acetic acid. For optimal results, NaOH and sodium chlorite/acetic acid were used in combination, which resulted in a hydrolysate containing mainly glucose. Glucose-rich hydroly-sates (75–80 % mol/L) support the growth of *Pseudomonas* strains, which led to the production of 20–35 wt.% of PHAs. In another study, alligator weeds were used as feedstock, and hydrolysate of weed produced particularly PHB in the presence of *C. necator* betaproteobacteria. Acid or enzymatic pre-treated alligator weed hydrolysate did not produce enough PHB, which was caused by an inhibitor that interfered with microbial growth. Maximum production of 5.0 g L^{-1} of PHB was obtained after 70 hours of fermentation under optimal conditions with enzymatic hydrolysate (Robak and Balcerek, 2018).

10.4.5 Electrode Materials from Green Waste

Green waste may be used as a feedstock producing affordable and ecological functional materials such as *electrodes*. Carbonization of green waste and subsequent use as an electrode material is a potential approach to functionalization. The electrodes serve in electro-biotechnological applications as microbial fuel cells and bioelectrosynthesis systems. A biobased system of biocomposites can help to promote greener chemistry and bioeconomy. One classic example of bioelectro methanogenesis, in which electroactive methanogens are used to convert the electrical source into methane fuel gas (Enzmann et al., 2019a). Electro-fermentation processes can be cited as another application of bioelectrochemical systems. In the electro-fermentation process, electricity is used to drive an electrochemical reaction in the desired direction in the presence of microorganisms such as *Clostridium acetobutylicum* to produce biofuels (biobutanol, bioethanol, etc.,). A significant barrier to the widespread adoption of electro-fermentation process technology has been high electrode prices, which make the technology economically unfeasible for the time being. Numerous electrode types analyzed for use in bioelectrochemical systems include carbonized materials, and underline the necessity of cost-effective designs (Enzmann et al., 2019b).

Several scientific papers have recently reviewed the carbonization of materials with lower lignin content, such as grasses, leaves, and green waste (Escobar et al., 2020). The hydrothermal carbonization (HTC) process was implemented to produce the biochar and subsequent utilization in the synthesis of electrodes. Low-lignin biomass was pretreated prior to the HTC process. The HTC process was carried out at different temperatures such as 200, 230, or 260°C depending on the production quality of biochar. Miscanthus and switchgrass yielded the largest biochar mass: 80 wt.% and 90 wt.% respectively at 200°C (Dąbkowska et al., 2019). However, it was observed that moisture content in the process significantly affected the quality of biochar and the length of the process. It was observed without the addition of water in the HTC process, the experiment lasted for one, two, or three hours at 300, 350, or 400°C respectively. In the presence of moisture content, the mass output of biochar declined from 82.6% to 35.2% with increasing temperature and time. In another study, the HTC process was performed using lawn grass. Lawn grass was dried, shredded, and carbonized at around 200 to 240°C for 30 minutes maintaining the liquid-to-solid ratio of 1:30. The total yield of hydrochar or biochar was between 30 wt.% to 50 wt.%. It was observed that long residence times of process yielded high quality of hydrochar from lawn grass (Guo et al., 2015). Various low-lignin green waste raw materials (wood chips, grass trimming, lawn grass, and elephant grass) are implemented to produce hydrochar and different parameters are considered to optimize the operating conditions to yield the maximum hydrochar.

Hydrophilic functional groups were present in the hydrochars synthesized from low-lignin green waste compared to the woody biomass. As result, carbonized hydrochar seems more suitable as an electrode material. The HTC processing of dead eucalyptus leaves was studied at different carbonization conditions such as 150–380°C, and liquid-to-solid ratio 1.10. It was observed that yields of biochar/hydrochar were reduced from 90 wt.% to 30 wt.% as the temperature increased from 150–380°C (Czajczyńska et al., 2017). Low-lignin green waste was also investigated for its conversion into coal as an energy source. Various woody biomass HTC processes were conducted and observed to produce a yield of 75 wt.% of coal from biochar using green waste. Some literature reports microwave HTC process of green waste containing leaves and deadwood that had fallen and decomposed. Prior to the microwave HTC process, grinding and drying treatment was performed at 110°C for 24 hours. The quality of hydrochar was assessed based on the calorific value and for different operating conditions, different calorific value hydrochar was produced. Reported operating conditions varied from temperature (120–200°C), residence time (0.5–2.5 h), and the liquid-to-solid ratio (5:1–10:1). Hydrochar yields ranged from 50.4 to 76.8 wt.%, depending on the source of green waste. Additionally it was observed that as the temperature increased, the carbon content of hydrochar increased with oxygen-deficient conditions (Mäkelä et al., 2015).

Large informative articles have already been published about carbonizing low-lignin green waste or herbaceous biomass. Nonetheless, electro-biotechnological applications do not examine carbonized materials as electrode material, hence little published research is available in the scientific community. Mainly, electrodes from low-lignin green waste or herbaceous biomass are for application as capacitors or in the semiconductors industry. Few articles report the use of carbonized alfalfa leaves as a cathodic catalyst in microbial fuel cells (Deng et al., 2017). It was observed, compared to a Pt/C cathode catalyst, a biomass-derived carbon material exhibited high current density and long-term stability. In conclusion, there is a substantial research opportunity that remains in examining electrode materials derived from low-lignin green waste and maybe finding an economical substitute for electro-biotechnological applications.

10.5   BIO-CIRCULAR ECONOMY IN A BIOREFINERY CONCEPT

The waste framework directive established the green waste hierarchy as the key waste management policy. The green waste hierarchy puts preventative measures before recycling, reuse, and disposal. Incineration with energy recovery is a common procedure for many countries when it comes to biomass waste. The ability to extract important bioproducts is lost when biomass waste is combusted for heat/energy recovery. In the incineration process, the high moisture content of biomass or green waste reduces destruction efficiency and increases the risk of pollution with persistent organic pollutants. Thus, in the case of green waste, the use of the incineration process for energy recovery should be kept to a minimum. The waste management policy should focus on recycling and reutilization of the green waste source for the production of bio-based value-added products.

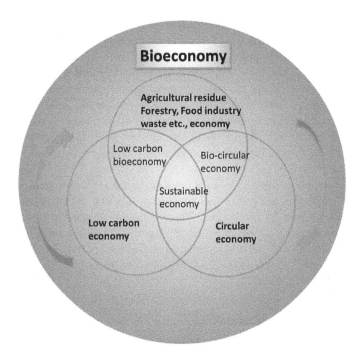

FIGURE 10.3 Bio-circular economy in a biorefinery concept for lignocellulosic biomass.

In the order to develop the utilization of biomass green waste, regulations and policies should be strong and supportive. In 2014, the linear economical model was converted to a circular economy strategy "take, make, and dispose of" to bring about a more circular approach to resource management. It was proposed in a circular economy that, about 70% of resources should be reused and recycled, with zero dumpings. The goal is to achieve a completely circular economy by 2030. In addition to implementing the waste hierarchy, the circular economy plan goes beyond regional, national, and international levels between industries. The circular economy strategy unites the waste management hierarchy and the bio-economy together, in order to focus on renewable carbon reserves from agricultural residue, forestry waste, and organic waste from various products. Figure 10.3, describes the circular economy from biomass waste in the biorefinery concept.

Transforming from a linear economy to a circular economy, the entire product value chain requires an alternative. An alternative to product design, identifying new zero-waste processes, waste to value-added products, and new business models depending on consumers' behavior. The socioeconomical, technological, and behavioral shifts will impact the circular economy.

10.5.1 THE BIOREFINERY CONCEPT

Integrated biobased product industries valorizing the lignocellulose green waste by the implementation of a range of diverse technologies are referred to as biorefineries.

Biorefineries are analogous to crude oil refineries. Biorefinery facilities sort biocrude oil into different streams such as fuels and raw materials used in petrochemical, pharmaceutical, and renewable sectors. The green waste biorefinery idea is based on mixed feedstock. The biorefinery concept can be implemented utilizing homogeneous feedstock from the food industry, and agriculture sector. As a result, the composition of green waste composition is a mixture of organic compounds including carbohydrates, cellulose, protein, fats, and lipids. Applying a biorefinery concept to green waste might help to deal with specific feedstock issues (Kumar and Verma, 2022). However, challenges may arise due to variable feedstock decomposition temperatures.

Very few green waste biorefineries at the industrial scale are implemented. Small- and medium-scale production facilities for food processing and farm waste biorefineries are also known to be limited. Three new value-added products such as bioplastics, biopesticides, and enzymes are found in the biorefinery concept (see Figure 10.1). Functional carbon electrodes and plasticizers have been formed as a result of biorefinery (Amore et al., 2016; Forster-Carneiro et al., 2013; Kamm et al., 2006; Rajesh Banu et al., 2020).

10.5.2 Enzymes from Green Waste

Enzymes are protein molecules synthesized from green waste that subsequently helps to catalyze biological complex reactions, including the breakdown of cellulose to glucose. Enzymes are more significant in a biorefinery process, as they influence the overall process efficiency, yields, and thermodynamic reaction kinetics (Bhardwaj and Verma, 2021). However, the cost of conventional enzyme production is high, and as a result the cost of the substrate, as the requirement is high. Substituting green waste as a substrate helps to produce the enzymes at a lower cost (Andler and Goddard, 2018). The utilization of various types of green waste as feedstock raw material for the synthesis of enzymes by microorganisms and fungi are shown in Table 10.3. Applying the zero-cost concept, mixed green waste especially food waste substrates are an efficient and inexpensive production method for the entire cellulase enzymes. It is identical to commercial enzyme preparation when it comes to the rate of sugar release from green waste.

10.5.3 Bioplastics from Green Waste

The fact is that petroleum-based plastics are not renewable and cause challenges in degradation in solid waste management. On the other hand, bioplastics are bio-based polymers include polyhydroxyalkanoates (PHAs) which are biodegradable. The proven industrial bioplastics are based on sugar molecules present in the feedstock and are highly valuable. The use of lower-value substrates such as green waste provides low production costs. To synthesis bioplastics, the requirement of organic acid from green waste would be a feasible solution for a low-cost bioeconomy. Many reviews have reported that the utilization of green waste as substrates is encouraging in bioplastic production. However, additional optimization studies are required (Sharif Hossain et al., 2016; Sharma et al., 2021; Shrestha et al., 2020).

In the bioplastic production process, studies have reported the fermentation of green waste first and then the addition of culture would provide high yields and better process efficiencies. Last but not least, the effluent from fermentation containing organic acids was coupled with microbial culture, which yields and stores bioplastics such as PHAs. The biopolymer polyhydroxyalkanoates (PHA) were formed along with polyhydroxyalkanoate (PHB) and polyhydroxy valerate (PHV) as co-polymers. However, the research findings reported were very different, with respect to the microbes used and the size of the culture (Hafuka et al., 2011).

Few studies report on a biorefinery concept for bioplastic manufacture with green waste as feedstock. Most of the research in current trends deals with improving the output of the main products such as biofuels. The valorization efficiency of bioplastics such as polylactic acid (PLA) has been successfully proven. It has been demonstrated that PLA was produced, ranging from 25–120 g/kg of zinc oxide nanoparticles used as a catalyst. Additionally, it was the first research that incorporated mixed green waste, including food waste, as a feedstock raw material in PLA synthesis. However, results are not similar to other investigations reported at smaller lab-scale studies.

10.5.4 Biopesticides from Green Waste

Bacillus thuringiensis (Bt) is a species of bacteria microbe that is well-known and extensively studied in the synthesis of biopesticide or as a biopesticide itself. Biopesticides have various applications in, public health sectors, the agriculture sector, and forest conservation (Miranda et al., 2021). Though biopesticides have wider application, production of biopesticides through conventional methods is expensive due to availability of raw materials, capital cost due to large equipment, complex operating procedures, and difficultly in selectivity (Vea et al., 2018). Thus, bacteria Bt cultured from green waste or natural soils used in biorefineries could be a competitive and promising substitute.

Studies have reported the culturing of Bt in semi-solid fermentation produces high yields of biopesticides compared to production in solid-state fermentation methods. They have also reported higher yields of Bt to have greater valorization efficiencies (CFU/kg of green waste) (Hölker and Lenz, 2005). The growth of microorganisms occurs on the solid substrate of green waste/organic waste without any free water, which was possible in solid-state fermentation. In the case of faster growth of Bt, a solid-state fermentation reactor has advantages over submerged fermentation. Increased fermentation productivity, product stability, lower pollutant discharge, and high valorization efficiency were also reported to be advantageous over submerged fermentation reactors. Still, the mass transfer efficiencies are low due to the substrate's high sugar concentration which subsequently inhibits the substrate (Ravindran et al., 2018). However, mass transfer efficiencies can be improved by increasing the moisture content of the substrate to make the culture medium a semi-solid state. Nevertheless, studies reported that semi-solid fermentation processes are performed at lab-scale, while solid-state fermentation processes are conducted at bench-scale. Thus, productivity, efficiencies, and yields are not directly comparable. Some studies have been carried out based on sterile and nonsterile conditions and observed no significant differences between the two conditions. Further studies are utilized to scale up the process to an industrial scale (McCreanor and Graves, 2017).

10.6 ECONOMICS OF GREEN WASTE AND FUTURE PERSPECTIVES

Global potential revenue of value-added chemical commodities such as enzymes, bioplastic, biopesticides, electrodes, and other bioproducts are presented in Figure 10.4. It can be seen that global estimates of all the bioproducts included in Figure 10.4 vary from 5–10%. Biopesticides (USD4.3 billion) and enzymes (USD4.46 billion) have the biggest potential revenue compared with other bioproducts. The market value of both enzymes and Bt is expected to increase at a rate of 7–10% CAGR, with an assumption base of activity that would not change or decrease during transportation. Globally, enzymes and Bt are more profitable due to their stability and activity which remains unchanged. Production of enzymes and Bt from green waste is an added advantage to the process, and significantly reduces the total operating cost while increasing the capital cost between new technologies.

It has been found that bioplastics showed just a minor increase in the market size fromUSD85 to 91 million as of March 2020. Other value-added building block compounds (includes 5-HMF, furfuraldehyde, succinic acid, and levulinic acid) produced using green waste have a total market value of USD755 million globally. Furfuraldehyde and succinic acid hold the major share among the other bioproducts. Well-established technology, applications in the chemical process industry, and global requirements as basic chemicals have supported the production of furfuraldehyde and succinic acid to become significant products. Bioproducts especially,

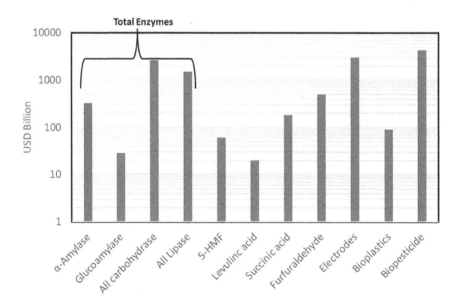

FIGURE 10.4 Potential revenue from value-added products and their global market size (2019–2020). Note: Values on Y-axis are taken on a logarithmic scale.

5-HMF, furfuraldehyde, succinic acid, and levulinic acid are necessarily produced from green waste with low-lignin content.

The market value of electrodes presented in Figure 10.4, is not completely from green waste. It is general market size, which represents total electrode production from different feedstock raw materials. It can be observed from the market research, that electrode-produced green waste has not yet been commercialized. The value presented here is encouraging research into production of electrodes from green waste hydrochar.

The large-scale development of green waste biorefineries is possible due to the industrial implementation of green waste valorization (Ghodke et al., 2015). Additional factors to examine include operational expenses, environmental implications, and the benefits derived from replacing fossil-based products. A summary of the economic and environmental value of the production and use of green waste can be characterized as follows: To provide an economic and environmental assessment of the cascade utilization of green waste and associated bioproducts, further research and development are needed and necessary.

10.7   CONCLUSION

The study describes recent advances in the development of a green waste feedstock that may be utilized in biorefineries to produce high-value products. Several different experiments produced bioplastics, enzymes, and Bt. The study evaluates the profitability of entering the market by estimating the valorization efficiency and possible revenue if utilizing waste as a feedstock. The results show that developing biorefineries that make biopesticides and enzymes from green waste have significantly higher revenues. As a result, the scale-up of biologically produced building blocks presents several difficulties and challenges. Bioplastics and other high-value chemicals, including enzymes, can be produced from cheap green waste. This study demonstrates that green waste with low-lignin is an acceptable alternative to traditional municipal solid waste. Thus, it is advisable to focus on technological advancement in the utilization of green waste in the future. It is also critical that future researchinvestigates the operating costs and environmental impacts as well as the impact of these emerging technologies.

REFERENCES

(CPCB), C.P.C.B., Ministry of Environment, F.& C.C., 2015. Assessment & characterisation of plastic waste generation in 60 major cities.

Abdel-Shafy, H.I., Mansour, M.S.M., 2018. Solid waste issue: Sources, composition, disposal, recycling, and valorization. *Egypt. J. Pet.* 27, 1275–1290. https://doi.org/10.1016/j.ejpe.2018.07.003

Agrawal, K., Verma, P. (2020) Production optimization of yellow laccase from *Stropharia* sp. ITCC 8422 and enzyme-mediated depolymerization and hydrolysis of lignocellulosic biomass for biorefinery application. *Biomass Convers. Biorefin.* 1–20. https://doi.org/10.1007/s13399-020-00869-w

Alriksson, B., Horvath, I.S., Sjöde, A., Nilvebrant, N.-O., Jönsson, L.J., n.d. Ammonium hydroxide detoxification of spruce acid hydrolysates. In: *Twenty-Sixth Symposium on Biotechnology for Fuels and Chemicals*, 911–922. Humana Press, Totowa, NJ. https://doi.org/10.1007/978-1-59259-991-2_78

Amore, A., Ciesielski, P.N., Lin, C.Y., Salvachuá, D., Nogué, V.S.I., 2016. Development of lignocellulosic biorefinery technologies: Recent advances and current challenges. *Aust. J. Chem.* 69, 1201–1218. https://doi.org/10.1071/CH16022

Andler, S.M., Goddard, J.M., 2018. Transforming food waste: How immobilized enzymes can valorize waste streams into revenue streams. *NPJ Sci. Food* 2, 19. https://doi.org/ 10.1038/s41538-018-0028-2

Antonetti, C., Bonari, E., Licursi, D., Di Nasso, N.N., Galletti, A.M.R., Cravotto, G., Chemat, F., 2015. Hydrothermal conversion of giant reed to furfural and levulinic acid: Optimization of the process under microwave irradiation and investigation of distinctive agronomic parameters. *Molecules* 20, 21232–21353. https://doi.org/10.3390/ molecules201219760

Anwar, Z., Gulfraz, M., Irshad, M., 2014. Agro-industrial lignocellulosic biomass a key to unlock the future bio-energy: A brief review. *J. Radiat. Res. Appl. Sci.* 7, 163–173. https://doi.org/10.1016/j.jrras.2014.02.003

Bhardwaj, N., Kumar, B., Agrawal, K., Verma, P., 2021. Current perspective on production and applications of microbial cellulases: a review. *Bioresour. Bioprocess*, 8, 1–34.

Bhardwaj, N., Verma, P., 2021. Xylanases: A helping module for the enzyme biorefinery platform. In: *Bioenergy Research: Revisiting Latest Developments*, 7, 161–179. Springer, Singapore.

Bioenergy, I., 2014. *IEA Bioenergy Take42 Biorefining*, International Energy Agency (IEA), Wageningen, the Netherlands.

Boldrin, A., Christensen, T.H., 2010. Seasonal generation and composition of garden waste in Aarhus (Denmark). *Waste Manag.* 30, 551–557. https://doi.org/10.1016/j. wasman.2009.11.031

Buschke, N., Schäfer, R., Becker, J., Wittmann, C., 2013. Metabolic engineering of industrial platform microorganisms for biorefinery applications - Optimization of substrate spectrum and process robustness by rational and evolutive strategies. *Bioresour. Technol.* 135, 544–554. https://doi.org/10.1016/J.BIORTECH.2012.11.047

Chojnacka, K., Moustakas, K., Witek-Krowiak, A., 2020. Bio-based fertilizers: A practical approach towards circular economy. *Bioresour. Technol.* 295, 122223. https://doi. org/10.1016/j.biortech.2019.122223

Cripwell, R.A., van Zyl, W.H., Viljoen-Bloom, M., 2021. Fungal biotechnology: Fungal amylases and their applications. In: *Encyclopedia of Mycology*, 326–336. Elsevier. https:// doi.org/10.1016/B978-0-12-809633-8.21082-0

Cunha, F.M., Esperança, M.N., Zangirolami, T.C., Badino, A.C., Farinas, C.S., 2012. Sequential solid-state and submerged cultivation of *Aspergillus niger* on sugarcane bagasse for the production of cellulase. *Bioresour. Technol.* 112, 270–274. https://doi. org/10.1016/J.BIORTECH.2012.02.082

Czajczyńska, D., Nannou, T., Anguilano, L., Krzyzyńska, R., Ghazal, H., Spencer, N., Jouhara, H., 2017. Potentials of pyrolysis processes in the waste management sector. *Energy Procedia* 123, 387–394. https://doi.org/10.1016/j.egypro.2017.07.275

Dąbkowska, K., Alvarado-Morales, M., Kuglarz, M., Angelidaki, I., 2019. Miscanthus straw as substrate for biosuccinic acid production: Focusing on pretreatment and downstream processing. *Bioresour. Technol.* 278, 82–91. https://doi.org/10.1016/j.biortech.2019. 01.051

Dahiya, S., Kumar, A.N., Shanthi Sravan, J., Chatterjee, S., Sarkar, O., Mohan, S.V., 2018. Food waste biorefinery: Sustainable strategy for circular bioeconomy. *Bioresour. Technol.* https://doi.org/10.1016/j.biortech.2017.07.176

Deng, L., Yuan, Y., Zhang, Y., Wang, Y., Chen, Y., Yuan, H., Chen, Y., 2017. Alfalfa leaf-derived porous heteroatom-doped carbon materials as efficient cathodic catalysts in microbial fuel cells. *ACS Sustain. Chem. Eng.* 5, 9766–9773. https://doi.org/10.1021/ acssuschemeng.7b01585

Dilipkumar, M., Rajasimman, M., Rajamohan, N., 2014. Utilization of copra waste for the solid state fermentatative production of inulinase in batch and packed bed reactors. *Carbohydr. Polym.* 102, 662–668. https://doi.org/10.1016/j.carbpol.2013.11.008

Du, Z., Mo, J., Zhang, Y., 2014. Risk assessment of population inhalation exposure to volatile organic compounds and carbonyls in urban China. *Environ. Int.* 73, 33–45. https://doi.org/10.1016/J.ENVINT.2014.06.014

Dussan, K., Girisuta, B., Haverty, D., Leahy, J.J., Hayes, M.H.B., 2013. Kinetics of levulinic acid and furfural production from *Miscanthus×giganteus. Bioresour. Technol.* 149, 216–224. https://doi.org/10.1016/j.biortech.2013.09.006

Enzmann, F., Mayer, F., Holtmann, D., 2019a. Process parameters influence the extracellular electron transfer mechanism in bioelectromethanogenesis. *Int J Hydrog. Energy* 44, 24450–24458. https://doi.org/10.1016/j.ijhydene.2019.07.039

Enzmann, F., Stöckl, M., Gronemeier, D., Holtmann, D.,2019b. Insights in the anode chamber influences on cathodic bioelectromethanogenesis—systematic comparison of anode materials and anolytes. *Eng. Life Sci.* 19, 795–804. https://doi.org/10.1002/elsc.201900126

Escobar, E.L.N., Da Silva, T.A., Pirich, C.L., Corazza, M.L. and Pereira Ramos, L., 2020. Supercritical fluids: A promising technique for biomass pretreatment and fractionation. *Front Bioeng. Biotechnol.* 8, 252. https://doi.org/10.3389/fbioe.2020.00252

Forster-Carneiro, T., Berni, M.D., Dorileo, I.L., Rostagno, M.A., 2013. Biorefinery study of availability of agriculture residues and wastes for integrated biorefineries in Brazil. *Resour. Conserv. Recycl.* 77, 78–88. https://doi.org/10.1016/j.resconrec.2013.05.007

Galletti, A.M.R., Antonetti, C., De Luise, V., Martinelli, M., 2012. A sustainable process for the production of γ-valerolactone by hydrogenation of biomass-derived levulinic acid. *Green Chem.* 14, 688. https://doi.org/10.1039/c2gc15872h

Ghodke, P., Ganesh, A., Mahajani, S., 2015. Stabilization of fast pyrolysis oil derived from wood through esterification. Int. J. Chem. React. Eng. 13, 323–334. https://doi.org/10.1515/ijcre-2014-0102

Ghodke, P.K., Ramanjaneylu, B., Kumar, S., 2021. Stabilization of bio-oil derived from macroalgae biomass using reactive chromatography. *Biomass Convers. Biorefinery.* https://doi.org/10.1007/s13399-021-01533-7

Girisuta, B., Danon, B., Manurung, R., Janssen, L.P.B.M., Heeres, H.J., 2008. Experimental and kinetic modelling studies on the acid-catalysed hydrolysis of the water hyacinth plant to levulinic acid. *Biores. Technol.* 99, 8367–8375. https://doi.org/10.1016/j.biortech.2008.02.045

Guo, S., Dong, X., Liu, K., Yu, H., Zhu, C., 2015. Chemical, energetic and structural characteristics of hydrothermal carbonization solid products for lawn grass. *BioResources* 10, 4613–4625. https://doi.org/10.15376/biores.10.3.4613-4625

Hafuka, A., Sakaida, K., Satoh, H., Takahashi, M., Watanabe, Y., Okabe, S., 2011. Effect of feeding regimens on polyhydroxybutyrate production from food wastes by *Cupriavidus necator. Bioresour. Technol.* 102, 3551–3553. https://doi.org/10.1016/j.biortech.2010.09.018

Hanc, A., Novak, P., Dvorak, M., Habart, J., Svehla, P., 2011. Composition and parameters of household bio-waste in four seasons. *Waste Manag.* 31, 1450–1460. https://doi.org/10.1016/j.wasman.2011.02.016

Herrmann, C., Prochnow, A., Heiermann, M., Idler, C., 2014. Biomass from landscape management of grassland used for biogas production: Effects of harvest date and silage additives on feedstock quality and methane yield. *Grass Forage Sci.* 69, 549–566. https://doi.org/10.1111/gfs.12086

Hinnemann, B., Moses, P.G., Bonde, J., Jørgensen, K.P., Nielsen, J.H., Horch, S., Chorkendorff, I., Nørskov, J.K., 2005. Biomimetic hydrogen evolution: MoS_2 nanoparticles as catalyst for hydrogen evolution. *J. Am. Chem. Soc.* 127, 5308–5309. https://doi.org/10.1021/ja0504690

Hitzl, M., Corma, A., Pomares, F., Renz, M., 2015. The hydrothermal carbonization (HTC) plant as a decentral biorefinery for wet biomass. *Catal. Today* 257, 154–159. https://doi.org/10.1016/j.cattod.2014.09.024

Hölker, U., Lenz, J., 2005. Solid-state fermentation - Are there any biotechnological advantages? Curr. Opin. Microbiol. 8, 301–306. https://doi.org/10.1016/J.MIB.2005.04.006

Hollins, O., Lee, P., Sims, E., Bertham, O., Symington, H., Bell, N., Sjögren, P., Lucie, P., 2017. Towards a circular economy - Waste management in the EU Study.

Hu, X., Gholizadeh, M., 2019. Biomass pyrolysis: A review of the process development and challenges from initial researches up to the commercialisation stage. *J. Energy Chem.* https://doi.org/10.1016/j.jechem.2019.01.024

Huang, C., Han, L., Liu, X., Ma, L., 2010. The rapid estimation of cellulose, hemicellulose, and lignin contents in rice straw by near infrared spectroscopy. *Energy Sources, Part A Recover. Util. Environ. Eff.* 33, 114–120. https://doi.org/10.1080/15567030902937127

Huber, G.W., Iborra, S., Corma, A., 2006. Synthesis of transportation fuels from biomass: Chemistry, catalysts, and engineering. *Chem. Rev.* 106, 4044–4098. https://doi.org/10.1021/cr068360d

Ito, K., Gomi, K., Kariyama, M., Miyake, T., 2013. Rapid enzyme production and mycelial growth in solid-state fermentation using the non-airflow box. *J. Biosci. Bioeng.* 116, 585–590. https://doi.org/10.1016/j.jbiosc.2013.04.024

Jiang, Y., Hebly, M., Kleerebezem, R., Muyzer, G., van Loosdrecht, M.C.M., 2011. Metabolic modeling of mixed substrate uptake for polyhydroxyalkanoate (PHA) production. *Water Res.* 45, 1309–1321. https://doi.org/10.1016/j.watres.2010.10.009

Jönsson, L.J., Martín, C., 2016. Pretreatment of lignocellulose: Formation of inhibitory by-products and strategies for minimizing their effects. *Bioresour. Technol.* 199, 103–112. https://doi.org/10.1016/j.biortech.2015.10.009

Juneja, A., Kumar, D., Williams, J. D., Wysocki, D. J., Murthy, G. S. 2011. Potential for etha-nol production from conservation reserve program lands in Oregon. *J Renew Sustain Energy* 3, 63102. https://doi.org/10.1063/1.3658399

Kamm, B., Gruber, P.R., Kamm, M., 2006. *Biorefinery industrial processes and products. Status and future direction*, vols. 1 and 2. WILEY-VCH Verlag GmbH & Co. KGaA, Weinheim. https://doi.org/10.1002/9783527619849

Keshav, P.K., Banoth, C., Kethavath, S.N., Bhukya, B., 2021. Lignocellulosic ethanol produc-tion from cotton stalk: An overview on pretreatment, saccharification and fermentation methods for improved bioconversion process. *Biomass Convers. Biorefinery.* https://doi.org/10.1007/s13399-021-01468-z

Kim, D., 2018. Physico-chemical conversion of lignocellulose: Inhibitor effects and detoxifica-tion strategies: A mini review. *Molecules.* https://doi.org/10.3390/molecules23020309

Kuglarz, M., Gunnarsson, I.B., Svensson, S.-E., Prade, T., Johansson, E., Angelidaki, I., 2014. Ethanol production from industrial hemp: Effect of combined dilute acid/steam pretreat-ment and economic aspects. *Bioresour. Technol.* 163, 236–243. https://doi.org/10.1016/j.biortech.2014.04.049

Kumar, B., Bhardwaj, N., Agrawal, K., Chaturvedi, V., Verma, P., 2020a. Current perspective on pretreatment technologies using lignocellulosic biomass: An emerging biorefinery concept. *Fuel Process. Technol.* 199, 106244.

Kumar, B., Bhardwaj, N., Agrawal, K., Verma, P., 2020b. Bioethanol production: Generation-based comparative status measurements. In: *Biofuel Production Technologies: Critical Analysis for Sustainability*, 155–201. Springer, Singapore.

Kumar, B., Bhardwaj, N., Verma, P., 2020c. Microwave assisted transition metal salt and orthophosphoric acid pretreatment systems: Generation of bioethanol and xylo-oligosaccharides. *Renewable Energy*, 158, 574–584.

Kumar, B., Verma, P., 2021. Techno-economic assessment of biomass-based integrated biorefinery for energy and value-added product. In: *Biorefineries: A Step Towards Renewable and Clean Energy*, 581–616. Springer, Singapore.

Kumar, B., Verma, P., 2022. *Pichia pastoris*: Multifaced fungal cell factory of biochemicals for biorefinery applications. In: *Fungal Biotechnology Prospects, and Avenues*, CRC Press. Boca Raton, Florida.

Langsdorf, A., Volkmar, M., Holtmann, D., Ulber, R., 2021. Material utilization of green waste: A review on potential valorization methods. *Bioresour. Bioprocess*. https://doi.org/10.1186/s40643-021-00367-5

Lee, J.Y., Yoo, C., Jun, S.Y., Ahn, C.Y., Oh, H.M., 2010. Comparison of several methods for effective lipid extraction from microalgae. *Bioresour. Technol.* 101, S75–S77. https://doi.org/10.1016/j.biortech.2009.03.058

Lin, Weiwei, Lin, M., Zhou, H., Wu, H., Li, Z., Lin, Wenxiong, 2019. The effects of chemical and organic fertilizer usage on rhizosphere soil in tea orchards. *PLoS One* 14, e0217018. https://doi.org/10.1371/journal.pone.0217018

Lyu, Y., Ishida, H., 2019. Natural-sourced benzoxazine resins, homopolymers, blends and composites: A review of their synthesis, manufacturing and applications. *Prog. Polym. Sci.* 99. https://doi.org/10.1016/J.PROGPOLYMSCI.2019.101168

Mäkelä, M., Benavente, V., Fullana, A., 2015. Hydrothermal carbonization of lignocellulosic biomass: Effect of process conditions on hydrochar properties. *Appl. Energy* 155, 576–584. https://doi.org/10.1016/j.apenergy.2015.06.022

Malinauskaite, J., Jouhara, H., Czajczyńska, D., Stanchev, P., Katsou, E., Rostkowski, P., Thorne, R.J., Colón, J., Ponsá, S., Al-Mansour, F., Anguilano, L., Krzyżyńska, R., López, I.C., A. Vlasopoulos, Spencer, N., 2017. Municipal solid waste management and waste-to-energy in the context of a circular economy and energy recycling in Europe. *Energy* 141, 2013–2044. https://doi.org/10.1016/j.energy.2017.11.128

Mamo, Z., Alemu, T., 2012. Evaluation and optimization of agro-industrial wastes for conidial production of Trichoderma isolates under solid state fermentation. *J. Appl. Biosci.* 54, 3880–3891.

Mandalika, A., Runge, T., 2012. Enabling integrated biorefineries through high-yield conversion of fractionated pentosans into furfural. *Green Chem.* 14, 3175. https://doi.org/10.1039/c2gc35759c

Maria, A., Galletti, R., Antonetti, C., De Luise, V., Licursi, D., Nassi, N., Nasso, D., 2012. Levulinic acid production from waste biomass. *BioResources* 7, 1824–1835. https://doi.org/10.15376/biores.7.2.1824-1835

Mashingo, M.S.H., Kellogg, D.W., Coblentz, W.K., Anschutz, K.S., 2008. Effect of harvest dates on yield nutritive value of eastern gamagrass. *Prof. Anim. Sci.* 24, 363–373. https://doi.org/10.15232/S1080-7446(15)30868-8

McCreanor, V., Graves, N., 2017. An economic analysis of the benefits of sterilizing medical instruments in low-temperature systems instead of steam. *Am. J. Infect. Control* 45, 756–760. https://doi.org/10.1016/j.ajic.2017.02.026

Menegol, D., Scholl, A.L., Dillon, A.J.P., Camassola, M., 2016. Influence of different chemical pretreatments of elephant grass (*Pennisetum purpureum*, Schum.) used as a substrate for cellulase and xylanase production in submerged cultivation. *Bioprocess Biosyst Eng* 39, 1455–1464. https://doi.org/10.1007/s00449-016-1623-8

Miranda, A.V.C., Espejo, Y. del C.B., Salas, J.L.T.F., Gonzales, H.H.S., Aguilera, J.G., Martínez, L.A., 2021. Biopesticides: Mechanisms of biocidal action in pest insects. *Res. Soc. Dev.* 10, e42010716893. https://doi.org/10.33448/rsd-v10i7.16893

Mishra, S., Roy, M., Mohanty, K., 2019. Microalgal bioenergy production under zero-waste biorefinery approach: Recent advances and future perspectives. *Bioresour. Technol.* https://doi.org/10.1016/j.biortech.2019.122008

Mittal, A., Decker, S.R., 2013. Special issue: Application of biotechnology for biofuels: Transforming biomass to biofuels. *3 Biotech* 3, 341–343. https://doi.org/10.1007/s13205-013-0122-8

Mohapatra, S., Mishra, C., Behera, S.S., Thatoi, H., 2017. Application of pretreatment, fermentation and molecular techniques for enhancing bioethanol production from grass biomass – A review. *Renew. Sustain. Energy Rev.* 78, 1007–1032. https://doi.org/10.1016/j.rser.2017.05.026

Murthy, P.S., Naidu, M.M., 2012. Production and application of Xylanase from *Penicillium* sp. utilizing coffee by-products. *Food Bioprocess Technol.* 5, 657–664. https://doi.org/10.1007/s11947-010-0331-7

Nascimento, R.P. do, Alves Junior, N., Coelho, R.R.R., 2011. Brewer's spent grain and corn steep liquor as alternative culture medium substrates for proteinase production by *Streptomyces malaysiensis* AMT-3. *Brazilian J. Microbiol.* 42, 1384–1389. https://doi.org/10.1590/S1517-83822011000400020

Pandey, A., Höfer, R., Taherzadeh, M., Nampoothiri, K.M., Larroche, C., 2015. Industrial biorefineries and white biotechnology. *Ind. Biorefineries White Biotechnol.* 1–710. https://doi.org/10.1016/C2013-0-19082-4

Peeters, J.R., Vanegas, P., Tange, L., Van Houwelingen, J., Duflou, J.R., 2014. Closed loop recycling of plastics containing Flame Retardants. *Resour. Conserv. Recycl.* 84, 35–43. https://doi.org/10.1016/j.resconrec.2013.12.006

Raheem, A., Sikarwar, V.S., He, J., Dastyar, W., Dionysiou, D.D., Wang, W., Zhao, M., 2018. Opportunities and challenges in sustainable treatment and resource reuse of sewage sludge: A review. *Chem. Eng. J.* 337, 616–641. https://doi.org/10.1016/j.cej.2017.12.149

Rajesh Banu, J., Kavitha, S., Yukesh Kannah, R., Dinesh Kumar, M., Preethi, Atabani, Kumar, G., 2020. Biorefinery of spent coffee grounds waste: Viable pathway towards circular bioeconomy. *Bioresour. Technol.* https://doi.org/10.1016/j.biortech.2020.122821

Ravindran, R., Hassan, S.S., Williams, G.A., Jaiswal, A.K., 2018. A review on bioconversion of agro-industrial wastes to industrially important enzymes. *Bioengineering* 5. https://doi.org/10.3390/BIOENGINEERING5040093

Ravindran, R. and Jaiswal, A.K., 2016. Microbial enzyme production using lignocellulosic food industry wastes as feedstock: A review. *Bioeng. (Basel, Switzerland)* 3. https://doi.org/10.3390/BIOENGINEERING3040030

Rivas, S., Vila, C., Alonso, J.L., Santos, V., Parajó, J.C., Leahy, J.J., 2019. Biorefinery processes for the valorization of Miscanthus polysaccharides: From constituent sugars to platform chemicals. *Ind. Crops Prod.* 134, 309–317. https://doi.org/10.1016/j.indcrop.2019.04.005

Robak, K., Balcerek, M., 2018. Review of second generation bioethanol production from residual biomass. *Food Technol. Biotechnol.* 56, 174–187. https://doi.org/10.17113/ftb.56.02.18.5428

Sadh, P.K., Duhan, S., Duhan, J.S., 2018. Agro-industrial wastes and their utilization using solid state fermentation: A review. *Bioresour. Bioprocess.* 5, 1–15. https://doi.org/10.1186/s40643-017-0187-z

Sajid, M., Zhao, X., Liu, D., 2018. Production of 2,5-furandicarboxylic acid (FDCA) from 5-hydroxymethylfurfural (HMF): Recent progress focusing on the chemical-catalytic routes. *Green Chem.* 20, 5427–5453. https://doi.org/10.1039/C8GC02680G

Sarkar, A., Praveen, G., 2017. Utilization of waste biomass into useful forms of energy. In: *Springer Proceeding in Energy*, 117–132. https://doi.org/10.1007/978-3-319-47257-7_12

Sharif Hossain, A.B.M., Ibrahim, N.A., AlEissa, M.S., 2016. Nano-cellulose derived bioplastic biomaterial data for vehicle bio-bumper from banana peel waste biomass. *Data Br.* 8, 286–294. http://dx.doi.org/10.1016/j.dib.2016.05.029

Sharma, P.K., Capalash, N., Kaur, J., 2007. An improved method for single step purification of metagenomic DNA. *Mol. Biotechnol.* 36, 61–63. https://doi.org/10.1007/s12033-007-0015-3

Sharma, P., Gaur, V.K., Sirohi, R., Varjani, S., Hyoun Kim, S., Wong, J.W.C., 2021. Sustainable processing of food waste for production of bio-based products for circular bioeconomy. *Bioresour. Technol.* https://doi.org/10.1016/j.biortech.2021.124684

Sharma, J., Sarmah, P. and Bishnoi, N.R. 2020. Market perspective of EPA and DHA production from microalgae. *Nutraceutical Fatty Acids from Oleaginous Microalgae.* Wiley. https://doi.org/10.1002/9781119631729.ch11

Shrestha, A., van-Eerten Jansen, M.C.A.A., Acharya, B., 2020. Biodegradation of bioplastic using anaerobic digestion at retention time as per industrial biogas plant and international norms. *Sustain.* 12. https://doi.org/10.3390/su12104231

Songulashvili, G., Spindler, D., Jimenéz-Tobón, G.A., Jaspers, C., Kerns, G., Penninckx, M.J., 2015. Production of a high level of laccase by submerged fermentation at 120-L scale of *Cerrena unicolor* C-139 grown on wheat bran. *C. R. Biol.* 338, 121–125. https://doi.org/10.1016/j.crvi.2014.12.001

Tahmoorian, F., Khabbaz, H., 2020. Performance comparison of a MSW settlement prediction model in Tehran landfill. *J. Environ. Manage.* 254. https://doi.org/10.1016/J.JENVMAN.2019.109809

Tyagi, V.K., Lo, S.L., 2013. Sludge: A waste or renewable source for energy and resources recovery? *Renew. Sustain. Energy Rev.* 25, 708–728. https://doi.org/10.1016/j.rser.2013.05.029

Ummalyma, S.B., Sahoo, D., Pandey, A., 2020. *Microalgal Biorefineries for Industrial Products, in: Microalgae Cultivation for Biofuels Production.* Elsevier, pp. 187–195. https://doi.org/10.1016/B978-0-12-817536-1.00012-6

Usman, M., Kavitha, S., Kannah, Y., Yogalakshmi, K.N., Sivashanmugam, P., Bhatnagar, A., Kumar, G., 2021. A critical review on limitations and enhancement strategies associated with biohydrogen production. *Int. J. Hydrogen Energy* 46, 16565–16590. https://doi.org/10.1016/j.ijhydene.2021.01.075

Vea, E.B., Romeo, D., Thomsen, M., Blikra, E., Romeo, D., Thomsen, M., 2018. biowaste valorisation in a future circular bioeconomy. *Procedia CIRP* 69, 591–596. https://doi.org/10.1016/j.procir.2017.11.062

Veana, F., Martínez-Hernández, J.L., Aguilar, C.N., Rodríguez-Herrera, R., Michelena, G., 2014. Utilization of molasses and sugar cane bagasse for production of fungal invertase in solid state fermentation using *Aspergillus niger* GH1. *Brazilian J. Microbiol.* 45, 373–377. https://doi.org/10.1590/S1517-83822014000200002

Ventorino, V., Robertiello, A., Cimini, D., Argenzio, O., Schiraldi, C., Montella, S., Faraco, V., Ambrosanio, A., Viscardi, S., Pepe, O., 2017. Bio-Based Succinate Production from *Arundo donax* Hydrolysate with the New Natural Succinic Acid-Producing Strain Basfia succiniciproducens BPP7. *BioEnergy Res.* 10, 488–498. https://doi.org/10.1007/s12155-017-9814-y

Vo, T.T.T., Chu, P.-M., Tuan, V.P., Te, J.S.-L., Lee, I.-T., 2020. The promising role of antioxidant phytochemicals in the prevention and treatment of periodontal disease via the inhibition of oxidative stress pathways: Updated insights. *Antioxidants* 9, 1211. https://doi.org/10.3390/antiox9121211

Yukesh Kannah, R., Sivashanmugham, P., Kavitha, S., Rajesh Banu, J., 2020. Valorization of food waste for bioethanol and biobutanol production. In: *Food Waste to Valuable Resources*, 39–73. Elsevier. https://doi.org/10.1016/B978-0-12-818353-3.00003-1

Zaher, U., Cheong, D.-Y., Wu, B., Chen, S., 2007. Producing energy and fertilizer from organic municipal solid waste. Department of Biological Systems Engineering. Washington State University.

Index